Analysis and Design of
PILE FOUNDATIONS

Proceedings of a
Symposium sponsored by the ASCE Geotechnical
Engineering Division and a
Session sponsored by the ASCE Technical Council
on Codes and Standards
in conjunction with the
ASCE National Convention
San Francisco, California
October 1-5, 1984

Joseph Ray Meyer, Editor

AMERICAN SOCIETY OF CIVIL ENGINEERS FOUNDED 1852

Published by the
American Society of Civil Engineers
345 East 47th Street
New York, New York 10017-2398

The material presented in this publication has been prepared in accordance with generally recognized engineering principles and practices, and is for general information only. This information should not be used without first securing competent advice with respect to its suitability for any general or specific application.

The contents of this publication are not intended to be and should not be construed to be a standard of the American Society of Civil Engineers (ASCE) and are not intended for use as a reference in purchase specifications, contracts, regulations, statutes, or any other legal document.

No reference made in this publication to any specific method, product, process, or service constitutes or implies an endorsement, recommendation, or warranty thereof by ASCE.

ASCE makes no representation or warranty of any kind, whether express or implied, concerning the accuracy, completeness, suitability or utility of any information, apparatus, product, or process discussed in this publication, and assumes no liability therefor.

Anyone utilizing this information assumes all liability arising from such use, including but not limited to infringement of any patent or patents.

Library of Congress Catalog Card No.: 84-72087
ISBN 0-87262-427-7
Manufactured in the United States of America.

Preface

Piles are one of the oldest type of foundation system known to man and as such have been previously studied in depth based on available modeling and monitoring techniques. An increased understanding of pile behavior coupled with the desire to develop cost effective foundation systems have resulted in a need to increase foundation efficiency. The increased knowledge of pile behavior has been spurred by the development of improved analytical techniques, technological advances in monitoring instrumentation, a better understanding of installation effects and the distinction between group and single pile behavior.

This special technical publication was assembled jointly by the Deep Foundation and Computer Applications and Numerical Methods Committees of the Geotechnical Engineering Division of ASCE to present information on recent developments in the analysis and design of pile foundations. The papers generally address axial and lateral capacity modeling, group effects and the effects of the installation process on pile behavior.

In addition to the articles on piles, there are five reviewer comments discussing a proposed standard on drilled shafts. The reviewers comments on the proposed drill shaft standard are related to work that has been underway for some time in developing a standard for the design and construction of those deep foundations. As a part of its responsibility, the National Bureau of Standards initiated work several years ago toward the revision of the A-56 standard on foundations and excavations (1952) of the American National Standards Institute. The NBS sponsored a series of background reports, one of which concerned drilled shafts. Subsequently, the Technical Council on Codes and Standards, A. Ivan Johnson, Chairman, of ASCE formed a committee to pursue the work. The Committee on Foundation and Excavation Standards, Felix Y. Yokel, Chairman, has several subcommittees at work in preparing documents for formal balloting by the full committee. The subcommittee on drilled shafts consists of Walter E. Hanson, James L. Parks, Lymon C. Reese, and Samuel D. Stone.

At the time of this writing the subcommittee, after many meetings, has proposed a draft of the standard on drilled shafts and the draft is being submitted to the full committee for balloting. The draft will have been available to the attendees of the San Francisco Convention and the papers at the convention, published here, will be directed toward a discussion of that draft. The points of view of various users of the standard are given: a structural engineer, a geotechnical engineer, a university professor, a user in governmental practice, and a contractor. A standard on drilled shafts is needed by engineers and contractors and the papers published herein are expected to make significant impact on the successful completion of such a standard.

The reviewers comments presented herein do not necessarily represent the views of the editor, the Deep Foundation and Computer Applications and Numerical Methods Committees or ASCE.

This special technical publication is composed of refereed papers reviewed in general accordance with ASCE Guidelines and carry the same status as any other ASCE Geotechnical Journal article. The papers are open for written discussion and comment the same as a journal paper and are also available for ASCE awards.

Most of the papers contained in this special technical publication were presented at four sessions of the 1984 ASCE Annual Convention in October, 1984 in San Francisco, California. The editor would like to thank Dr. Mike Holloway, Dr. Amr Azzouz and Dr. Lymon Reese for their assistance in developing this special technical publication, Dr. Harry Coyle, Pile Subcommittee chairman and Dr. Wayne Clough, Computer Applications and Numerical Methods Committees Chairman, for their general administrative guidance and Dr. Mike O'Neil, Deep Foundations Committee Chairman for his unselfish assistance in developing the ideas and organization necessary to initiate this special technical publication.

The Deep Foundations and Computer Practices Committees hopes that the information contained in this Special Technical Publication will benefit the readers and serve to initiate further investigations into the behavior of pile foundations.

Joseph Ray Meyer, P. E.
Editor

TABLE OF CONTENTS

Finite-Strain Axial Analysis of Piles in Clay

By Gustav A. Nystrom[1], M. ASCE

ABSTRACT

In the late 1970's, a series of research projects developed simple effective-stress methods to predict the axial capacity of piles in normally-consolidated clay. During the last project, known as ESACC, four different effective-stress methods were developed. Capacities predicted by many different methods were also compared with a high-quality pile data base. The only effective-stress method which contained no empirical factors was found to hindcast the test data as well as any empirically-based method. That method, known as ESM2 was built around a simplified, one-dimensional, cavity expansion solution.

The ESM2 boundary-value approach has now been extended significantly. The first extension was to use the same axisymmetric model to analyse pile installation, consolidation, and axial loading. This effective-stress cavity expansion model was used to evaluate the effects of both ends of the pile, the effect of pile loading direction, and the effect of pile loading rate. Other improvements over earlier work include use of a full finite-strain theory, and an extension of the modified-Cam-clay constitutive relations which accounts for the effect of the third stress invariant.

The paper compares the results from one-dimensional and axisymmetric models. Then it summarizes some new insights regarding the axial behavior of piles in normally-consolidated clay.

INTRODUCTION

Most analysis and design procedures for evaluating the axial behavior of piles are empirical. For such empirical methods, a characteristic soil parameter (such as undrained shear strength) is measured at the site of a pile test. Then a relation is sought between the axial pile performance and the soil parameter. Since empirical methods provide no insight about the mechanism by which soil resists the pile load, application of these methods to conditions different than those supported by pile tests involves the exercise of considerable engineering judgement.

Around 1975, a series of research programs was initiated to develop predictions of axial pile capacity based on theoretical considerations. The resulting pile capacity methods employ the effective

[1]Research Specialist, Exxon Preduction Research Co., Houston, Texas

stress principle, which states that the total stress at a point is made up of the sum of effective stress carried by the solid phase and pore pressure carried by the fluid phase. These effective stress methods explicitly track effective stresses and pore pressures in the soil from in-situ conditions, through pile installation and set-up to axial loading to failure.

In 1980, this series of research projects culminated in the conclusion of the ESACC (Effective Stress Axial pile Capacity Cooperative) project. The ESACC project consisted of a) developing theoretical methods for the prediction of axial capacity in clay, b) performing several series of pile tests, and c) comparing the predictions to the measurements.

The ESACC project developed four different effective stress methods for predicting axial pile capacity. Three of the methods employed a combination of theoretical concepts and empirical factors. The fourth method, known as ESM2 (Effective Stress Method 2) was strictly theoretical and contained no empirical factors.

Table 1 summarizes ESACC's hindcasts of pile test data (Ref. 4). The hindcasts employ empirical methods (API and lambda), hybrid methods containing both theory and empirical factors (ESM1, ESM3, and ESM4), and the strictly theoretical method (ESM2). We see that ESM2 hindcasts the pile test data as well as any other method. It even matched the data better than methods ESM3 and ESM4, which contained empirical factors based on the ESACC pile test data. Reference 3 contains a discussion of a hybrid effective stress method.

The remainder of this paper discusses only strictly theoretical methods. Their superior theoretical content is likely to lead to a better understanding of the behavior of soil around an axially-loaded pile. These methods are also more easily modified to treat other materials, geometries, or loading conditions.

The first section will summarize an implementation of ESM2. ESM2 will be shown to be simple enough for use as a design tool, since the most complicated calculation involves only one spatial dimension.

The majority of this paper will discuss axisymmetric finite-strain calculations. One objective of performing these calculations was to obtain a better understanding of ESM2's success in matching pile test data. The second objective was to evaluate effects related to the axial pile direction (which is neglected for ESM2 calculations). These axial dimension effects include the effects of pile tip and pile top; the effect of pile loading direction; and the relations among pile installation, surface heave, and ensuing pile capacity.

The complexity of these axi-symmetric calculations rules out their use for routine calculations. But the insight gained from them should lead to a wiser use of simpler design methods.

TABLE 1

ESACC COMPARISON: RATIO OF PREDICTED TO MEASURED FRICTION

Method	Test	Range	Mean	Standard Deviation of the Mean
API	Cambridge	.79 to 1.10	.91	.14
	Hamilton	1.57 to 2.56	2.12	*
	Empire	.84 to 1.05	.96	.11
Lambda	Cambridge			
	Hamilton	1.30 to 2.12	1.75	*
	Empire	1.17 to 1.44	1.31	.11
ESM1	Cambridge	.95 to 1.90	1.42	.40
	Hamilton	.87 to 1.82	1.35	*
	Empire	.95 to 1.26	1.11	.15
ESM2	Cambridge	.91 to 1.49	1.20	.24
	Hamilton	.41 to .86	.64	*
	Empire	1.06 to 1.18	1.13	.06
ESM3	Cambridge	1.09 to 2.00	1.55	.38
	Hamilton	1.11 to 1.45	1.28	*
	Empire	.83 to .97	.89	.07
ESM4	Cambridge	1.12 to 1.61	1.31	.22
	Hamilton	1.22 to 1.62	1.42	*
	Empire	.83 to .92	.88	.05

*Only two data points, so standard deviation not computed.

ONE-DIMENSIONAL ANALYSIS

The ESACC project developed ESM2, a purely analytical effective stress method for predicting axial pile capacity in clay. ESM2 consists of two distinct steps: 1) A one-dimensional finite-element analysis models pile installation and subsequent reconsolidation, and 2) A simple formula, called LOD2, predicts the maximum skin friction in terms of the stresses at the pile-soil interface (calculated in the first step) and the soil's mechanical properties.

ESM2 is a very simplified model of pile behavior. But it matches pile test data fairly well; provides insight about pile-soil interaction; and represents a starting point for more realistic effective stress models. We will now discuss this one-dimensional model and provide the groundwork for subsequent discussions of more realistic and complicated analyses.

Pile Installation - In general, the pile problem involves three dimensions. ESM2's first step involves only one spatial dimension because it assumes both axi-symmetry and plane strain conditions. Assuming axi-symmetry is very reasonable, as long as no lateral pile loads are considered. On the other hand, the assumption that there are no strains in the vertical direction is mainly justified by the resulting computational simplicity.

At first sight, it might appear that since a pile is installed vertically, to neglect vertical displacements in the model of installation is inconsistent. Fig. 1 should help to clarify the situation. It separates all soil surrounding a pile into 4 zones. Zone D is the soil modelled by ESM2. In this zone, the displacements are mainly radial. Zones A and B are affected by the pile's end conditions. ESM2 ignores these zones (but the axisymmetric analysis of Section 3 does not). Zone C is the material adjacent to the pile, which undergoes tremendous strain during pile installation. A basic assumption made for all the analyses in this report is that installation strains right adjacent to the pile have a negligible effect on the pile's behavior. The soil in Zone C just shears at constant volume. Within the context of critical state soil behavior (modified-Cam-clay is one of several critical state constitutive relations), such an assumption appears well justified.

For a closed-ended pile of radius r_o, the cavity radius is increased from a_o to $2\,a_o$, where $r_o = \sqrt{3}\,a_o$. This specific choice of radii increases the cavity volume by the volume of the pile. It also doubles the initial cavity size. Modified-Cam-clay constitutive relations are used to solve the cylindrical cavity expansion problem.

Let us now apply ESM2 to a specific pile problem. The same pile problem will be analyzed later using a more realistic model.

The soil parameters are: shear modulus G = 3520 kPa, initial void ratio e_o = 1.116, elastic logarithmic bulk modulus κ = .03, plastic logarithmic bulk modulus λ = .15, critical state ratio M = 1.2, void ratio at intersection of critical state line and unit pressure e_{cs} = 1.75, effective stress friction angle in triaxial compression ϕ = 30 degrees, permeability k = 4×10^{-5} m/day, and buoyant submerged unit weight of solid = 10 kN/m^3. Initially, the vertical stress is -138. kPa, the lateral stresses are -76.4 kPa, pore pressures are at equilibrium, and the soil is normally-consolidated.

Before pile installation, we assume there is a cylindrical cavity of radius = .45 m = actual pile radius/$\sqrt{3}$. The stresses correspond to the in-situ stresses for a penetration of 13.8 m below the mudline. The insitu vertical effective stress, σ_{vo} will be used later to scale calculated stresses.

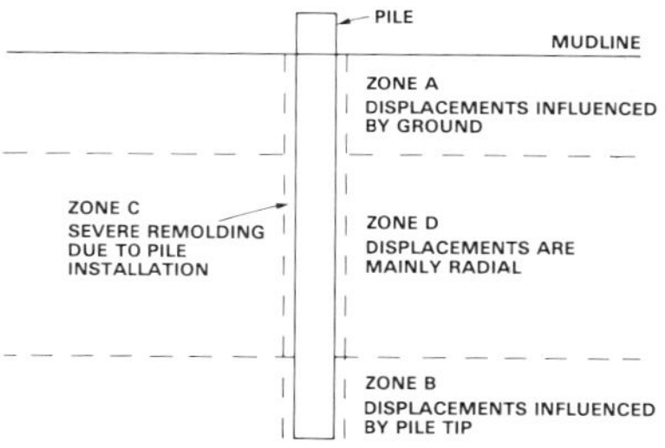

FIG. 1. SOIL DISPLACEMENT DUE TO PILE INSTALLION

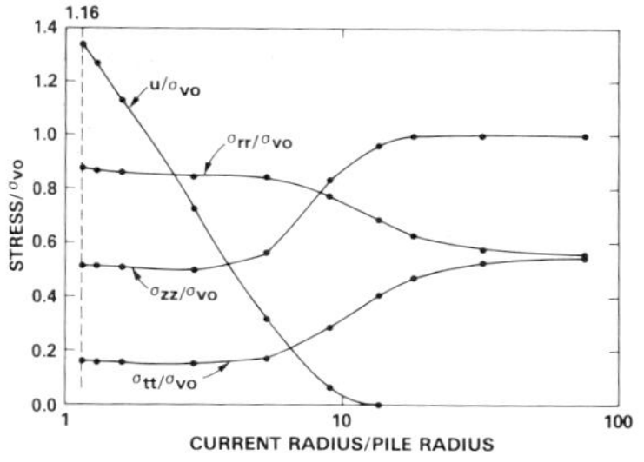

FIG. 2. ESM2 STRESSES AFTER PILE INSTALLATION

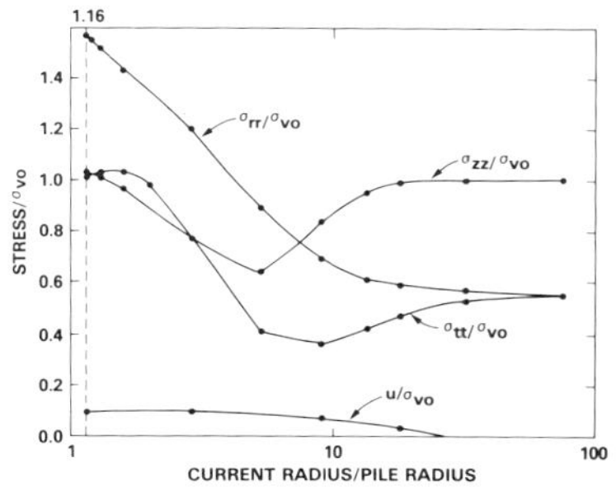

FIG. 3. ESM2 STRESSES 1100 DAYS AFTER PILE INSTALLATION

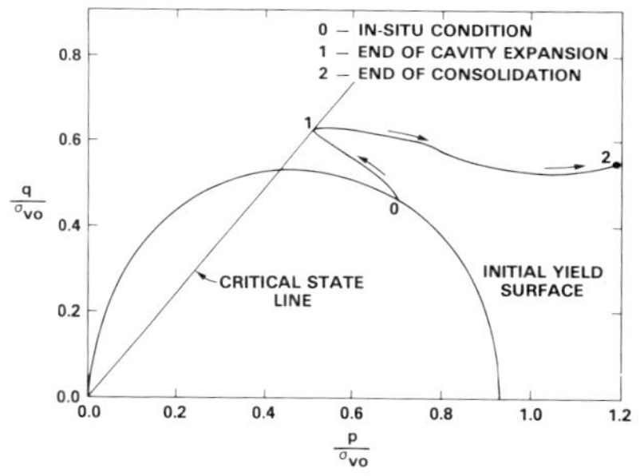

FIG. 4. ESM2 STRESS PATH AT PILE-SOIL INTERFACE

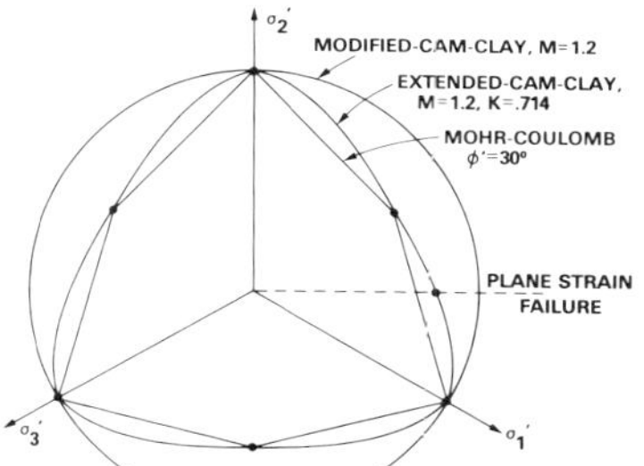

FIG. 5. COMPARISON OF THREE DIFFERENT FAILURE SURFACES IN THE π PLANE

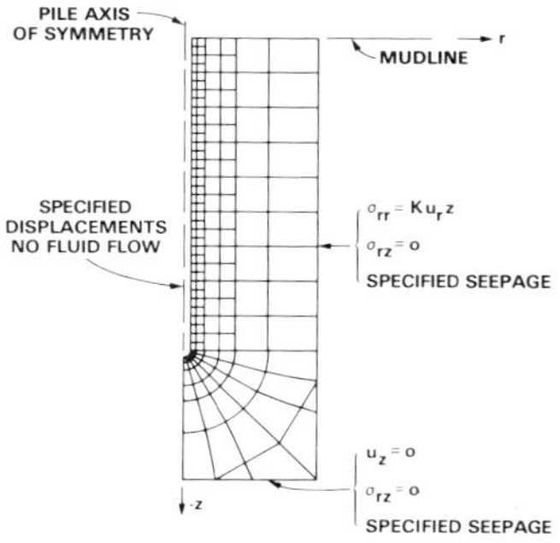

FIG. 6. AXISYMMETRIC FINITE ELEMENT MESH

Pile installation is modelled by a doubling of the cavity size over a short period of time (.03 days). Fig. 2 plots the soil effective stresses right after the cavity expansion. Note that the expansion has introduced sizable excess pore pressures.

Pile set-up is modelled by holding the cavity size constant and allowing the pore pressures to dissipate. After 80 days, the maximum pore pressure has been halved; and after 1000 days, it has been reduced to 10% of the initial value. Fig. 3 plots the stresses after 1,100 days of consolidation. As a result of pile installation (i.e. cavity expansion) the nearby soil is under a larger confining stress than in-situ.

We are mainly interested in the stresses near the pile-soil interface. It is this soil which is assumed to fail when the ultimate axial pile load is applied. Fig. 4 plots the stress path at the pile-soil interface during pile installation and consolidation. The stress path is shown in terms of effective stress invariants p and q, which are the mean normal stress and Mises equivalent shear stress. Fig. 4 shows that the size of the yield surface has increased about 50% due to the effect of pile installation and consolidation.

Pile Loading - According to method ESM2, the skin friction is calculated by substituting the results of the cavity expansion calculations into a formula known as LOD2. LOD2 states

$$\tau = \exp\left(\frac{e_{cs} - e_f}{\lambda}\right) \sin\phi \cos\phi \qquad (1)$$

where τ is the skin friction at axial pile failure, e_{cs} and λ are input soil parameters used for the cavity expansion calculation, e_f is the final void ratio at the pile-soil interface calculated via the cavity expansion, and ϕ is the effective-stress friction angle observed under triaxial compression conditions. The units on τ are the same as the units used to define e_{cs}.

For the example problem, after 1000 days of consolidation, the calculated void ratio near the pile-soil interface is $e_f = 1.050$. Substituting into the LOD2 formula, the predicted ultimate skin friction is

$$\tau = \exp\left(\frac{1.75 - 1.050}{.15}\right) \times .5 \times .866 = 46.0 \text{ kPa.} \qquad (2)$$

For normally consolidated material, the skin friction should be linear with depth. And the pile capacity consists of the skin friction integrated over the pile surface area plus the end bearing component.

Strengths and Weaknesses - Having described the ESM2 method, we can summarize its strengths and weaknesses.

Strengths:

1. based on physics and simple assumptions (contains no site-specific empirical factors).
2. good match with pile test data base (Table 1).
3. accounts for the effects of partial set-up on axial capacity.
4. input soil parameters are obtained from a few standard laboratory tests.
5. calculations are moderately easy to perform. The cavity expansion calculation is one-dimensional.

Weaknesses:

1. no information about the vertical dimension: tip effects, difference between tension and compression, etc.
2. use one model for pile installation (one-dimensional finite-element calculation using modified-Cam-clay constitutive relations) and another model (the LOD2 formula, derived from Mohr-Coulomb yield criteria) for pile loading.
3. only information regarding the pile loading event is the skin friction.
4. constitutive relations do not account for softening associated with very large strains.

References 1, 4, 7, and 8 contain a fuller introduction to the method, and several cavity-expansion solutions. Note that several different versions of ESM2 have been implemented by different individuals. All past versions have used an infinitesimal-strain theory with corrections for some finite-strain effects.

AXISYMMETRIC ANALYSIS

This section describes axisymmetric analyses performed to provide more insight about axial pile behavior than is possible using ESM2. Then it will discuss some of the resulting insights regarding pile behavior and the analysis of pile behavior. In terms of the ESM2 strengths/weaknesses listed above the axisymmetric analysis eliminates the first three weaknesses at the expense of the last strength.

Axisymmetric effective stress analyses of pile behavior have been reported earlier. As part of the ESACC project, Randolph and Wroth made some calculations employing a very rough mesh and a simple solution scheme (Reference 8). Then in 1983, Potts and Martins (Reference 6) reported on more realistic calculations. However, these axisymmetric calculations modelled special laboratory pile tests, for which pile installation did not affect the surrounding soil stresses.

Finite-Element Model - The finite-element model used to perform the axisymmetric analyses will now be described.

kinematics:
axisymmetric displacements and fluid flow allowed in the radial and vertical directions.

field equations:
force equilibrium enforced at each node. For each increment, the program iterates until a preset tolerance is met at all nodes.

Darcy's dissipation relation governs pore fluid flow

a full finite-strain theory is used. Such a theory correctly accounts for finite strains and rotations. But there is a computational limit to the amount of strain any one element may experience.

constitutive relations:
the extended-Cam-clay relations are used. They represent an improvement over the well-known modified-Cam-clay constitutive relations. Fig. 5 shows the modified-Cam-clay, extended-Cam-clay, and generalized-Mohr-Coulomb yield surfaces plotted on the π plane. The π, or deviatoric plane shows principal stresses on a plane of constant mean normal stress, p. The figure shows that the extended-Cam-clay yield surface represents a compromise between the smooth and simple shape of modified-Cam-clay and the more realistic shape of generalized-Mohr-Coulomb. References 2 and 5 contain more detailed information about the extended-Cam-clay constitutive relations.

body forces:
the buoyant weight of soil was included

initial stresses:
same as the in-situ stresses

finite-element mesh:
166 isoparametric soil elements with linear strain variation. See Fig. 6

boundary conditions:
Figure 6 illustrates the boundary conditions.

Note that the value of seepage coefficient was chosen to grossly represent the far-field resistance to fluid flow. The boundary condition is stated as: normal pore fluid velocity = seepage coefficient x (local pore pressure - reference pore pressure). Setting the seepage coefficient to zero would correspond to no fluid flow, while setting it to a very large value would correspond to specifying the boundary pore pressure.

Loading Condition - The only loading condition which varied over time was the displacement at the pile-soil interface. During the cavity expansion period, the cavity radius was doubled over a period of

.03 days. Along the cavity's cylindrical surface, the expansion was cylindrical; while along the hemispherical pile tip, the cavity expansion was spherical. Then during the set-up consolidation stage, no pile displacements took place. Finally, during the pile loading stage the pile was displaced axially as a rigid body. The direction and rate of displacement were varied from case to case. But for the baseline case the pile was pulled in tension over a period of .02 days.

Pile Installation - To arrive at the solution for the baseline case the computer solution was broken up into a number of sequential steps. Each step contained a number of increments and iterations. The table below shows that steps 2-4 made up the cavity expansion phase; steps 5-7 made up the consolidation phase, and steps 8-9 made up the pile loading phase.

step	specification	cavity radius (m)	pile elevation (m)	time span for step (days)
1	in-situ conditions	.45	0.	n.a.
2	expand cavity to .50 m	.50	0.	.01
3	expand cavity to .75 m	.75	0.	.01
4	expand cavity to .90 m	.90	0.	.01
5	consolidate to .1 days	.90	0.	.1
6	consolidate to 100. days	.90	0.	100.
7	consolidate to 1000. days	.90	0.	900.
8	displace pile up to .02 m	.90	.02	.01
9	displace pile up to .10 m	.90	.10	.01

Figs. 7 and 8 give a global view of the solution prior to loading by plotting lines of equal stress intensity at different stages of the solution. Those figures show qualitatively how the stress state varies with location and loading phase.

The excess pore pressure plots are particularly interesting. They show that right after cavity expansion (Fig. 7), the highest pore pressures are around the pile tip. But, with time, the pile tip pore pressures dissipate faster (the flow there is spherical, whereas elsewhere it is cylindrical). After 1000 days of consolidation (Fig. 8), the highest excess pore pressures are one quarter of the way down the pile - even though they are quite small.

Pile Loading - Fig. 9 plots the stress path in p-q space for the point very close to the pile wall and half-way down the pile. We see that the cavity expansion and consolidation path is similar to the path obtained using the one-dimensional ESM2 model (Compare Figs. 4 and 9).

The stress paths for cavity expansion and pile loading are similar because they are both "undrained". For undrained conditions, the yield surface does not change significantly, so the stress path follows it up to the critical state condition (where q is a maximum).

During the ESACC program, intermediate theoretical predictions were compared to the corresponding pile measurements. The worst match was that predicted pore pressures generated during pile loading were significantly larger than the measured values (Reference 4). Pore pressure generation predicted by the axisymmetric model agreed with ESM2-based predictions. Thus, this poor match remains unexplained.

Downdrag - This paper has presented a one-dimensional ESM2 solution, and introduced the corresponding two-dimensional axisymmetric solution. Overall, the two solutions for the pile installation phase are similar. An obvious difference deals with downdrag. The one-dimensional cavity expansion model assumes that there are no vertical strains associated with cavity expansion or consolidation. On the other hand, the axisymmetric solution calculates those strains.

The calculated results show that radial cavity expansion induces vertical motion of the surrounding soil. During cavity expansion, the soil is nearly incompressible. As the cavity expands laterally, the surrounding soil is pushed upwards. For example, for the base case the radial displacement is .45 m while the maximum surface heave is .30 m. Such an upward soil displacement introduces a compressive pile load right after the cavity expansion.

Likewise, when the soil is allowed to consolidate, the soil moves radially towards the pile and downwards. The maximum surface subsidence associated with consolidation for the baseline case is .092 m. This downward soil movement leaves the pile in tension before the pile loading is applied.

The rest of this paper will show other consequences of the downdrag associated with a cavity-expansion model of pile installation. But we will confirm that this downdrag has no effect on ultimate pile capacity, which is the quantity of most practical importance. (Indeed, if it were otherwise, the cavity expansion model would not predict capacity, since it neglects vertical pile displacement and associated residual pile stresses during installation).

Skin Friction at a Typical Point - Consider a typical point along the pile-soil interface, say a point half-way up the pile. Since the axial capacity of a long pile in clay is obtained via skin friction, there is a great deal of interest in evaluating the ultimate skin friction at such a point. (In this context, skin friction refers to the ultimate shear stress available to resist a quick, axial pile load).

The Introduction stated that the ESM2 one-dimensional analysis seems to predict the skin friction as well as any other method. This section will compare the skin friction predicted from that one-dimensional model to predictions for the axisymmetric model. For convenience, skin friction is normalized by the in-situ vertical effective stress σ_{vo}.

An earlier section reported on an application of the ESM2 method. Employing the CAMFE program to model pile installation, and the LOD2 formula for pile loading, we calculated τ/σ_{vo} = 46.0 kPa ÷ 138. kPa = .33. For the baseline finite-strain axisymmetric model, the result is τ/σ_{vo} = 62.8 kPa ÷ 136. kPa = .46. Thus, the axisymmetric skin friction is higher by 33%.

In order to clarify the 33% difference, we performed a third set of pile installation calculations, and a third set of pile loading calculations. For pile installation, we solved the one-dimensional cavity expansion problem with the same finite-strain program used to solve the axisymmetric problem (ABAQUS). This set of calculations had characteristics intermediate between the CAMFE one-dimensional calculations and the ABAQUS axisymmetric calculations.

The LOD2 formula was developed by ESACC based on Mohr-Coulomb and Cam-clay concepts. An analogous formula may be obtained by calculating the undrained shear strength for an extended-Cam-clay material under simple shear conditions. To underscore the analogy with the LOD2 formula, we have named this new formula LODB. LODB states that the skin friction is given by

$$\tau = \frac{M}{\sqrt{3}} \left(\frac{2\,K}{1 + K} \right) \; a \left(\frac{p}{a} \right)^{\kappa/\lambda} \tag{3}$$

where M, K, κ, and λ are input soil parameters, and a and p are the yield surface semi-radius and mean normal stress, calculated using a cavity-expansion model.

TABLE 2: SKIN FRICTION AT TYPICAL POINT

This table gives the value of skin friction divided by in-situ vertical stress. The pile load is applied 1000 days after installation.

		Pile Installation Model		
		CAMFE one-dimensional	ABAQUS one-dimensional finite strain	ABAQUS axisymmetric finite strain
Pile Loading Model	LOD2 formula	.33	.31	.36
	LODB formula	.45	.41	.45
	ABAQUS axisymmetric	n.a.	n.a.	.46

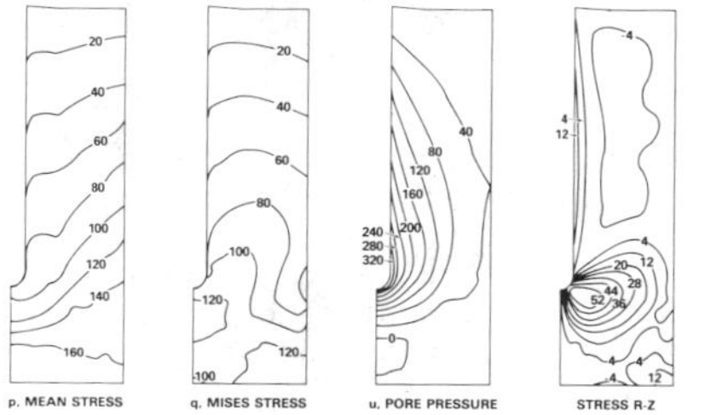

FIG. 7. STRESS CONTOURS AFTER CAVITY EXPANSION

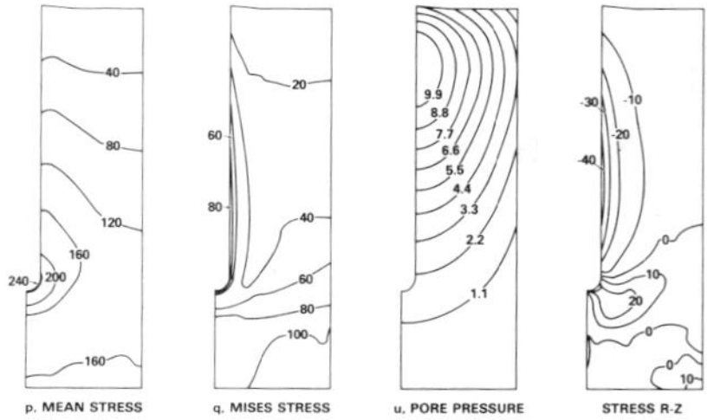

FIG. 8. STRESS CONTOURS AFTER 1000 DAYS OF CONSOLIDATION

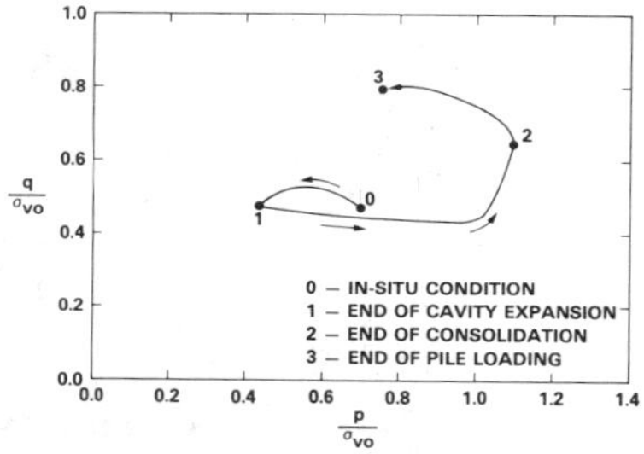

FIG. 9. STRESS PATH AT TYPICAL POINT ADJOINING PILE

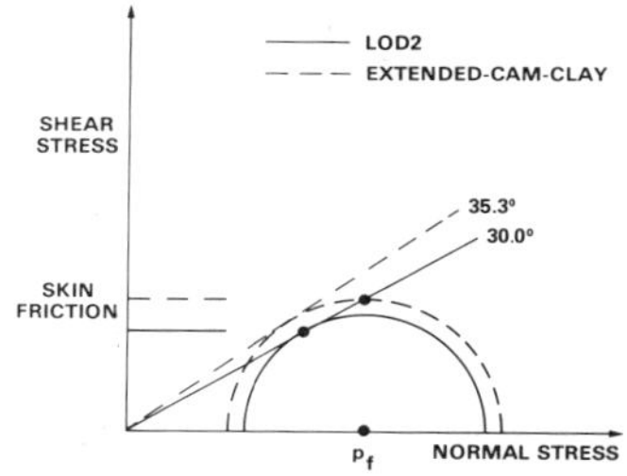

FIG. 10. SKIN FRICTION DIFFERENCES

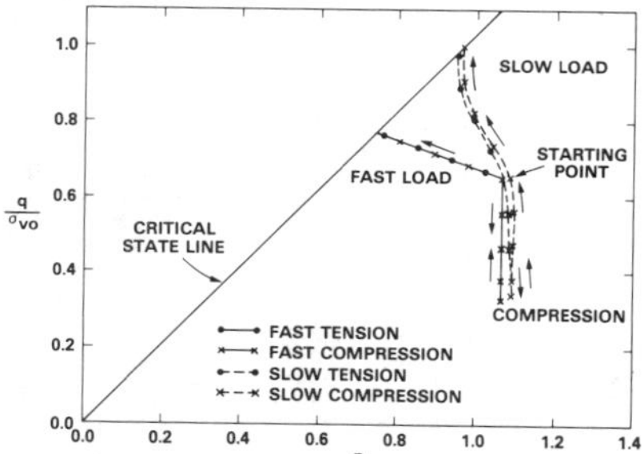

FIG. 11. STRESS PATHS FOR DIFFERENT PILE LOADING DIRECTIONS AND TIME

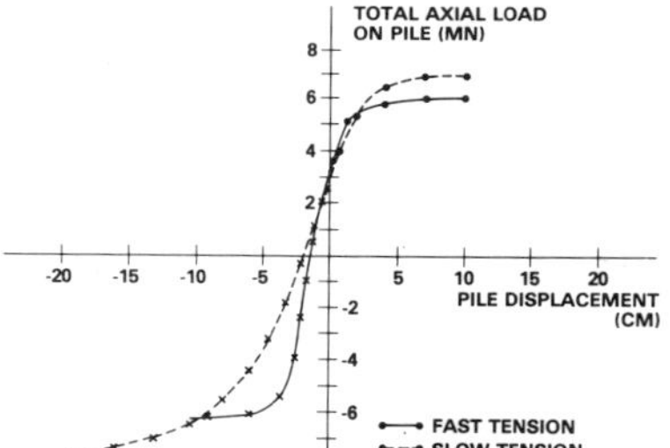

FIG. 12. TOTAL AXIAL LOAD VS. AXIAL DISPLACEMENT

Table 2 presents the skin friction at the typical point 1,000 days after pile installation. The predictions come from various combinations of three different pile installation models and three pile loading models. Table 2 clearly shows that: 1) Calculated skin friction is insensitive to the choice of an installation model, and 2) Loading models which employ the extended-Cam-clay relations predict about 33% higher skin friction than LOD2 (Cam-clay and Mohr-Coulomb).

Fig. 10 shows why the extended-Cam-clay skin friction predictions are 33% higher than the corresponding LOD2 predictions. The 33% difference is due to two effects. The larger effect is due to the fact that under plane strain conditions (applicable at the time of failure near the pile), the chosen extended-Cam-clay material is stronger than the generalized-Mohr-Coulomb material used to derive the LOD2 formula. Qualitatively, this is shown on Fig. 5. Quantitatively, Van Eekelen (Reference 9) has shown that if the friction angle for a generalized-Mohr-Coulomb material is 30.0 degrees, and one chooses an extended-Cam-clay material as shown in Fig. 5, its friction angle under plane strain conditions will be 35.3 degrees.

The smaller effect deals with orientation of stress components in space. For extended-Cam-clay, the maximum shear stress is σ_{rz}, so the skin friction corresponds to the radius of the Mohr's circle. On the other hand, for LOD2 calculations, the skin friction at the pile wall is assumed to correspond to the point where the circle is tangent to the Mohr's envelope.

Applying some trigonometry to Fig. 10, the result is

$$\text{LOD2:} \quad \frac{\tau}{p_f} = \sin(30°) \times \cos(30°) = .433$$

$$\text{extended-Cam-clay:} \quad \frac{\tau}{p_f} = \sin(35.3°) = .577 \qquad (4)$$

where p_f is the value of p at the failure condition. The ratio between the two skin frictions is 1.33. Thus the 33% discrepancy is fully explained by the effects illustrated in Fig. 10.

It should be noted that with extended-Cam-clay one can match Mohr-Coulomb at any two failure conditions by appropriate selection of the input soil parameters M and K. But at this point, the objective is not to match the earlier ESM2 method, but to understand how the choice of a computational model affects the calculated skin friction.

For performing the cavity expansion calculation, a finite - strain solution (such as the ABAQUS one-dimensional solution) is preferred over a CAMFE solution which only contains some finite-strain terms. Finite-strain solutions (axisymmetric or one-dimensional ABAQUS) show the soil strength at the pile wall increasing and then decreasing. (See Fig. 9.) On the other hand, all calculations lacking the full finite-strain theory (axisymmetric ABAQUS with small strain theory, one-dimensional ABAQUS with small strain theory, and one-dimensional CAMFE with some finite-strain terms) show the soil strength increasing monotonically. See Fig. 4.

Quantitatively, the CAMFE solution seems to overestimate the shear strength at the pile wall by 15% at the end of cavity expansion. At this stage, the strains near the pile are 60%, so that it is not surprising that omitting some finite strain terms would introduce some error. Fortunately, the calculated error disappears as the pore pressures are allowed to dissipate with time.

In summary, to calculate undrained skin friction for a long pile: the first step is to perform the cavity expansion - consolidation calculation; the second step is to calculate the mean stress at the pile surface at failure; and the third step is to calculate the skin friction. The first two steps are fairly insensitive to the choice of a computational procedure. And the results of the third step depend in a predictable manner on the choice of soil constitutive relation.

Pile End Effects - The previous section discussed skin friction at a typical point in the middle of the pile. This section will illustrate how the stresses at the pile-soil interface vary with depth. In particular, the calculated effect of the pile's two ends will be examined.

At many key stages along the solution, the calculated stresses along the majority of the pile-soil interface are proportional to depth. For example, at the time of axial pile failure, the calculated skin friction is nearly linear with depth. The one-dimensional ESM2 method would imply such a variation -- but of course it gives no guidance on the extent of pile end effects. The axisymmetric model results confirm that for a long and rigid pile in normally - consolidated clay, neglecting the ends' effect on skin friction is quite appropriate.

At other stages along a pile's life, the stresses are not linear with depth. For example, Fig. 8 shows that near the end of the consolidation stage, the excess pore pressure is not linear with depth.

Let us now examine the extent of pile end effects. If we limit ourselves to calculated friction along the pile, the axisymmetric results indicate that:

1. at the pile bottom, the skin friction deviates by more than 25% from the linear variation over a distance of about 1 pile radius.

2. at the pile top, the axisymmetric model neglected the strength of the top 1 m (1.2 radii) of soil. This was done to avoid numerical problems associated with soupy soils having a strength very close to zero. That top meter of material simply applied a surcharge to the underlying soil. For the soil modelled as having strength, the skin friction

deviates by more than 25% from the linear variation over a distance of about 1 pile radius.

No great significance should be attached to the specific distances quoted above, since they are of the order of element dimensions and they vary somewhat depending on loading rate and loading direction. The main point is that calculated end effects are significant over a relatively small distance.

Effects of Load Duration and Time - Up to this point, most of the discussion has dealt with the baseline loading conditions. That corresponds to the application of a fast tensile failure load 1000 days after pile installation. This section will discuss four different loading conditions: fast tensile load, fast compressive load, slow tensile load, and slow compressive load.

All four calculations discussed in this section have identical cavity expansion and set-up histories. Then the pile loading stage is defined by specifying the following pile movement:

loading condition	vertical displacement (m)	time interval for displacement (days)
fast tension	.1	.02
fast compression	-.1	.02
slow tension	.1	2000.
slow compression	-.2	2000.

Fig. 11 presents the stress paths for this series of analyses. It shows that all four calculations started at the same condition -- the point shown in Figure 9 corresponding to stress conditions after 1,000 days of set-up consolidation. The two cases corresponding to a "fast load" end up at the same failure condition. And the two cases corresponding to a "slow load" end up at another failure condition corresponding to 20% larger skin friction.

The implications are:

1) for undrained, fast pile loading, skin friction is independent of load direction. This is in line with observed pile performance.

2) for drained, slow pile loading, the skin friction is about 20% larger and independent of load direction. That a slow compression load should encounter a greater skin friction is not surprising. But what is surprising is that a pile loaded slowly in tension would also encounter a greater skin friction.

The slow tension result illustrates how effective-stress pile analysis can help insight. Before seeing these results, one's intuition is that a slow compressive load would involve a higher mean normal stress (p) than a slow tension loading. But Fig. 11 shows that this effect accounts for only a 2% difference in skin friction -- almost imperceptible. The figure shows that the 2% effect is swamped by the 20% effect associated with a reduction in p for undrained loading (due to the constraint that for undrained loading, the stress path must stay close to the elliptical yield surface).

Fig. 11, shows that downdrag effects lead to a stress-path detour for the two compression cases. After pile installation, downdrag has left the pile in tension. For the two compression cases, the result of specifying a compressive displacement is to first unload the pile and then reload it in compression. On Fig. 11, this appears as a "downdrag detour." But we note that after the detour, the stress invariants return to the initial condition. Thus, elastic downdrag has no effect on calculated pile capacity. For the two tension cases, the stress path goes directly from the "starting point" to failure.

After discussing skin friction, let us examine the end bearing contribution to pile capacity. Table 3 presents both components of pile capacity, as calculated using the axisymmetric model.

The first observation is that for the current conditions, the skin friction component does dominate the total capacity, as expected. The next observation is that the end bearing subtotal depends both on load direction and rate. Its value ranges from a negligible value (for slow tension) to 2021 kN (for slow compression). By comparison, the end bearing component is normally calculated as 9 x pile cross-sectional area x undrained shear strength. For current conditions this would be $9 \times \pi \times (.7794\ m)^2 \times 94\ kPa = 1600\ kN$. This is very close to the fast compression subtotal indicated on Table 3.

TABLE 3: PILE LOADS AT FAILURE (kN)

LOAD ON PILE	LOADING CONDITION			
	fast tension	fast compression	slow tension	slow compression
skin friction component	5003	-5049	6263	-6296
end bearing component due effective stress	-44	-764	60	-1964
end bearing component due excess pore pressure	612	-916	14	-57
weight of displaced soil	508	508	508	508
total	6080	-6221	6846	-7810

Fig. 12 plots the calculated load-displacement curves for the four pile loading conditions. That curve shows how downdrag leaves the pile in tension at the end of installation. It also shows that a relatively large displacement is required to fail a compression pile slowly.

The results presented in this section have a direct application to tension leg platform piles. Those piles are exposed to a significant long-term tension load during the life of the structure. Thus, there is a great interest in evaluating the pile capacity after the pile has seen years of continuous tension load.

These results indicate that for a long, rigid pile in clay, the effects of the long-term tension are beneficial. To obtain a quantitative estimate of the TLP pile's capacity, a specific load-time history is needed. But, one can use Fig. 11 for a "ballpark" estimate. If the long-term load is 1/4 of the pile's capacity, then years after installation the stress state might be 1/4 of the way up the slow-tension curve. If at that point a very large tensile load is applied, the stress path will be similar to the "fast load" stress paths on Fig. 11. The resulting failure load will be in the order of 4% higher than if no long-term load had been present. This is a small increase. But the important point is that the long-term load does not reduce the final capacity, as is generally believed.

Initial Vertical Displacement - Fig. 12 suggests that a downward displacement of .018 m might leave the installed pile at the no-load condition. For one calculation, that vertical displacement was superposed on the standard cavity expansion to model pile installation. The calculated pile load was reduced about 10% during the installation stage. But for subsequent stages, there was no effect on calculated pile load.

The conclusion is that the pile's response to applied axial loads is insensitive to vertical displacements during installation. This result is very important. It confirms a basic assumption of the cavity-expansion method: that vertical pile movement during installation has a negligible effect on the pile's subsequent response to loads.

CLOSING

Unanswered Questions - The effective stress results presented in this paper appear reasonable. And, as stated in the Introduction, this kind of analysis has been found to hindcast pile test data about as well as any other method.

But a number of key questions remain to be resolved and will remain so until some of the theoretical assumptions are checked out in the field. Such field measurements must include high-quality pile test data local to the pile-soil interface for a site with homogeneous and throughly - tested soil conditions.

Some specific questions that have not been settled yet are:

1) What effect do the large vertical displacements associated with pile installation have on the behavior of soil right at the interface (Region C in Fig. 1)?
2) How important is the anisotropic nature of soil plasticity (Bauschinger effect)?
3) How important is the strain-softening sometimes observed in the laboratory?
4) For a long pile at failure, is the radial normal effective stress proportional to the local skin friction?

CONCLUSIONS

Conclusions regarding the axial behavior of a deep pile installed in normally consolidated clay are summarized below, in the order discussed in the text.

1. The ESACC project showed that a simple cavity-expansion calculation hindcasts pile test data as well as any other method. That method, called ESM2, employs no empirical parameters.

2. For an axisymmetric analysis, cavity expansion induces ground heave, and consolidation induces settlement. But these vertical motions do not affect pile capacity.

3. Undrained skin friction was calculated using a variety of computational procedures. Calculated differences were found to reside in the calculation of skin friction given the mean normal stress at the pile surface at failure. They depend in a predictable manner on the choice of a soil constitutive relation.

4. Calculated end effects are significant over a relatively small distance. Otherwise, calculated skin friction is quite linear with depth.

5. Skin friction is independent of loading direction (tension versus compression).

6. Very slow pile loading develops about 20% larger skin friction than fast pile loading.

7. The long-term tensile load applied to a tension-leg-platform pile is not detrimental to ultimate pile capacity.

8. The pile's response to applied axial loads is insensitive to vertical displacements during installation.

9. For piles in normally consolidated clay, purely-analytical results seem reasonable. But a number of key questions remain to be resolved.

ACKNOWLEDGEMENTS

This work may be considered an extension of the ESACC work, so it is fitting that the ESACC project contributors be acknowledged. The ESACC project spanned the years 1978-80 and employed three geotechnical contractors (the University of Cambridge Engineering Department, McClelland Engineers, and Woodward - Clyde, Inc.). The project was administered by Amoco Production Company and sponsored by 17 firms (including Brown & Root, Building Research Establishment, Exxon Production Research, Gulf Oil Corporation, the Norwegian Geotechnical Institute, Pennzoil Company, and Phillips Petroleum Company).

The computer program ABAQUS is the property of Hibbitt, Karlsson and Sorensen, Inc. The ABAQUS soils capabilities, which are described in Reference 5, were developed as a joint effort between Hibbitt, Karlsson and Sorensen, Inc. and Exxon Production Research Co. This successful development effort has benefitted from the efforts of many, but particularly E. P. Sorensen, J. D. Murff, S. S. Wilder, H. D. Hibbitt, and A. B. Potvin.

Appendix I. - REFERENCES

1. Carter, J. P., Randolph, M. F., and Wroth, C. P., "Stress and Pore Pressure Changes in Clay during and after the Expansion of a Cylindrical Cavity", International Journal on Numerical and Analytical Methods in Geomechanics, Vol. 3, No. 4, 1979, pp. 305-322.

2. Hibbitt, Karlsson and Sorensen, Inc., "Theory Manual, ABAQUS General-Purpose Finite Element Program, Providence R. I., 1983.

3. Kirby, R. C., Esrig, M. I., and Murphy, B. S., "General Effective Stress Method for Piles in Clay, Part I - Theory," Proceedings of the Conference on Geotechnical Practice in Offshore Engineering, edited by S. G. Wright, Austin, Texas April 1983.

4. Kraft, L. M., Kirby, R. C., and Murphy, B. S., "ESACC Project Summary Report," prepared for Amoco Production Company, July 1980.

5. Nystrom, G. A., "The Use of Soil Mechanics Capabilities in a General Purpose Finite Element Program," Proceedings of ASCE Symposium on Prediction and Validation of Consolidation, San Francisco, October 1, 1984.

6. Potts, D. M. and Martins, J. P., "The Shaft Resistance of Axially Loaded Piles in Clay," Geotechnique, Vol. 32, No. 4, 1982.

7. Randolph, M. F., Carter, J. P., and Wroth, C. P., "Driven Piles in Clay - the Effect of Installation and Subsequent Consolidation," Geotechnique, Vol. 29, No. 4, 1979, pp. 361-393.

8. Randolph, M. F., and Wroth, C. P., "Report on the Development of a Load Module (LOD2) for Estimating the Shaft Capacity of a Driven Pile," prepared for the ESACC Project, 1980.

9. Van Eekelen, H. A. M., "Isotropic Yield Surfaces in Three Dimensions for Use in Soil Mechanics," International Journal for Numerical and Analytical Methods in Geomechanics, Vol. 4, 1980, pp. 89-101.

10. Zienkiewicz, O. C., and Pande, G. N., "Some Useful Forms of Isotropic Yield Surfaces in Soil and Rock Mechanics," Finite Elements in Geomechanics, eidted by G. Gudehus, John Wiley and Sons, 1977.

Appendix II - NOTATION

a yield surface semi-radius

e_{cs} void ratio at intersection of critical state line and p=1

e_o initial void ratio

G shear modulus

k permeability

K third stress invariant soil parameter

M critical state ratio q/p at critical state

p mean normal stress; first stress invariant

p_f value of p at the failure condition

q Mises shear stress intensity; second stress invariant.

r third stress invariant or radial coordinate, depending on context

r_o pile radius

u excess pore pressure

z vertical coordinate

κ elastic logarithmic bulk modulus

λ plastic logarithmic bulk modulus

γ_B submerged unit weight of solid

γ_w unit weight of water

σ_{vo} in-situ vertical effective stress

τ skin friction at axial pile failure

ϕ effective stress friction angle

Centrifugal Modeling of Pile Foundations

Hon-Yim Ko[1], R.H. Atkinson[2], G.G. Goble[3], Carl D. Ealy[4]

Abstract

Modeling of pile foundations in a centrifuge, where the self weight induced stresses in the soil are simulated by the elevated gravity loading, is described in this paper. The primary purpose of the test program was to simulate two full scale prototype experiments from which load test data are available for comparison with centrifuge test results.

Hollow pipe piles in a stiff clay were tested in simulation of the tests at the University of Houston site. Tapered timber piles in sand were tested in simulation of the tests at the Locks and Dam No. 26 site. Both single piles and pile groups were tested under axial loads.

The validity of centrifugal modeling of pile foundations is demonstrated by testing different scaled models at gravity levels necessary to simulate the prototype. Effects of inflight installation, soil properties and soil saturation on pile behavior are studied. Load transfer data are obtained on both types of tests and compared with the prototype.

Results of this program show that it is feasible to duplicate prototype data by testing models in a centrifuge. Since it is possible to construct centrifuge specimens having uniform and known soil properties, it can be a useful tool for studying fundamental aspects of pile behavior and for testing theoretical explanations of performance. In many projects it may be advisable to include centrifuge

[1] Prof. & Chairman, Dept. of Civil Engineering, University of Colorado, Boulder, CO 80309
[2] Principal, Atkinson-Noland & Assoc., 2619 Spruce St., Boulder, CO 80302
[3] Prof., Dept. of Civil Engineering, University of Colorado, Boulder, CO 80301
[4] Turner-Fairbank Highway Research Center, Federal Highway Admin., 6300 Georgetown Pike, McLean, VA 22101

model testing as part of the preconstruction design and planning because of the low costs involved and the possibility of performing a parametric design study.

Introduction

Full scale testing of geotechnical structures such as retaining walls, dams, foundations, etc. is rarely performed due to the great expense and difficulty of loading these structures to meaningful levels. When full-scale field tests are conducted, they are usually of one design at a specific site and thus do not yield information on the influence of variables such as structure design or soil properties. Furthermore, the soil properties and their variability are usually not well known and controlled.

Model tests on scaled structures in the laboratory can overcome many of these problems. For structures located within or loaded by a soil mass, proper simulation of field (prototype) conditions requires that the proper magnitude and distribution of soil interaction forces be modeled. The stress-strain-deformation properties of soil are directly dependent on the magnitude and nature of the applied stress state which is, for most structures, usually generated by the self-weight or body forces of the soil under normal gravity conditions. When testing small-scale models in laboratory situations, however, proper modeling of the soil stress field and the resulting soil properties is usually not possible under 1 g conditions.

A method of achieving the required soil stress field is to conduct a model test in an increased gravity field, created by using a geotechnical centrifuge. The acceleration generated by the centrifuge increases the gravity induced body forces on the soil particles and, hence, the apparent unit weight of the soil. Thus, the in-situ stress field in a soil deposit can be correctly modeled.

The theory and application of centrifugal testing of soil-structure models which has been discussed in detail by Schofield (7), is briefly reviewed. If λ denotes the scaling factor between model and prototype

$$L_p = L_m \lambda, \tag{1}$$

then scaling relationships for quantities of importance in geotechnical engineering can be derived as listed in Table 1.

Table 1. Scaling Factors for Various Quantities

Length	λ
Stress	1
Time	λ^2
Strain	1
Force	λ^2
Area	λ^2

Volume	λ^3
Specific Weight	$1/\lambda$
Gravitational Acceleration	$1/\lambda$
Mass Density	1
Mass	λ^3

These relationships are based on the assumption that the same soil material is used for both the model and prototype. Since, as discussed above, body forces cannot be neglected, the required scaling relation between model and prototype specific weights can be achieved by centrifugal acceleration. This modeling approach yields scaling factors of unity for both stress and strain and permits a direct comparison between prototype and model behavior at equivalent locations.

The difficulties of full scale field testing apply to single piles and especially to pile groups where the magnitude of loads required to produce group failure become extremely difficult to generate. The deformation and failure of pile foundations is primarily influenced by the properties of the surrounding soil and its in-situ stress state. For these reasons, results from the few model pile group studies at 1 g conditions (Whitaker, (10); Saffery and Tate, (6) and Sowers, (9)) have been largely qualitative due to the lack of adequate scaling relationships between prototype and model (Rocha, (5)).

After a limited application in the 1930's, centrifugal modeling of geotechnical structures was abandoned in the West until the late 1960's. The technique was first used at Cambridge University with other geotechnical centrifuge facilities following in Japan and the United States. An excellent discussion of centrifuge testing is provided by Schofield's 20th Ranking Lecture, 1980.

Research on the behavior of pile foundations using the centrifuge has begun only fairly recently. Scott's (8) research on pile groups in silt subjected to cyclic lateral loads produced results which were internally consistent and demonstrated the feasibility of conducting pile load tests in a centrifuge. The lack of a prototype with which to compare results and thus verify the similitude relationships was a shortcoming of the study.

Axially loaded piles in sand were investigated by Hougnon (3) at the University of Colorado. Although this was largely a feasibility study, some useful data regarding the effect of taper and soil density on bearing capacity was obtained. Problems encountered with the preparation of soil specimens of uniform properties and with the loading apparatus limited the usefulness of Hougnon's work.

Purpose of Research Program

The research reported herein had as its principal objective the testing, in a centrifuge, of single piles and pile groups that modeled full scale field tests. One of the specimen test programs was conducted on single piles and a 9-pile group at the University of Houston campus (O'Neill, et al. (4)). This test was conducted in an overconsolidated clay deposit. The other specimen test to be modeled was conducted at Lock and Dam No. 26 on the Mississippi River near Alton, Illinois by Woodward-Clyde Consultants (11). Single pile tests and tests on an 8-pile group in a cohesionless alluvial sand deposit were to be modeled.

In conducting the research program experimental techniques were developed for, 1) soil preparation and placement, 2) fabrication and instrumentation of model piles and pile groups, 3) pile installation equipment and techniques and 4) pile load testing equipment and techniques.

A series of model pile tests were conducted in both sand and clay to evaluate both qualitatively and quantitatively the method of centrifugal testing. The method of modeling of models as described below was the approach used to provide quantitative verification of centrifugal model testing of piles.

Description of Experimental Program

The research program consisted of two separate parts; tests of model piles in clay and tests of model piles in sand. These two studies have in common the use of the same centrifuge, the same pile loading mechanisms and the same data acquisition system which are discussed below. The soil preparation, model pile design, instrumentation, test procedure and results for the sand and clay studies are discussed separately below.

Tests were conducted using the 10 g-ton geotechnical centrifuge located at the University of Colorado, Figure 1. This machine can generate an acceleration of 100 g on a 200 lb (90.8 kg) package. This centrifuge is equipped with swinging "baskets" mounted at the ends of a symmetrical 42-inch (1.1 m) radius arm. The hinged baskets permit the resultant of gravity and centrifugal forces to act at all times perpendicular to the bse of the model foundation. The centrifuge is equipped with 56 electrical slip rings for power and electrical signal transmission and two hydraulic slip rings for control of the loading mechanisms.

The soil container for both the sand and clay tests was a 15-inch (38.1 cm) diameter aluminum cylinder 12 inches (30.5 cm) deep, Figure 2. The driving and loading of the piles were performed in-flight using a Bellofram cylinder. A LVDT with a 5-inch (12.7 cm) range was used to monitor pile installation while a more sensitive unit with a 0.50 inch (1.27 cm) range was used to monitor pile motion during load teting. Applied loads were measured using special strain-gage type load cells. Data was recorded using either analog plotters or a digital data logging system.

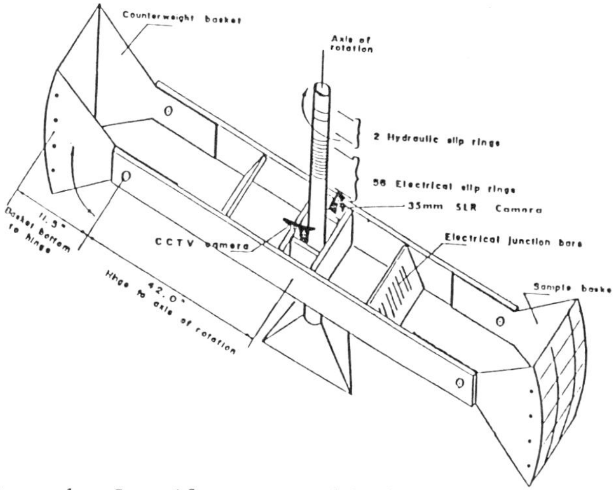

Figure 1. Centrifuge arms and baskets

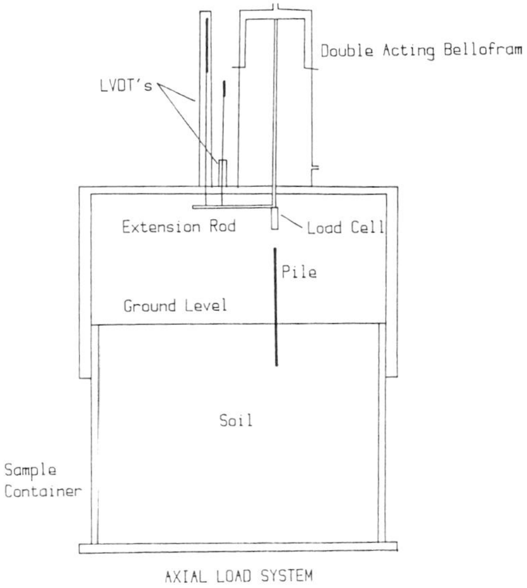

Figure 2. Mechanism for loading model piles in the centrifuge.

Pile Tests in Clay

Soil Preparation and Pile Design

A Georgia kaolin was used to simulate the Beaumont clay from the University of Houston site as the prototype soil was unavailable in block samples. The laboratory soil was prepared at an initial water content of 60% (L_w = 45%), placed in the soil container equipped with an extension tube and consolidated under a series of increasing static loads. The preconsolidation stress was determined using procedures developed by Atkinson and Bransby (1). After the static consolidation pressure was relieved, the soil container was installed in the centrifuge and brought up to the test g level to produce the desired vertical distribution of undrained shear strength, Figure 3. Shear strength, S_u, of the clay soil was measured in-flight using a specially developed vane shear device.

Model piles were fabricated to 1/50, 1/70, and 1/100 scale factors. The 1/50 and 1/70 scale piles were fabricated from split aluminum tubing which permitted installation of internal strain gages before the two halves were epoxied together. Strain gages were installed at diametrically opposite positions at 5 levels to permit determination of both axial load and bending moment.

Test Program and Results

Considerable effort was expended in developing soil preparation and consolidation procedures to obtain the desired magnitude and profile of S_u versus depth. A total of 15 tests conducted in this effort yielded shear strengths at pile mid-depth and ultimate capacities for the single pile models. The desired S_u was 2.4 ksf (115 kPa) at pile mid-depth but the actual values ranged from 1.6 ksf (91 kPa) to 2.9 ksf (139 kPa). Figure 4 presents ultimate pile capacity of the model piles (at prototype scale) versus S_u. The correctness of the similitude relations assumed can be verified internally by comparing single pile results from the 50 g, 70 g and 100 g tests. These results, when plotted to prototype scale, yield consistent results. The prototype results also fall along the same curve, thus further verifying the similitude relations.

Pile group tests were conducted at 70 g and 100 g scales (Table 2). The measured group factor from the four groups (two at 70 g and two at 100 g) range from 0.96 to 1.04. The prototype had a measured group factor of 0.98. The greater factor measured in Test No. 21 may have resulted from underground model pile deviations from the driven pile grouping. Upon excavation of this model pile group, several piles were observed to have deviated laterally several pile diameters at the pile tip depth.

Axial load transfer data obtained from the strain gage output was consistent from all tests, both on single piles and pile groups, with the tip taking from 5 to 15 percent of the applied load with the

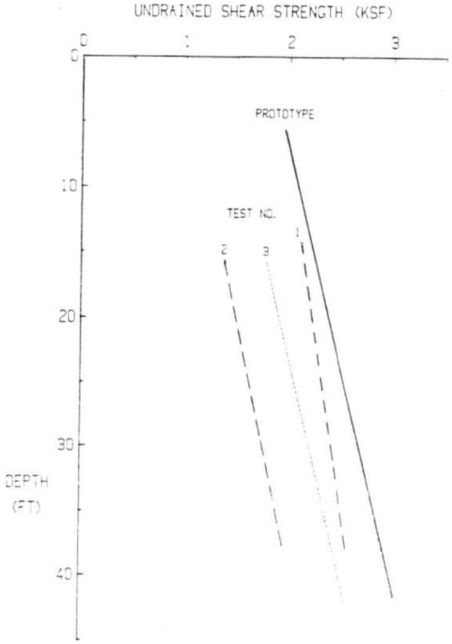

Figure 3. S_u profiles for initial tests (1 ksf = 47.9 kPa; 1 ft = 0.305m).

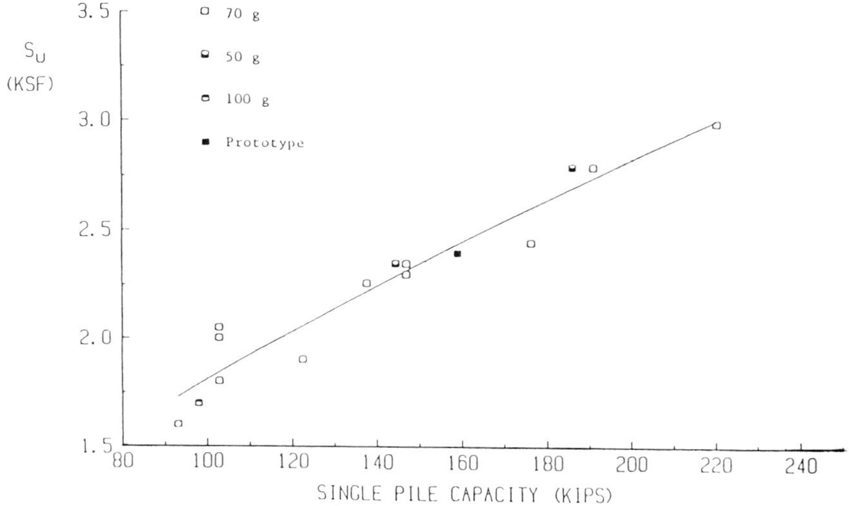

Figure 4. S_u vs. single pile capacity (1 KSF = 47.9 KPa; 1 KIP = 4.45 KN)

Table 2. GROUP TEST DATA, CLAY

Test No. g-level	17 70	18 100	21 100	22 70	Prototype 1
S_u at pile middepth, KSF	2.45	1.7	3.2	1.8	2.4
Single pile capacity, KIPS	176	98	330	104	160
Pile group capacity, KIPS	1520	850	3100	940	1419
Group factor	0.96	0.96	1.04	1.00	0.98

Conversion factors: 1 KSF = 47.9 KPa, 1 KIP = 4.45 KN

remainder of the load transferred fairly uniformly along the pile length. Figure 5 shows a typical model load transfer curve and the load transfer curve from the prototype single pile test. The difference in the two curves at the pile top may reflect the fact that the prototype piles were driven in holes which had been preaugered for a depth of 10 feet (3 m). Attempts to auger holes at the model scale were unsuccessful.

Load-displacement curves are presented in Figure 6 for single piles and in Figure 7 for group tests. The single pile model results show slightly stiffer behavior which may reflect the effect of not driving in preaugered holes as well as a slightly greater axial pile stiffness due to sizing the aluminum model piles to model the bending stiffness of the steel prototype pile. The influence of S_u on capacity is evident in the group load-displacement curves.

The pile prediction program PILGP1 developed by Ha and O'Neill (2) was used to predict model test results based on pile geometry and properties, soil properties and loading conditions. PILGP1 uses a finite difference approach to model single pile behavior and the elastic equations of Mindlin to compute group behavior.

Soil behavior for single piles is defined by the f-z curve which relates unit shear stress versus shear deformation along the pile sides and the Q-z curve which relates tip load to tip deformation. The load transfer data obtained at the 5 strain gage locations and the

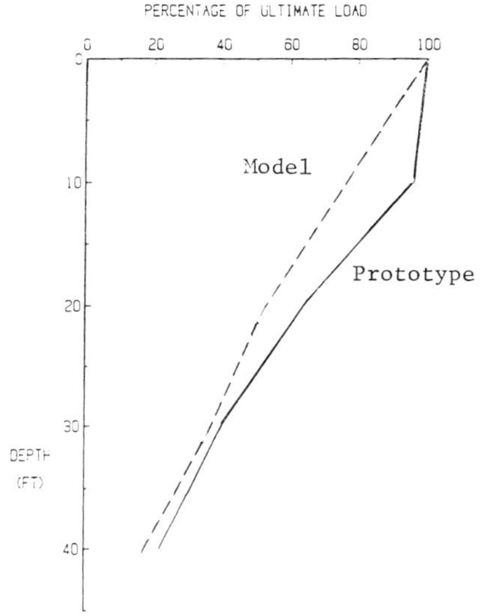

Figure 5. Load transfer curves from model and prototype tests in clay (1 kip = 4.45 KN; 1 ft = 0.305 m).

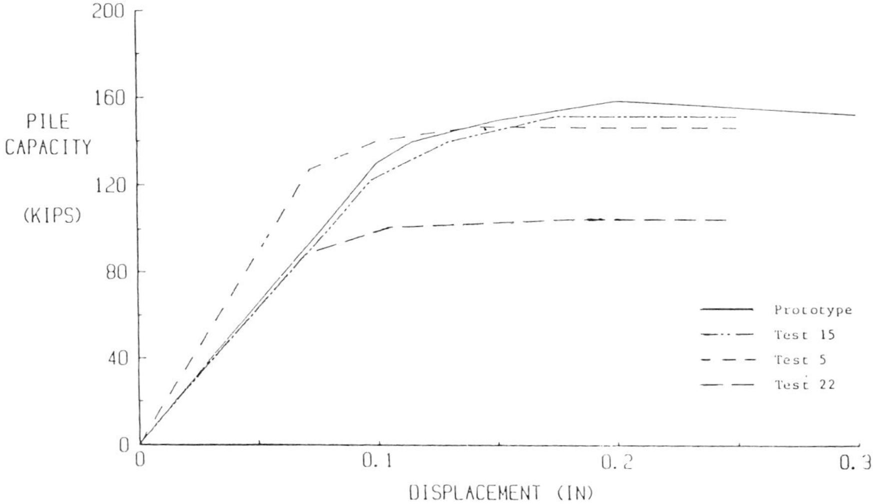

Figure 6. Load deflection curves, single pile tests in clay (1 kip = 4.45 KN; 1 in. = 2.54 cm).

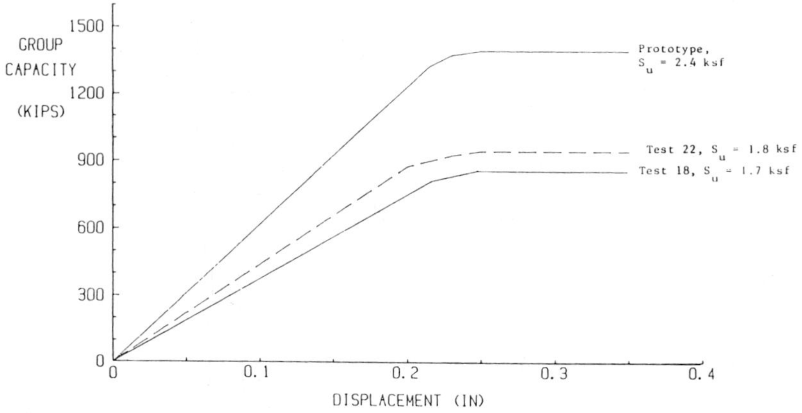

Figure 7. Load-deflection curves, group tests in clay, (1 kip = 4.45 KN; 1 in. = 2.54 cm)

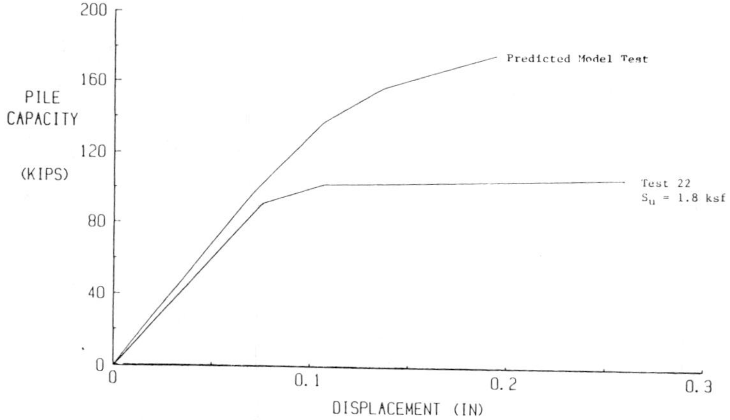

Figure 8. Single pile load-displacement as predicted by PILGP1 (1 KIP = 4.45 KN; 1 in. = 2.54 cm.).

load-deformation data at the pile top were used to determine the f-z and Q-z curves for the model pile tests.

Predicted results using model pile data for input to PILGP1 are presented in Figure 8 for single pile test and in Figure 9 for the pile group test.

Pile Tests in Sand

Soil Preparation and Pile Design

Previous experience with centrifugal testing of model piles in sand at the University of Colorado reported by Hougnon (3) showed a strong sensitivity of pile performance to soil preparation methods and soil uniformity. Since the prototype soil profile contained 5 different layers, (Figure 10), obtaining a representative sample was not possible. Another difficulty with using the prototype soil in the model tests was the need to remove coarser fractions prior to use in centrifuge model testing.

After reviewing of the prototype soil profile data, it was decided to fabricate a uniform cohesionless model soil with a dry unit weight of 100 pcf (15.89 kN/m^3) and an angle of internal friction, ϕ, equal to 40° from a commercially bagged, poorly graded, medium sand obtained from the FHWA Soil Laboratory, McLean, VA. The "raining" method was used to place the soil in the test container. The required density was obtained by properly controlling the rate of sand placement and the height of drop.

The prototype piles were naturally tapered wood piles, 40 ft (12.2 m) in length with a top diameter of 14 inches (35.6 cm) and a tip diameter of 10 inches (25.4 cm). Model piles were fabricated to a correct dimensional scale for 50 g, 70 g and 100 g scales using wood having the same modulus as the prototypes.

The top portion of the tapered wood model piles were strain-gaged to measure the applied load. Each pile was calibrated individually. An aluminum pile at a 70 g scale was fabricated with five levels of strain gages mounted internally. This pile, which had a uniform section with depth, was used to determine the load-transfer and tip loading of a pile in the test sand.

Single piles and the piles in the group tests were installed in-flight under full scale gravity conditions. A template on the soil surface was used to guide the piles and maintain proper spacing for the pile groups. After all piles in the 2 x 4 pile group had been separately driven, a rigid aluminum pile cap was installed to tie them together. This cap was made in three parts and was clamped to the model piles. After installation of the cap, the model piles were pushed in 0.2 inch (0.5 cm) as a group to minimize any disturbance produced by the pile cap installation.

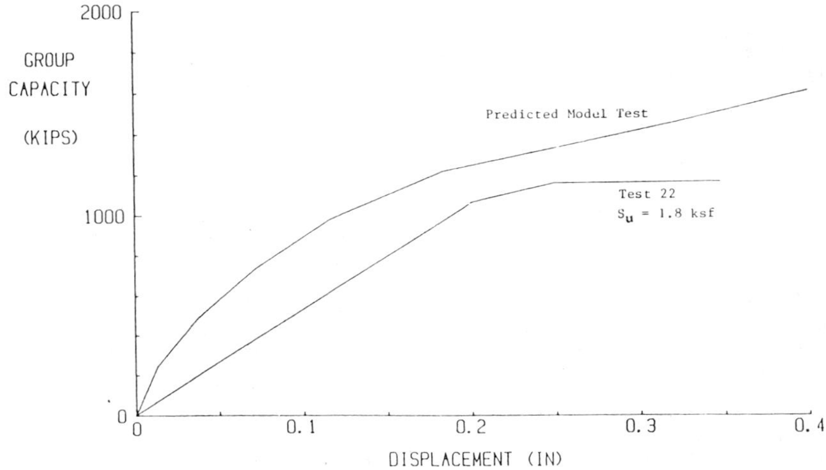

Figure 9. Pile group load-displacement as predicted by PILGP1 (1 KIP = 4.45 KN; 1 in. = 2.54 cm.).

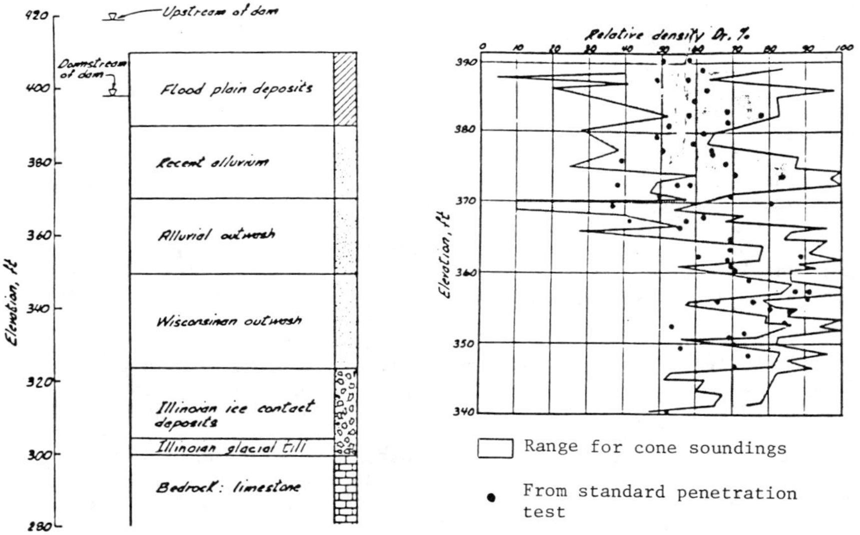

Figure 10. Site conditions, Lock & Dam No. 26 (From Woodward-Clyde Report).

Test Program and Results

A total of 28 centrifuge tests were conducted in the test program in sand. In addition to modeling the full scale test conditions, tests were conducted to study the effects of soil density, model scale, pile installation techniques, pile shape, and soil saturation.

The effect of in-flight installation vs. 1 g installation on single model piles was investigated at both 50 g and 70 g scale levels. Figure 11 shows the results of the 70 g test. The 50 g test showed similar behavior. These results show that it is necessary to have the proper simulation of the soil stress field both during the driving and testing of model piles. As it was necessary to stop the centrifuge to reposition the driving mechanisms for each pile in the pile group, a test series was conducted which showed that interrupting the centrifuge test had no effect on subsequent pile performance as long as both driving and testing were carried out at the elevated gravity levels.

To verify the internal consistency of the similitude relations over a range of model scales, pile tests were conducted at 50 g, 70 g and 100 g model scales. The load-settlement curves from these tests, converted to prototype scale, are shown in Figure 12. Five of the six curves are in good agreement, thus verifying the similitude assumptions.

Four group tests were successfully conducted. Driving load records from the strain gages at the top of the pile were obtained from two tests. Figure 13 shows one of these records where the influence of the driving order is clearly seen.

After installation of all piles and the pile cap, the individual pile loads were monitored as the pile group was loaded. Table 3 presents the individual capacities as well as the total group load. Although in all tests Pile No. 1 took less load than other piles, the influence of driving order is otherwise not pronounced. The group efficiencies computed for the four groups are presented in Table 3. Except for the test in the lowest density soil which indicated a group efficiency of 1.32, the tests all showed an efficiency of 1.15.

One single pile and one group test were conducted to investigate the effect of soil saturation on pile performance. Because of test problems, the measured load from the group tests were not reduced by the ratio of $\gamma_{buoyant}/\gamma_{dry}$. These problems, which included swelling effects on the wooden pile, were corrected and the test rerun on a single pile, Figure 14. The reduction of both capacity and stiffness is clearly seen.

The aluminum model pile was used to obtain the load transfer curve shown in Figure 15. Approximately 73 percent of the total resistance was provided by tip resistance. The load-transfer along the pile length is seen to be uniform.

The finite difference program PILGP1 developed by Ha and O'Neill

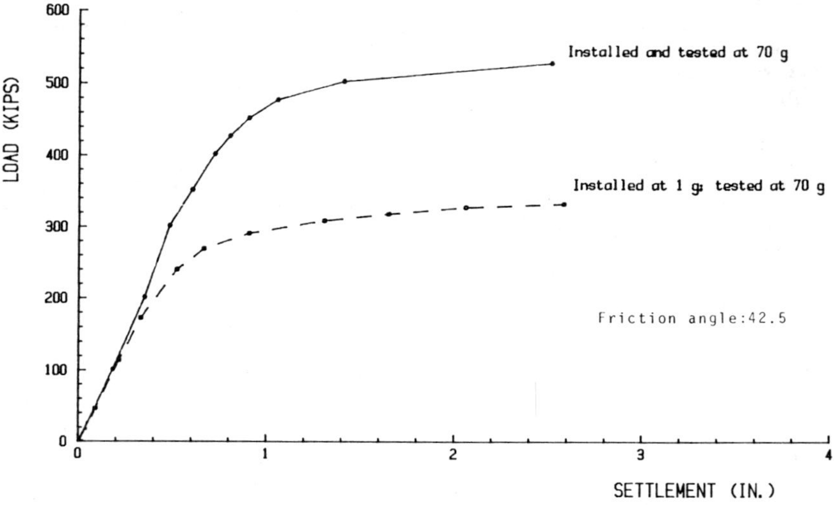

Figure 11. In-flight vs. 1 g installation in sand.

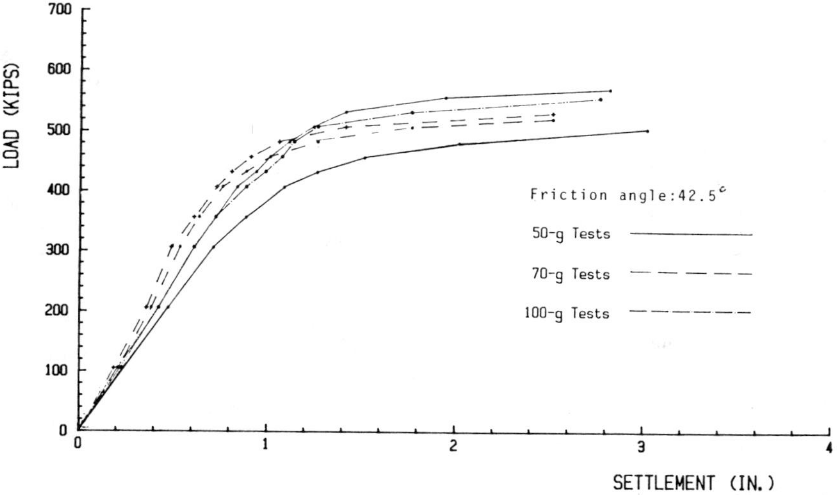

Figure 12. Load-seattlement curves in sand.

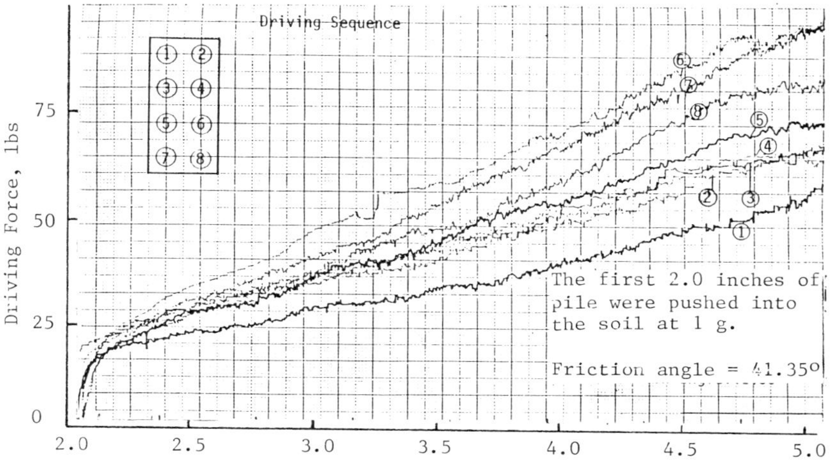

Figure 13. Driving Records for Test 5.2

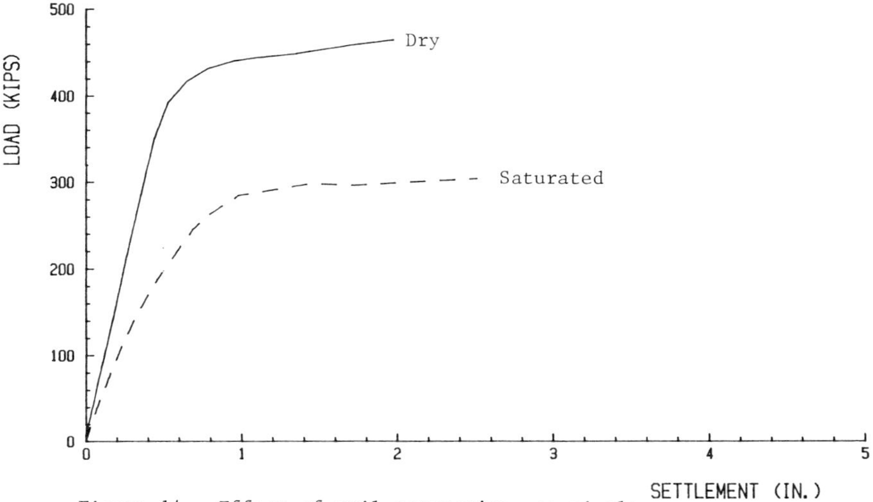

Figure 14. Effect of soil saturation on single pile load-settlement curve.

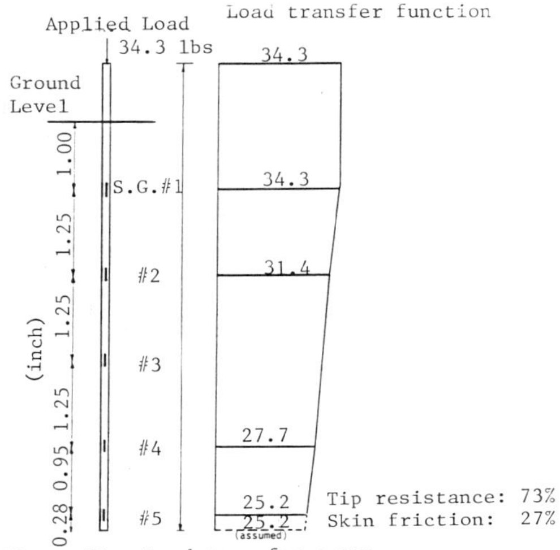

Figure 15. Load-transfer curve, aluminum pile in sand. (1 lb=4.448 N, 1" = 2.54 cm)

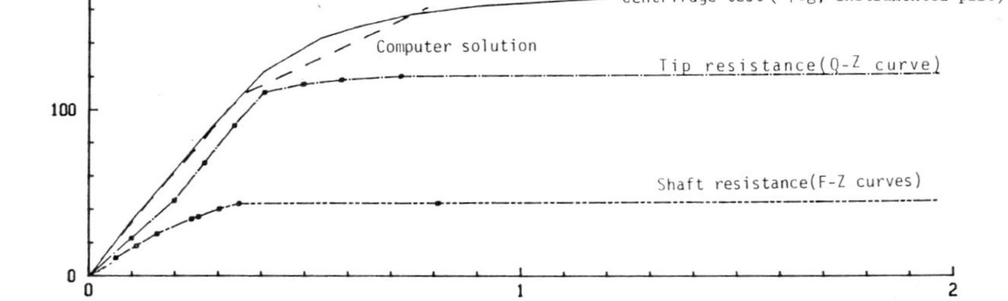

Figure 16. Measured model vs predicted single pile response in sand (prototype scale)

Table 3 RESULTS OF THE PILE GROUPS TESTS

	Individual Pile Capacities (lbs)			
Pile No	Test 5.1 ϕ=40.8°	Test 7.1 ϕ=40.8°	Test 5.2 ϕ=41.35°	Test 5.4 ϕ=42.5°
1	54.7	57.6	57.1	102.4
2	67.1	59.1	72.8	103.0
3	69.2	60.3	74.4	114.9
4	70.8	*	73.6	114.9
5	*	58.4	75.6	144.5
6	76.6	63.5	82.4	114.9
7	70.8	58.6	72.8	142.6
8	77.0	64.0	80.8	113.5
Total load from individual piles	-	-	590	952
Total load from load cell	551	482	600	1017
Computed group efficiency	1.32	1.19	1.15	1.15

* Gages did not work 1 lbf = 4.448 N

(2) was used to predict model pile performance. Because the available data from the aluminum pile test provided only ultimate values for side shear (f) and tip loads (Q), it was necessary to make a number of computer runs in which both the shape of f-z and Q-z curves and the magnitude of displacement at ultimate value, z_c, were varied. The best fit of predicted load and deformation to measured was obtained when z_c for the side shear stress (f) was 0.35 inch (prototype scale) while the value for the tip load (Q) was 0.65 inch (prototype

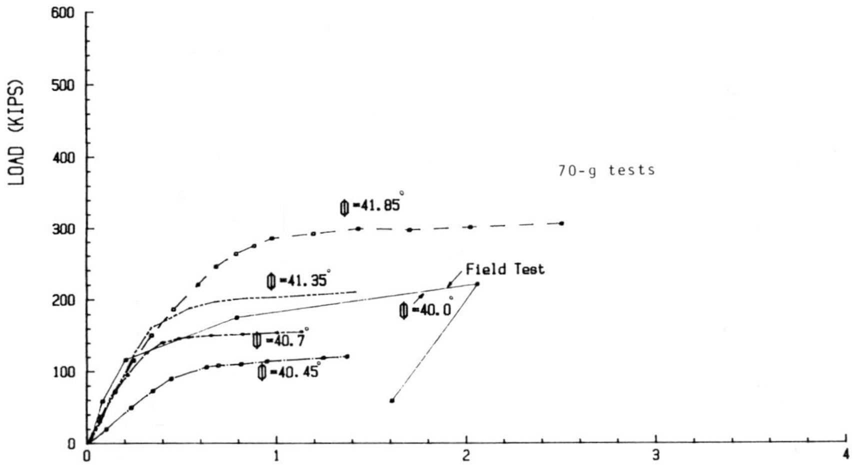

Figure 17. Comparison of extrapolated model response vs. prototype pile response in sand.

scale) or 6 percent of the pile tip diameter. The shape of the Q-z curve was modified from that recommended by Ha and O'Neill to obtain a closer fit. The same f-z curve was used for the full length of the pile except for the top 5.8 feet (prototype scale) for which zero shear strength was assigned. The measured model vs. predicted pile response is shown in Figure 16 where the Q-z and total shear force versus z curves are also shown. Good agreement with the initial pile response is achieved; however, the curves begin to diverge after the side shear has reached ultimate value.

The field test was conducted in saturated conditions in a soil mass with variable properties as shown in Figure 10. The measured ratio of 1.61 for dry to saturated load from the saturated single pile test was applied to the results of tests from the parametric series to obtain an estimate of saturated performance. These results are shown plotted in Figure 17 where the best fit is obtained for $\phi = 41.35°$. The extreme sensitivity of model pile capacity to the ϕ value is evident. The field pile was driven by a conventional pile driving hammer while the model pile was installed by a steady force or jacking procedure. It is speculated that this difference in pile installation may produce different types and magnitudes of soil disturbance in the cohesionless soil that may significantly affect pile behavior.

Conclusions

The results from both the sand and clay tests verify the assumed similitude relations as evidenced by the modeling of models results. This important conclusion gives the required credibility to the results obtained on pile behavior of the influence of various soil and pile factors. The modeling of models results when scaled to prototype scale gave a reasonable prediction of field test results considering the variable nature of the actual soil being modeled.

A pile behavior computer program was used to predict test pile behavior using load transfer data generated from the instrumented test piles. Only fair prediction of the full measured load-displacement curve was achieved even after several computer runs in which input quantities were varied to improve the prediction. It is the opinion of the investigators that given soil property data and soil profile but without field test results, predictions from centrifuge model tests would give results as accurate as, and very likely much more accurate than, those generated from computer predictions.

The experimental procedures developed and the verification established in this program open the door to investigating in controlled laboratory setting the many factors that influence pile performance. The centrifuge technique should also be useful and very cost-effective in establishing the predicted behavior and the sensitivity to design changes of pile foundation for large projects.

Acknowledgement

This research was supported by the Federal Highway Administration, Department of Transportation under Contract DTFH 61-81-R-00034.

References

1. Atkinson, J.H. and Bransby, P.L. (1978), The Mechanics of Soils, McGraw-Hill Book Company (UK), Ltd., London.

2. Ha, H.B. and O'Neill, M.W. (1981), "Field Study of Pile Group Action (Appendices), FHWA Report RD-81/03 and RD-81/004.

3. Hougnon, J.H. (1980), "Centrifugal Modeling of Axially Loaded Piles," M.S. Thesis, Department of Civil Engineering, University of Colorado, Boulder, Colorado.

4. O'Neill, M.W., Hawkins, R.A. and Mahar, L.J. (1980), "Field Study of Pile Group Action (Final Report)," FHWA Report No. RD-81/002.

5. Rocha, M. (1957), "The Possibility of Solving Soil Mechanics Problems by the Use of Models," 4th International Conference on Soil Mechanics and Foundation Engineering, 1, pp. 183-188.

6. Saffery, M.R. and Tate, A.P.K. (1961), "Model Tests on Pile Groups in a Clay Soil with Particular Reference to the Behavior of the Group when it is Loaded Eccentrically," 5th International Conference on Soil Mechanics and Foundation Engineering, 2, pp. 227-268.

7. Schofield, A.N. (1980), "Cambridge Geotechnical Centrifuge Operations," 20th Rankine Lecture, Geotechnique, 30, pp. 227-268.

8. Scott, R.F. (1979), "Centrifuge Studies of Cyclic Lateral Load-Displacement Behavior of Single Piles," Report to American Petroleum Institute, OSAPR Project 8.

9. Sowers, G.F. Martin, C.B., Wilson, L.L. and Faucold, M., "The Bearing Capacity of Friction Pile Groups in Homogeneous Clay from Model Studies," 5th International Conference on Soil Mechanics and Foundation Engineering, 2, pp. 155-157.

10. Whitaker, T. (1957), "Experiments With Model Piles in Groups," Geotechnique, 7, pp. 147-167.

11. Woodward-Clyde Consulting Engineers (1979), Phase 4, Vol. 3, Results and Interpretation of Pile Driving Effects Test Program, Existing Lock and Dam No. 26, Alton, IL.

PREDICTION OF SETTLEMENT OF SINGLE PILES IN CLAY

By Timothy B. Aschenbrener,[1] A.M. ASCE, and Roy E. Olson,[2] F. ASCE

ABSTRACT

Analyses by trial were used to fit theoretical load-settlement curves to measured curves for driven piles.

For the "elastic" method, analyses of load tests of best quality indicated that designers should use $E/c_u = 750$ for analyses in the working stress range and set limits on the local side shear in accord with the best currently available method for analysis of capacity.

For the "load-transfer" method, the scatter in data was such that no effort was made to include effects of load-shedding (reduction in local pile/soil shearing stress after failure) nor to use generally non-linear shapes for f-x curves. Instead, a simple bilinear relationship was selected (elasto-plastic response). The best fits of measured and computed curves in the working stress range occurred when the tip settlement at failure was about one percent of the tip diameter and the side shear was fully developed at a movement of about 0.1 inch.

INTRODUCTION

Prediction of the response of a single friction pile to axial loading involves an analysis of a soil-structure interaction problem. Under ideal circumstances, the designer would like to predict the entire load-settlement curve, as well as the rate of load transfer from the pile to the soil as a function of depth. Under the conditions of existing technology the designer will generally be satisfied with a prediction of the failure load and an estimate of the settlement under working loads. We are here concerned only with the prediction of settlements.

Although simple rules of thumb can be developed for estimating settlement under working loads (23), a more general analysis must be of the soil-structure interaction type. Such analyses have been of three types: (1) elastic, (2) load transfer, and (3) finite element. As a result of their comparative simplicity, and the availability of appropriate computer programs, the elastic and load transfer methods have been used widely. Finite element methods, although potentially more general, have not been used extensively and will not be considered here.

---

1 Graduate Student, Department of Civil Engineering, University of Texas, Austin, Texas

2 Professor, Department of Civil Engineering, University of Texas, Austin, Texas

In this paper, the elastic and load transfer methods will be described and then data from load tests will be used to evaluate parameters needed in the analyses as well as to develop measures of the scatter between measured and predicted settlements.

METHODS OF ANALYSIS

Elastic Method

As the term is used here, the "elastic method" refers to a group of methods in which the pile is subdivided into elements, the load transferred to the soil by each element is idealized, the soil is treated as a homogeneous, isotropic, weightless, linearly elastic, half space, and stresses and displacements in the soil are calculated using Mindlin's (10) equation. Displacements of the pile and soil depend on the depth-dependent rate of load transfer from the pile to the soil. A correct solution is assumed to have been attained when the vertical displacements of the pile, calculated for a particular variation of pile load with depth, match the calculated displacements of the soil at the same depths, where the soil is loaded at various depths by forces transferred from the pile.

The idealization used for the loads transferred from the pile to the soil include single point loads at the center of each element (2, 16, 20), line loads applied around the circumference of each pile element at its mid-depth (11), and stresses distributed over the surface of each pile element (9, 14, 15).

We assumed uniform stresses over the surface of each element and performed the calculations using a program TAPILE written by H.G. Poulos. A limit can be set on the load transferred by any element (14). A limiting load is reached at any depth, application of more load to the pile causes loads to be transferred to the soil at greater depths. When every element has transferred its peak load the pile plunges. There is no provision for load shedding (reduction in load transfer after failure in any one element). For layered systems, Young's modulus for the equivalent homogeneous soil was taken as the average of the value where the load is transferred and where the deformation is calculated (13) and thus the values of E used in the analysis are depth dependent. A simplified procedure based on a method originated by Steinbrenner (18), is used to account for the presence of a layer whose upper boundary is below the pile tip. Tapered piles are assumed to be step tapered and the load transferred on the bottom of each element is taken into account separately from the load transferred in side shear.

The elastic method, as it was used here, clearly involves substantial simplification of actual conditions. The major source of error, however, may be less in the simplifying assumptions of the analysis than in the problem of defining soil properties (Young's modulus and limiting side shear between the pile elements and soil).

Load Transfer Method

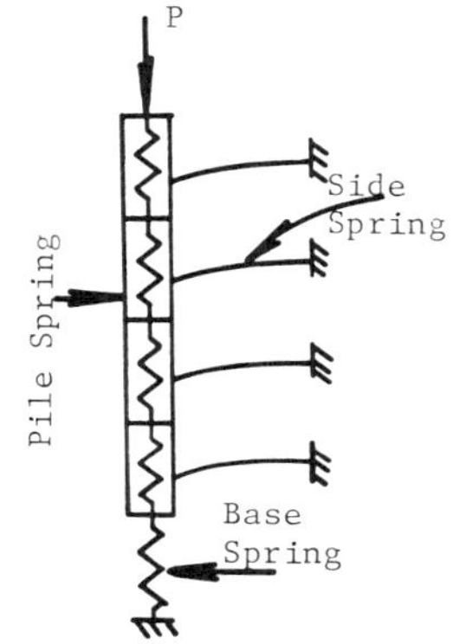

Fig. 1 Pile-Soil Model for the f-x Approach

In the load-transfer method, the pile is idealized as a series of elements, each of which is an elastic spring (Fig. 1). Each pile spring element interacts with pile spring elements above and below, and with a side spring, which represents the soil. The soil spring applies a vertical force, F (or stress f), to the pile element which is a function only of the vertical displacement, x, of the element. The F-x curve can be non-linear. The side soil springs are assumed to be fixed at their outer extremity, thus making each soil layer independent of all other layers. The soil under the pile tip is also represented by a non-linear spring.

The analysis is performed in a series of steps: (1) Assume a tip displacement and calculate the resulting tip force. (2) Assume the pile element (initially the bottom element) is rigid and calculate the side shear on it from the the displacement and the side F-x curve. (3) Calculate a force in the top of the pile element as the sum of the tip and side forces. (4) Calculate a new average displacement as the sum of the tip displacement and half of the elastic compression of the element. (5) Calculate a new side force corresponding to the revised displacement. (6) Return to step 3 and cycle through steps 3 through 5 until only a negligible change occurs for any iteration. (7) Move up to the next pile element above the base and return to step 2. (8) When the top element has been reached and iterations have ceased, calculate the top load and displacement and return to step 1. (9) When a complete load-settlement curve has been calculated at the pile top, the analysis has been completed.

The load-transfer method was originated by Seed and Reese (17) and extended by Coyle and Reese (1). It has the advantages of simplicity, rapid execution of computer programs, and ease in accounting for non-linear responses and layered systems. The major defects in the method are that it is empirical and does not involve direct use of measured stress-strain properties of the soil, and that displacements in upper soil layers due to movements in lower layers are not taken into account.

For this study, analyses were performed using a computer program PILSET which allows for f-x curves to be input as hyperbolas, or input point by point, with separate f-x curves for each soil layer. In the discussion to follow, the designation "f-x" will be used as a general name for this method; the definition of the axes of the f-x curves will vary.

DATA SET

A data set of pile load test results was set up by Olson and Dennis (12) and subsequently corrected and updated. From this set of 102 references and 1000 load tests, a subset was selected involving nine references (fifteen load tests) on instrumented piles, in compression, in clay. For each reference, one load test was selected initially, and a trial-and-error procedure was used to find the soil properties that would yield the best correlation between measured and computed load-settlement curves. An effort was made to express these soil properties in a general form and the resulting parameters were then used with a larger set composed of 68 tests at 26 sites, from 24 references, for piles in compression, in clay, to obtain an idea of the probable accuracy of the methods.

STATISTICAL METHODS

Scatter in a single parameter can be represented by relatively simple parameters which depend on the distribution of data, e.g., a coefficient of variation may be used for a normal distribution. When the scatter in the relationship between two variables is desired, parameters like a standard error of estimate or correlation coefficient (linear relationship) may be appropriate. What is needed here is a parameter that gives a single valued measure of the scatter between two non-linear relationships. For continuous functions, f(x) and g(x), such a measure of scatter is the "metric" (also termed the Euclidean norm, L2 norm, and H_o norm). The metric M is defined as (6):

$$M = [\int \gamma(x) \left| f(x) - g(x) \right|^2 dx]^{\frac{1}{2}} \qquad (1)$$

where $\gamma(x)$ is a weighting function.

In the present case, f(x) is the measured load-settlement curve and g(x) is the calculated curve. The weighting function was defined as $(1/f(x))^2$. The metric then represents the square of the area between the measured and computed load-settlement curves divided by the square of the area under the measured curve.

A value for the metric can be defined from the origin out to any load. As a matter of convenience, we defined metrics at different factors of safety, and defined failure as occurring when the gross butt settlement of a pile, S, is given by (3):

$$S = \frac{PL}{AE} + C + 0.01D \qquad (2)$$

where P is the applied load, L is the total pile length, A is the average pile cross-sectional area, E is Young's modulus of the pile, C is the settlement needed to develop side shear (taken as 0.15 inch), and D is the tip diameter.

In attempting to predict load-settlement curves at design time, there are two coupled sources of error, one is in the prediction of the appropriate dimensionless relationship between stress and strain (the actual variables depend on the selected method of settlement analysis), and the other is in predicting the failure load. For example, if the F-x method is used, the engineer may first select a curve of c_d/c_a vs x, where c_d is developed pile/soil adhesion, c_a is the adhesion at failure, and x is movement, and then select the value of c_a. It is convenient to uncouple the problem here by selecting values of c_a by trial so the predicted and measured failure loads coincide, and attention can be concentrated on the settlement prediction alone (see later more specific comments for each method).

Metrics can be evaluated up to any desired load. Values were calculated at failure and at factors of safety of two and three. To obtain a feel for the relationship between the numerical value of the metric and the scatter between measured and computed load-settlement curves, curves with a range in scatter are presented in Fig. 2. Qualitatively, correlations between measured and computed load-settlement curves appears to be excellent for metrics less than 0.5, very good up to 1.0, good up to 1.5, fair up to 2.0, and poor beyond.

A simpler measure of scatter would be the ratio of calculated to measured settlement, S_c/S_m, but this value gives less evidence of overall scatter. To obtain an idea of the relationship of metrics to scatter in S_c/S_m, values of $|S_c-S_m|/S_m$ at a factor of safety of two were summarized in terms of the metric (Table 1).

Table 1 Comparison of Measured and Computed Settlements at a Factor of Safety of Two, for a Range in Values of Metrics

Range in Metric	|Sc-Sm|/Sm, % Average	Range	Standard Deviation	Number of Observations
0 - .50	5	0-16	5	12
0.51- 1.00	16	0-33	9	29
1.01- 2.00	30	4-83	22	20
2.01-12.00	66	29-96	20	15

ANALYSES USING ELASTIC METHOD

A set of trial analyses was first performed using various numbers of pile elements. It was found that use of elements with lengths equal to five times the diameter led to stable solutions and acceptable execution times. For the pile elements, Young's modulus (E) was assigned the following values: 29×10^6 psi (200 G Pa) for steel, 4.5×10^6 psi (31 G Pa) for reinforced concrete, 3.0×10^6 psi (20 G Pa) for unreinforced concrete, and 1.7×10^6 psi (12 G Pa) for timber. The source documents did not provide information that would yield test-specific values of E. An equivalent E was used for concrete-filled steel pipe piles. Raymond step-tapered piles were considered unreinforced. By fixing the values of E for the piles, the differences between measured and computed settlements was attributed to errors in E of the soil. This assumption is not entirely realistic but data did not exist to allow differentiation of the pile and soil values of E and the soil values seemed more uncertain.

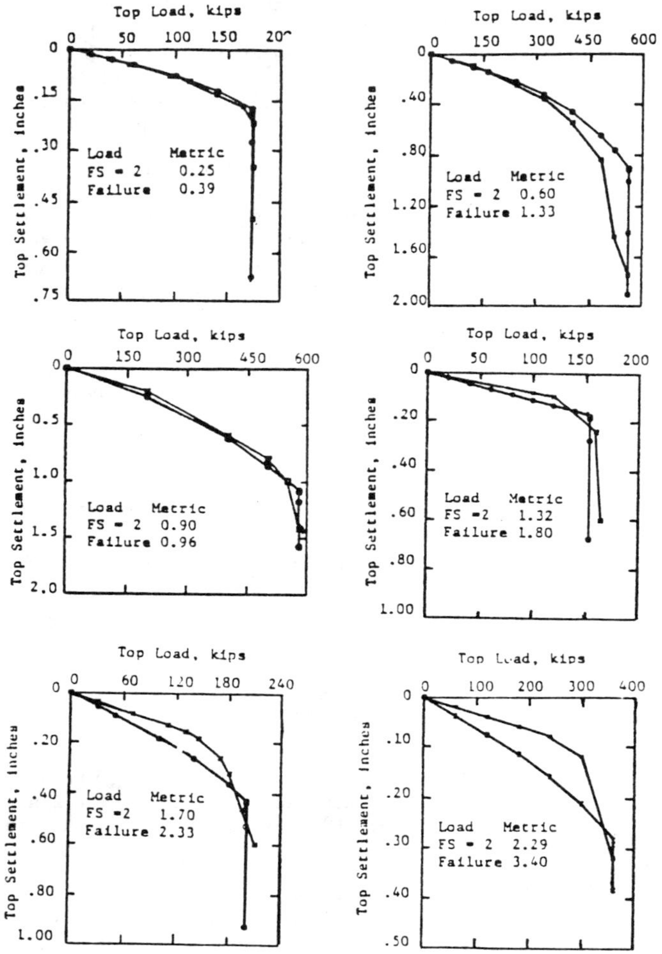

Fig. 2 Measured and Calculated Butt Load Versus Butt Settlement Curves with Various Metrics (Note: 1 inch = 25.4 mm, 1 kip = 4.4 kN)

Of the 67 tests analyzed, 44 (66%) were for steel pipe piles, 12 (18%) were for cast-in-place concrete piles, 7 (10%) were for precast concrete piles, 3 (4%) were for timber piles, and 2 (3%) were H piles.

The required soil properties are Poisson's ratio and Young's modulus. In all analyses, Poisson's ratio was set at 0.45. No values for Young's modulus were available from the original reports. Accordingly, for each load test, an attempt was made to vary E for each layer until the best fit (smallest metric) was obtained between the measured and computed load-settlement curves in the working stress range (around a factor of safety of two). However, there were insufficient data, in general, to find a unique set of values of E. To simplify the analyses, the decision was made to assume a constant value of E/c_u (dimensionless) for all layers for any one test and thus reduce the problem to a trial analysis for the appropriate E/c_u value.

Layers of sand were encountered along the side and at the tip of the pile in some cases. When layers of sand along the side of the pile were encountered, a test was included only if the calculated capacity due to the sand in side shear was less than 25% of the total calculated capacity, and the thickness of the sand above the pile tip was less than 30% of the depth of penetration of the pile. Side shear stresses in the sand were approximated using (12):

$$f_s = 0.58 \log N$$

where f_s is the side shear at failure in units of kips per square foot, and N is the standard penetration resistance in blows per foot (not corrected for overburden pressure). The values for E for the layers of sand was determined from E/f_s equal to a constant (same constant as for clays).

There are three sites where one or more piles encountered sand at the pile tip (Alsancak, Morganza, Donaldsonville). The limiting stress in the tip, q_p, was estimated using:

$$q_p = F_D \bar{\sigma}_{vo} N_q$$

where F_D is a depth correction factor (12), $\bar{\sigma}_{vo}$ is the free-field vertical effective stress at the depth of the pile tip, and N_q is a bearing capacity factor (40 for dense siliceous sand). The values of E for the sand at the tip of the pile were estimated based on correlations published by Sykora and Stokoe (19). For the three sites with sand at the tip of the pile, the standard penetration resistances were 40 to 50 blows/foot (130-160 blows/meter) and the average E was taken to be 6000 ksf (300 MN/m^2).

The values of E/c_u that led to the smallest metric at a factor of safety of two were converted back to values of E, and plotted versus c_u (Fig. 3). The undrained strengths were all measured using unconfined compressive, or unconsolidated-undrained triaxial, tests and no factors were used to correct the strengths to one standard. In cases involving multilayered soils, a weighted (by layer thickness) average E

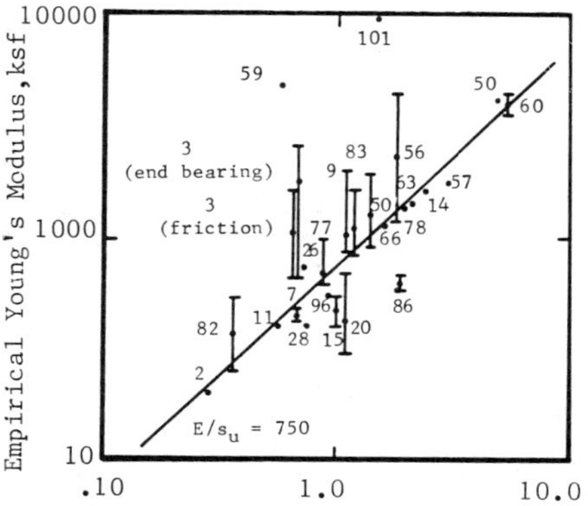

Fig. 3 Correlations Between Empirical Values of Young's Modulus and Undrained Shearing Strength
(Note: 1 ksf = 50 kN/m^2)

and c_u was plotted to avoid having multiple points for a single test. In cases where there were multiple load tests in a single profile, the range in data is shown by a vertical line with end tick marks and a symbol at the average E. The values of E/c_u range from 300 to 8500.

The quality of the data incorporated into the data set ranged widely. We assigned each test a data quality factor (DQF) which ranged from 1 to 5. Tests with low values of DQF often lacked adequate soil or pile data, had set-up times of a week or less, had been tested to failure before, or had other known defects. Tests in Fig. 3 with low values of DQF included 59 and 101 (uncertain wall thickness and set-up time), and 86 (pile tilted during driving and was pulled back into a vertical position after much of the pile set-up had occurred). If data for these three sets of test are eliminated, the range in E/c_u is 300 to 4000.

Some of the other tests were unusual as well. The tests at Empire, La. (No. 56) involve 14-inch (360 mm) diameter pipe piles threaded through 18-inch (460 mm) pipe piles which had been driven and cleaned out. The set-up time was more than a year. The piles were all loaded to failure several times in both compression and tension prior to the tests used here. The large scatter in values of E/c_u may have resulted from the unusual testing procedure.

Three tests were performed at the site marked No. 83 (location confidential) on pipe piles driven open ended and filled with concrete down to the soil plug. For two tests where the depth to the plug was known, E/c_u = 700. For the third test the depth was uncertain. If the depth included in the source was used then E/c_u = 1375 but if the plug was deeper, the E/c_u would approach 700.

For the site at Donaldsonville, La. (No. 9) a pile with its tip in a soft clay, just above a dense sand, had less settlement in the working stress range than another pile which was driven into the supposed supporting layer (dense sand). This unlikely circumstance implies the presence of an uncertain stratigraphy.

Fourteen load tests were performed at Morganza (No. 3). The data showed considerable scatter. Piles were driven with their tips either just above a supporting sand layer or with the tips in the sand but some piles with tips in the upper soft clay compressed less than piles with tips in sand. These tests were unusual in that the load test was performed by successively loading to higher loads but completely unloading in between the increments of load.

If data for cases 3, 56, 59, 86, and 101 are eliminated, and only the average E and c_u are used for each case, the average E/c_u is 740 with a standard deviation of 170 and range from 380 to 1100. A higher value of E/c_u is obtained if all tests are included.

It appears that use of an E/c_u of about 750 is in general accord with the best data. The 68 tests were reanalyzed with $E/c_u = 750$ and the range in the ratio of measured to calculated settlements at a factor of safety of two was 0.38 to 3.8 with an average of 1.06. These data tend to be dominated by a few sites where multiple tests were performed. If average data for each site are used, the range is 0.53 to 2.20 with an average of 1.07 (Table 2). Use of $E/c_u = 800$ would lead to an average ratio of measured to calculated settlements of 1.00.

The metrics (Table 2) tend to be rather large, in part because of use of a value of E/c_u that was smaller than the mean value for the whole group, because of the tendency of the elastic method to produce a bilinear Q-S curve whereas measured relationships are curved, and because some piles underwent much less settlement than would be predicted using E/c_u of even say 800 (Morganza, Terrebonne, Texas Gulf Coast) or much more settlement (Alsancak, Toronto, Confidential).

A frequency curve using site-average values of the ratio of measured settlement (S_m) to the settlement calculated using $E/c_u = 750$ (S_c), at a factor of safety of two, is shown in Fig. 4. Use of averages for each site has the advantage of not weighting the data heavily toward a few sites where multiple tests were performed but it diminishes the apparent scatter and precludes estimating probable errors for individual piles.

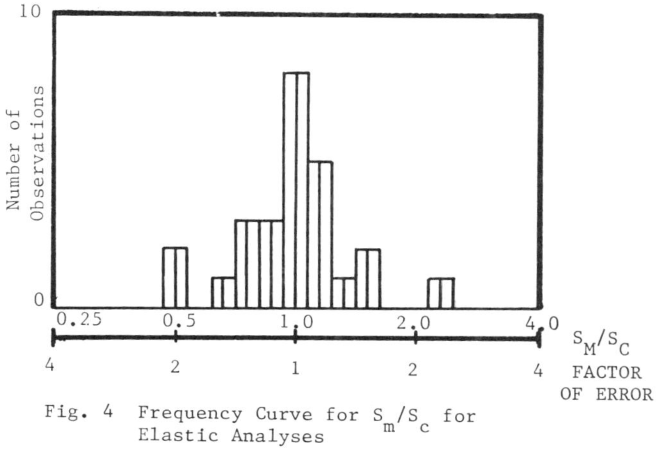

Fig. 4 Frequency Curve for S_m/S_c for Elastic Analyses

Table 2 Summary of Average Parameters Calculated using the Elastic Method

Site	Ref.	N	E/c_u	S_m/S_c at $F_s = 2$	Metric @ $F_s = 2$	Metric @ Failure
San Francisco	2	1	710	1.03	0.31	0.72
Morganza, La *	3	7	3910	0.80	1.34	1.69
Morganza, La **	3	7	5830	0.76	2.62	2.62
New Orleans, La	7	2	660	0.92	0.67	1.45
Donaldsonville, La	9	4	950	1.02	1.10	1.55
Toyko, Japan	11	1	700	1.00	0.79	1.30
Cleveland, Ohio	14	2	700	1.35	0.93	1.30
Alsancak, Turkey	15	2	480	1.28	0.56	2.89
Toronto, Ontario	20	6	380	2.20	3.88	5.69
New Orleans, La	26	1	1100	0.67	1.70	2.33
Omaha, Ne	28	1	530	1.14	0.73	1.11
Burnside, La	50	3	910	1.55	1.16	2.37
Argonne, Il	50	1	820	1.00	1.00	2.47
Empire, La	56	4	1300	0.80	1.42	1.79
Kontich, Belgium	57	1	600	1.15	0.66	2.07
Terrebonne, La	59	4	8020	0.49	3.96	5.54
Fulton, Ms	60	4	750	1.07	0.84	1.41
San Francisco, Ca	63	2	690	0.91	0.74	1.10
South Dakota	66	1	700	1.05	0.28	1.29
Houston, Tx	78	1	700	1.09	0.25	0.39
Hamilton AFB, Ca	82	3	1030	0.77	1.54	2.28
Confidential	83	3	930	1.10	0.71	1.23
Confidential	86	2	320	1.66	1.72	2.25
Alexandria, Va	96	1	700	0.83	0.31	0.74
Texas Gulf Coast	101	1	8000	0.53	3.34	4.92
Memphis, Tn	77	3	775	1.27	0.89	3.02

* piles with tips in clay
** piles with tips in sand

ANALYSES USING THE LOAD-TRANSFER METHOD

Tip Load-Settlement Curve

Data could be found for only five sites (Fulton, Houston, Omaha, San Francisco, Tokyo) where reasonably complete tip load-settlement curves could be determined and four more sites where tip loads and settlements at failure were known (Table 3). Dimensionless load-settlement data are presented in Fig. 5. Major sources of scatter are believed to include: (1) local variation in soil properties, (2) variable degrees of prestress of the pile tips, (3) presence of some open-ended pipe piles, and (4) instrumentation errors. There is, perhaps, some evidence that smaller tip deformations occurred for the open-ended pipe piles but calculations indicated that these piles should have been plugged. Considering the lack of data, the scatter, and the fact that the tip stress represented only about 7% of the pile capacity, the f-x curve for the pile tip was approximated as bilinear with a settlement at failure of 1% of the tip diameter.

Table 3 Data on Tip Loads for Instrumented Piles

Site	Ref. No.	N_c	c_u ksf	Percentage of Total Load in the Tip	Tip Settlement at Failure as a Percentage of Pile Diameter
Confidential	86	16	2.50	6.7	2.4
Empire	56	16	1.00	7.5	0.6
Fulton	60	20	10.00	6.9	1.5
Hamilton AFB	82	12	0.95	5.3	7.7
Houston	78	23	2.00	9.8	1.0
New Orleans	7	0	0.60	0.0	1.2
Omaha	28	12	0.95	5.1	1.3
San Francisco	2	16	0.28	13.0	1.0
Tokyo	11	4	0.51	3.7	1.6

Note: 1 ksf = 50 kN/m^2

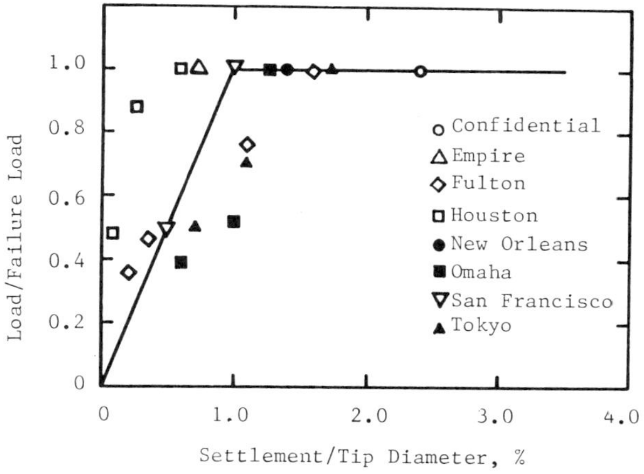

Fig. 5 Dimensionless Tip Load Versus Tip Settlement Curves

Side Load-Settlement Curves

Because the load-transfer method utilizes curves for each layer independently, it made sense to try to develop curves of side shear versus vertical movement for each layer for the tests on instrumented piles. It is convenient to use the name "f-x" for these curves where f may be the side force per pile element, side shearing stress, or a dimensionless force or stress, and x may be either settlement or some form of dimensionless movement. In developing f-x curves, the curves of load in the pile (Q) versus depth (z), can be used to determine the local pile/clay developed adhesion (c_d):

$$c_d = \frac{1}{C}\frac{dQ}{dz}$$

where C is pile circumference. Curves of Q versus z generally exhibit a zone of low load transfer near the top (Fig. 6), apparently caused by poor contact between the pile and the clay due to cyclic lateral deflection of the pile during driving. For the five tests using well instrumented piles, this top zone had a depth of about 1.2 c_u/γ where c_u and γ are the undrained shearing strength and total unit weight of the upper soil, respectively. In this zone, the side shear at failure was only about 50% of the side shear based on an average value of α where:

$$\alpha = c_a/c_u$$

and c_a and c_u are the average pile/soil adhesion, at failure and soil undrained shearing strength, respectively. However, attempts to use such results with the larger data set were unsuccessful, and the decision was made to treat all soil layers the same regardless of their location along the side of the pile.

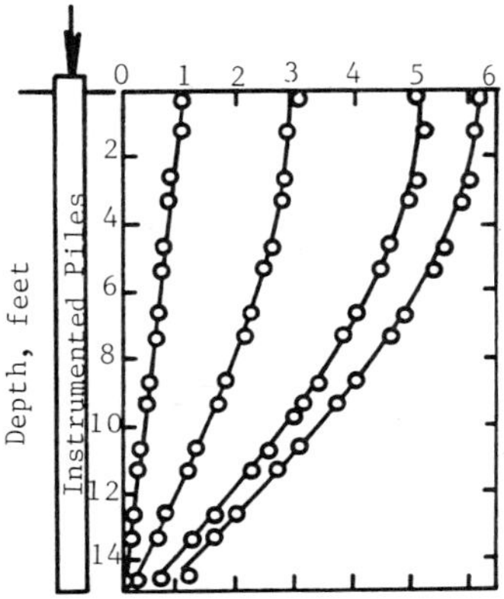

Fig. 6 A Set of Load Distribution Curves Obtained from a Pile in San Francisco (Ref. 2) (Note: 1 kip = 4.4 KN)

Curves of load in the pile versus depth were first used to calculate curves of developed adhesion (c_d) versus vertical movement (Δz), the "f-x" curves. The f-x curves were then used in a computer program, PILSET, to calculate curves of butt load (Q) versus butt settlement (S). Small differences between computed and measured Q-S curves were minimized by adjusting the f-x curves. It should be noted that some of the data sets included only one or two Q-depth curves, thus yielding only one or two points on the f-x curves.

It is convenient to non-dimensionalize the developed adhesion. When an α method of analysis is used (12, 21, 22), the dimensionless shear stress is logically $c_d/\alpha c_u$, where c_d is the developed soil/pile adhesion at any given deformation, c_u is undrained shearing strength, and α is here taken as the value for this particular load test at failure:

$$\alpha = \frac{Q_m - Q_p}{\Sigma c_u A_s}$$

where Q_m is the measured pile capacity, Q_p is measured (if available) or computed tip capacity, A_s is pile surface area in any layer, and the sum is performed over the depth of contact between the pile and the soil. When Q_p was calculated, we followed convention and used $N_c = 9$ rather than the generally higher values indicated in Table 3.

It next became necessary to select a shape for the f-x curves. In the program PILSET, the f-x curves can be input as any non-linear or linear relationship, or can be input as a hyperbolic function. The fact that the Q-S relationship is usually distinctly curved, is an indication that the f-x relationships are also curved. However, the Q-S curves come in a variety of shapes and substantial scatter was observed in all plots based on attempts to apply any simple relationship to a large collection of data. Consequently, the decision was made to use bilinear f-x curves and reduce the problem to finding the deformation at failure, Δz_f. Consideration was given to providing for a moderate decrease in side shear after failure (7). Most deformation-controlled load tests indicate that such "load shedding" (reduction in local c_d after the local peak) occurs. However, the amount of supporting data is small, some data indicate no load shedding, and the scatter in data is large enough that load shedding becomes a minor effect overall, and thus it was not included.

Trial and error analyses were performed with the set of 68 load tests until the value of Δz_f was found that yielded the smallest metric at a factor of safety of two. Values of Δz_f ranged from 0.05 inch (1 mm) to .27 inch (7 mm) (Fig. 7) and averaged 0.088 inch (2.2 mm). Values of Δz_f tended to be the mirror image of the values of E found in the elastic analyses. Some outlying values were partially from tests of low data quality factor. Some of the data may have scattered because of the presence of stiff underlying layers, the effect of which is not taken into account in the f-z approach. Some scatter is probably due to variable depths of lateral shaking of the pile during driving. Some of the data may be in error due to problems in data collection, publication, or interpretation. Numerous other sources of scatter can be imagined. No dominant source of scatter could be isolated. Consequently, analyses were performed using $\Delta z_f = 0.1$ inch, a value in keeping with other applications as well, e.g., wave equation analyses (4, 5, 8).

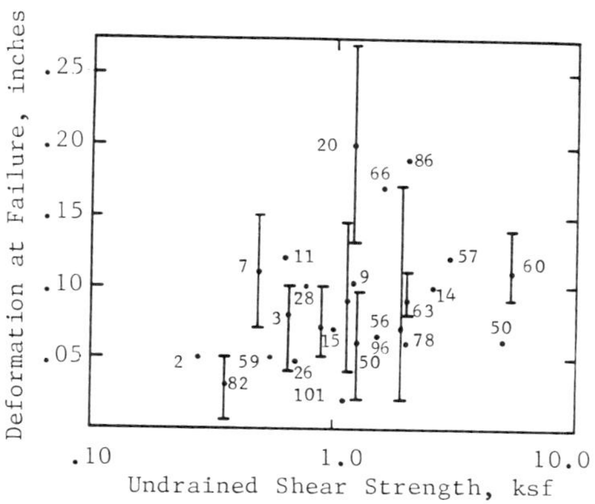

Fig. 7 Correlation Between the Deformation at Failure Versus the Undrained Shearing Strength.

Metrics and settlement ratios are presented in Table 4. The metrics are generally higher for the f-x approach than when the

Table 4 Summary of Average Parameters Calculated using the Load Transfer Method

Site	N	Metric FS = 2	Metric Failure	S_m/S_c
San Francisco, Ca	1	2.16	3.10	.53
Morganza, La *	7	1.30	1.85	.81
New Orleans, La	2	1.32	1.47	1.02
Donaldsonville, La	2	2.60	1.84	.81
Tokyo, Japan	1	1.83	2.33	1.23
Cleveland, Ohio	2	0.78	1.08	2.12
Alsancak, Turkey *	1	1.23	2.03	.87
Toronto, Ontario	6	2.20	4.33	1.48
New Orleans, La	1	.64	1.06	0.80
Omaha, Nebraska	1	.66	.99	1.20
Burnside, La	3	1.58	2.57	1.13
Argonne, Il	1	1.09	2.34	.73
Empire, La	4	2.20	2.60	.93
Kontich, Belgium	1	.85	1.06	1.26
Terrebonne, La	4	3.61	4.51	.47
Fulton, Ms	4	0.92	1.15	.89
San Francisco, Ca	2	0.86	1.03	1.02
South Dakota	1	1.75	3.12	1.30
Memphis, Tn	3	.70	2.60	1.29
Houston, Tx	1	1.51	1.96	.75
Hamilton AFB, Ca	3	3.19	5.23	.35
Confidential	3	1.79	2.22	.90
Confidential	2	2.02	2.54	1.49
Alexandria, Va	1	.32	.80	.84
Southeast Texas	1	3.00	3.83	.49

* piles which were driven into sand at these sites were not analyzed with the f-x method

elastic method is used. The frequency curve of average settlement ratios (one value of S_m/S_c at a factor of safety of 2), shown in Fig. 8, also indicates more scatter in the f-x predictions.

ESTIMATION OF SETTLEMENTS WITHOUT ANALYSES

For those who may lack the time, funds, or inclination, to perform computer based analyses, it may be of interest to know what the distribution of actual settlements was. Unfortunately, we did not evaluate and save the settlements for points in a working stress range but we did store settlements at a load of 95% of the plunging failure load. These settlements were selected for each pile type separately for piles that had tips in clay and more than 80% of the side shear in clay. The data are summarized in Table 5. The data are perturbed by a few tests where piles were loaded to quite large settlements without actually plunging. Hcwever, the weighted average mean settlement is only 0.7 inch (20 mm).

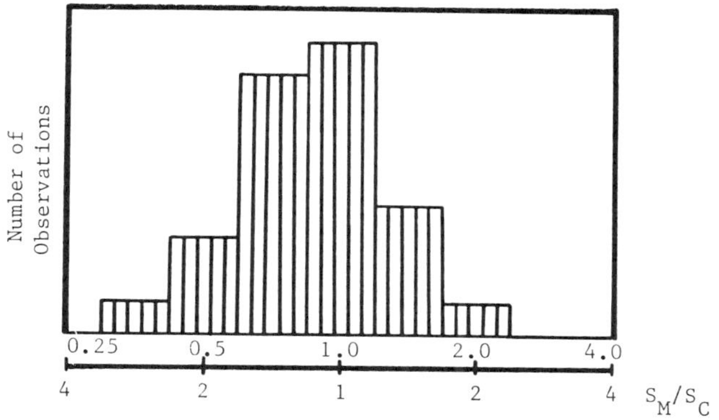

Fig. 8 Frequency Curve for S_m/S_c for Load-Transfer Analyses

FACTOR OF ERROR

Table 5 Summary of Statistical Data Relating to Settlements at a Load Equal to 95% of the Plunging Failure Load, for Piles with their Tips in Clay and More than 80% of the Side Shear from Clay

	Settlement (inches)				
		Cast-in-Place	Precast	H	Steel
	Timber	Concrete	Concrete	Pile	Pipe
Average					
Standard Deviation	0.52	0.55	0.75	0.38	0.94
Minimum	0.41	0.40	0.84	0.11	1.22
Maximum	0.09	0.10	0.10	0.15	0.05
Number of Tests	1.60	2.00	3.00	0.52	8.00
	30	27	28	12	51

Note: 1 inch = 25.4 mm

SUMMARY AND CONCLUSIONS

Data from axial load tests with instrumented piles in predominately cohesive soils, were used to find the best-fit parameters to use with both the elastic and load-transfer approaches to predicting load-settlement curves. The elastic approach produces a soil response that is bilinear in the sense that the soil is linearly elastic but an upper limit is placed on the stress between the pile and the soil. The f-x curves were made bilinear because the scatter in data did not justify an attempt to take into account nonlinear response and load shedding.

Because the goal was to consider only the settlement behavior, the limiting side stresses were adjusted to ensure that the calculated and measured pile capacities were the same.

For the elastic method, the most accurate predictions were based on a value of about 750 for E/c_u, with c_u measured using unconfined compression or unconsolidated-undrained triaxial compression tests. For the load-transfer method, a bilinear t-z curve was assumed and the deformation at failure was 0.1 inch (2.5 mm) for side shear and 1% of the tip capacity for end bearing.

In the cases studied here, the elastic method was more accurate than the load-transfer method but the differences were not major. The elastic method had the advantage that the presence of comparatively rigid layers at shallow depth below a pile tip can be taken into account. Although this advantage was significant in this study, it should be noted that it would be unusual to stop a pile tip just above a stiffer layer. The ability of the load-transfer method to utilize nonlinear load-deformation relationships is a significant advantage in attempting to predict the whole Q-S curve but appears to be of limited value in predicting response at factors of safety around two. The ease of accounting for load shedding is another advantage of the load-transfer approach but it appears that this feature could easily be incorporated into the elastic approach as well. In any case, the small amount of data on this effect, and the large scatter in available data, led to the decision not to include effects of load-shedding.

In trying to predict pile response under working loads, the major source of error seems to involve the prediction of the pile capacity, particularly when the factor of safety used in design is small and the method of analysis overpredicts the capacity.

The data presented here indicate that if we know the pile capacity we should be able to calculate settlements under working loads with a factor of two most of the time.

ACKNOWLEDGEMENTS

The data base used in this study was developed with financial support from the American Petroleum Institute under the guidance of a committee chaired by J.D. Murff.

REFERENCES

1. Coyle, H.M., and L.C. Reese (1966), "Load Transfer for Axially Loaded Piles in Clay," Journal of the Soil Mechanics and Foundation Engineering Division, ASCE, Vol. 92, No. SM2, pp. 1-26.

2. D'Appolonio, E., and J.P. Romauldi (1963), "Load Transfer in End Bearing Steel H-Piles," Journal of the Soil Mechanics and Foundations Division, ASCE, Vol. 89, SM2, pp. 1-25.

3. Davisson, M.T. (1973), "High Capacity Piles," Innovations in Foundation Construction, Soil Mech. Div., Illinois Section, ASCE, Chicago, pp. 81-112.

4. Goble, G.G., and F. Rausche (1976), "Wave Equation Analysis of Pile Driving, WEAP Program," Implementation Package 76-13, Federal Highway Administration.

5. Hirsch, T.J., L. Carr, L.L. Lowrey, Jr. (1976), "Pile Driving Analysis - Wave Equation User's Manual," Implementation Package 76-13, Federal Highway Administration.

6. Korn, G.A., and T.M. Korn (1961), Mathematics Handbook for Scientists and Engineers, McGraw Hill Book Company, pp. 424.

7. Kraft, L.M., R.P. Ray, and T. Kagawa (1981), "Theoretical t-z Curves," Jour., Geot. Engr. Div., ASCE, Vol. 107, No. 11, pp. 1543-1561.

8. Lang, G.R., Jr. (1980), "Predicting the Drivability of Large Diameter Offshore Piling," Offshore Technology Conference, Paper 3827, Vol. 3, pp. 301-312.

9. Mattes, N.S., and H.G. Poulos (1969), "Settlement of Single Compressible Pile," Journal of the Soil Mechanics and Foundation Division, ASCE, Vol. 95, SM1, pp. 189-207.

10. Mindlin, R.D. (1936), "Force at a Point in the Interior of a Semi-Infinite Solid," Physics, Vol. 7, pp. 195-202.

11. Nair, K. (1967), "Load-Settlement and Load Transfer Characteristics of a Friction Pile Subject to a Vertical Load," Proceedings of the Third Pan-American Conference on Soil Mechanics and Foundation Engineering, Vol. 1, pp. 565-590.

12. Olson, R.E., and N.D. Dennis (1982), "Review and Compilation of Pile Test Results, Axial Pile Capacity," report to Amer. Petroleum Institute for Project PRAC 81-29, Geotech. Engr. Report CR 83-4, Geotech. Eng. Center, Dept. of Civil Engr., Univ. of Texas, Austin, Texas 78712.

13. Poulos, H.G. (1979), "Settlement of Single Piles in Non-homogeneous Soil," Journal of the Geotechnical Engineering Division, ASCE, Vol. 105, No. GT5, pp. 627-641.

14. Poulos, H.G., and E.H. Davis (1968), "The Settlement Behavoir of of Single Axially-Loaded Incompressible Piles and Piers," Geotechnique, Vol. 18, pp. 351-371.

15. Poulos, H.G., and N.S. Mattes (1969), "The Behavior of Axially-Loaded End-Bearing Piles," Geotechnique, Vol. 19, pp. 285-300.

16. Salas, J.J. and J.A. Belzunce (1965), "Resolution Theorique de la Distribution des Forces dans les Pieux," Proceedings of the Sixth International Conference on Soil Mechanics and Foundation Engineering, Vol. 2, pp. 309-313.

17. Seed, H.B., and L.C. Reese (1957), "The Action of Soft Clay Along Friction Piles," Transactions, ASCE, Vol. 122, pp. 731-754.

18. Steinbrenner, W. (1934), "Tafeln Zur Setzungberechnung," Die Strasse, Vol. 1, pp. 221.

19. Sykora, D.W., and K.H. Stokoe, II (1983), "Correlations of In Situ Measurements in Sand of Shear Wave Velocity, Soil Characteristics, and Site Conditions," Masters Thesis, Geotechnical Engineering Report GR 83-33, The University of Texas at Austin.

20. Thurman, A.G., and E. D'Appolonia (1965), "Computed Movement of Friction and End-Bearing Piles Embedded in Uniform and Stratified Soils," Proceedings of the Sixth International Conference on Soil Mechanics and Foundation Engineering, Vol. 2, pp. 323-327.

21. Tomlinson, M.J. (1957), "The Adhesion of Piles in Clay Soils," Proc., Fourth Int. Conf. on Soil Mech. and Found. Engr., Vol. 2, pp. 66-71.

22. Tomlinson, M.J. (1971), "Some Effects of Pile Driving on Skin Friction," Behavior of Piles, Instn. of Civil Engrs., London, pp. 107-114, 149-152.

23. Vesic, A.S. (1970), "Tests on Instrumented Piles, Ogeechee River Site," Jour., Soil Mech. and Found. Div., ASCE, Vol. 96, No. SM2, pp. 561-584.

SHAFT CAPACITY OF DRIVEN PIPE PILES IN CLAY

by Robert M. Semple[1], M.ASCE and W. John Rigden[2]

Abstract: A compilation of axial load test records for driven pipe piles in clay published by the American Petroleum Institute is analysed to yield new criteria for skin friction and pile scale effects. The concept is directed towards design of large offshore foundations. Peak skin friction α values obtained from results of load tests on relatively short, rigid piles are related to the degree of overconsolidation of the soil. A scale factor depends on pile length-to-diameter ratio which is a convenient simplification of pile-soil stiffness characteristics that are believed to control scale effects. Computed pile shaft capacities correlate well with values measured for a range of pile sizes and cohesive soil conditions. Predicted capacities for offshore piles are generally greater than given by the current API methods. The reduction in required pile length varies from zero to about 15 percent depending on the soil-pile conditions. Success in representing existing load test data by peak skin friction and a scale factor increases prospects for improving pile design by site specific in situ measurement of skin friction on pile or model pile elements.

Introduction

The demands of offshore construction continue to focus attention on improving axial capacity prediction for driven pipe piles. Considerable research efforts have been made over the last 10 to 15 years involving soil mechanics theories, model and field scale pile testing. Recently, the American Petroleum Institute (API) sponsored a two year project, undertaken by Olson and his co-workers at the University of Texas at Austin, to establish a data bank of static pile load test records. Existing data were thoroughly examined and comprehensively documented in that well conceived and executed study (33).

Application of the test data in offshore design requires extrapolation to the pile sizes and to some of the soil conditions encountered. Accordingly, capacity criteria derived from pile load test results should reflect sound physical principles. This paper presents new criteria for skin friction in clay interpreted from the API data base.

Analytical Developments

Simple effective stress analyses are well represented by the proposals of Burland (6) and Meyerhof (31) wherein skin friction is related to the effective overburden pressure, $\bar{\sigma}_v$, by a parameter β that incorporates the frictional characteristics of the soil and a coefficient of earth pressure at failure, K_s. As

[1] Director, McClelland Limited, McClelland House, Chantry Place, Headstone Lane, Harrow, Middlesex.
[2] Manager, Civil and Geotechnical Branch, Central Engineering Department, BP International Limited, London.

discussed by Randolph (42), Meyerhof's recommendation for K_s, derived from soil mechanics theory and pile load test data, are well supported by results of recent high quality model tests (14).

More sophisticated effective stress methods based on critical state concepts and cavity expansion theory (20, 22) have contributed significantly to understanding while not fully explaining test observations (e.g.14). Recent analytical developments (3) have included detailed consideration of likely strain paths around the pile tip during installation, and this may increase the reliability of theoretical prediction.

Initial attempts to generalise pile load test information used the most obvious soil characteristic, undrained shear strength, s_u. The α coefficient, defined as the fraction of s_u mobilized as skin friction, has generally been correlated with undrained shear strength. However, McClelland (28) noted that a decision was taken in early Gulf of Mexico offshore practice to relate α to the degree of overconsolidation of the soil. This decision recognised that in existing $\alpha - s_u$ relationships, which showed α decreasing with increasing s_u, the shear strength values had a close positive correspondence with the overconsolidation ratio, OCR, of the soils. Hence α was taken as unity for normally consolidated clays, regardless of s_u, and smaller α values were used for overconsolidated clays as independently suggested by Wroth (55). For many areas of offshore development outside the Gulf of Mexico, this concept was effectively lost in the mid-1970's when the American Petroleum Institute incorporated an $\alpha - s_u$ correlation into RP2A (1).

Semple (48) estimated OCR values for some pile load test sites and was able to show that measured α values could be related to OCR. Randolph and Wroth (41) converted Meyerhof's semi-empirical effective stress procedure (31) into curves of α versus the strength ratio $s_u/\bar{\sigma}_v$ which is related to OCR but can be deduced more directly from site investigation data. Comparison of these $\alpha - s_u/\bar{\sigma}_v$ curves with pile test data (16) indicated reasonable agreement for relatively short piles but that scale effects due to pile length and flexibility (21, 42) should also be considered. The relationship of α to $s_u/\bar{\sigma}_v$ and pile scale is considered further herein.

A true representation of offshore pile response should recognise time varying load effects on soil resistance (5) and performance criteria (29). However, this paper considers only the static capacity of pipe piles as it is assessed for conventional steel framed offshore structures.

Recent Pile Testing

Research in Europe and the USA has tended to focus on cyclic load response (15, 25, 27, 39). The most recent static tests performed on relatively large pile sections, at Empire, La. and Long Beach, Calif. (7, 37), provided important new information for data base interpretation. The industry-sponsored "ESACC" project (20, 22) stimulated static testing of instrumented piles at a University of California test site (19), and at model scale at Cambridge University (14, 50).

Recent Data Interpretations

Working with the data base they compiled for API, Dennis and Olson (9) developed an $\alpha - s_u$ correlation in which α reduced from 1 to 0.3 with increasing shear strength. A length-related correction factor was formulated to account for the capacity of long piles in (stiff) normally consolidated clays being underestimated by this $\alpha - s_u$ correlation.

Randolph (43) found that the α values correlated well with the strength ratio $s_u/\bar{\sigma}_v$ in general accordance with the concept published earlier by Randolph and Wroth (41). For strength ratios up to unity, Randolph found that $\alpha = k(s_u/\bar{\sigma}_v)^{-\frac{1}{2}}$ with $k \approx 0.5$ provided a good average fit to the test data. Randolph further generalised the relationship by defining k as the square root of the strength ratio for normally consolidated soil. The relationship was later expanded to include strength ratios greater than unity (44). Randolph (43) also noted a possible effect of pile length, with a tendency for longer piles to have lower capacities than computed from the average $\alpha - s_u/\bar{\sigma}_v$ correlation. He recommended that the pile length effect be assessed separately, considering pile compressibility and likely residual values of skin friction, as described by Randolph (42).

New Approach

In assessing prospects for improving pile capacity predictions for long offshore piles, Focht and Kraft (12) recommended that the problem be conceived has having two components. The first component is the load transfer response of a soil element as characterised by its local peak and residual skin friction and associated displacements, commonly termed t-z curves. The second component involves the integration of this local soil response over the pile length with due regard to pile compressibility and the resulting variation in load transfer between peak and residual values at various points on the pile shaft. Focht and Kraft suggested as an objective the development of an analysis based on t-z soil response curves in which the value of peak skin friction is predicted by applying soil mechanics theory to assessing stress changes in soil elements.

Peak skin friction for a soil profile is the value that would be deduced from a load test on an incompressible pile. Accordingly, one approach to interpreting the API data base is first to distinguish peak skin friction as measured on relatively rigid piles and then to identify appropriate adjustments for longer, less stiff piles.

Pile Test Results

Although the 1982 API data base (33) contains results from over 1000 pile load tests, only a fraction of these are from driven steel pipe piles in predominantly cohesive soil profiles. The data were quality graded using a five point scale and the lowest quality data excluded from further consideration by the compilers in developing correlations (9). The same approach has been taken in this study.

Table 1 lists data from 24 sites that we have selected for analysis. The data are given in order of increasing strength ratio, $s_u/\bar{\sigma}_v$, which are averages for the embedded pile lengths. The identifying Load Test Numbers (LTN) from the API listing are indicated. Several of the data lines in Table 1 are averages of more than one test pile at a site. This averaging was performed to avoid clutter on data plots and giving undue weight to essentially repetitive results from a soil-pile condition.

TABLE 1. – Summary of Pile Load Test Results

Load Test Number (1)	L, in metres (2)	D, in millimetres (3)	L/D (4)	s_u, in kilopascals (5)	$\bar{\sigma}_v$, in kilopascals (6)	$s_u/\bar{\sigma}_v$ (7)	α (8)
8	20.4	762	27	30	144	0.21	0.92
6,20	21.6	457	47	31	147	0.21	0.93
3,7,17	19.2	610	32	31	142	0.22	0.99
478,489	15.2	356	43	104	448	0.23	1.05
491,493	12.2	356	34	162	718	0.23	1.00
854,855	43.9	305	144	38	162	0.23	0.79
868	96.0	610	158	80	354	0.23	0.55
869	73.8	610	121	67	273	0.25	0.71
873	22.6	767	30	170	651	0.26	1.13
451	66.4	325	205	60	223	0.27	0.52
42	30.5	325	94	45	153	0.29	0.65
444,450	45.7	325	142	52	148	0.35	0.42
507,508	29.0	330	88	39	105	0.37	1.02
30	13.7	325	42	45	112	0.40	0.94
150	18.3	325	57	33	51	0.43	0.97
45	48.2	610	79	64	152	0.43	0.59
844,846 848,851	11.6	114	101	21	44	0.47	0.64
856	12.2	168	72	16	33	0.50	0.62
325	14.0	351	40	30	59	0.51	0.78
67	39.6	274	145	165	297	0.56	0.49
43	30.5	610	50	52	91	0.57	0.59
443,449	22.9	325	71	52	91	0.57	0.52
368,369	25.9	325	80	61	99	0.62	0.56
435,436 437,438	25.3	274	92	185	244	0.76	0.48
70	14.9	528	28	53	66	0.79	0.52
998	32.0	274	117	115	141	0.82	0.52
106	12.8	325	40	96	110	0.87	0.57
547,549	16.8	610	28	100	87	1.15	0.55
31,32	13.7	325	42	137	112	1.22	0.47
829,830	13.1	274	48	110	80	1.37	0.49
495,497	20.4	610	34	208	105	1.98	0.44
*	9.1	450	20	144	54	2.65	0.51
23,24	18.3	762	24	335	115	2.90	0.46

Note: L = Embedment length; D = Outside diameter
1 m = 3.28 ft; 1 mm = 0.039 in; 1 kPa = 20.9 psf

The second from last data line refers to LTN 860 in the API listing. However the pile data on Table 1 were taken from Rigden et al (46) who presented results for a driven pipe pile which are considered more relevant than the jacked pile data from the same site given in the API listing. All the piles represented in Table 1 were installed by driving. Most α values are from compression tests after allowing for end bearing which was culculated to be small for these friction piles. Unit end bearing was taken as 9 x s_u at the pile tip.

TABLE 2. – Sources of Pile Test Data

Load Test Number (1)	Shear Strength Test (2)	Source Reference (3)
8	U	Mansur and Focht (24)
6,20	U	Mansur and Focht (24)
3,7,17	U	Mansur and Focht (24)
487,489	U	Cox, Kraft and Verner (7)
491,493	U	Cox, Kraft and Verner (7)
854,855	V	Confidential
868	U	Confidential
869	U	Confidential
873	U	Pelletier and Doyle (37)
451	U	Peck (36)
42	U	Darragh and Bell (8)
444,450	U	Peck (36)
507,508	U	Raymond (45)
30	U	Woodward, Lungren and Boitano (54)
150	U	A.R.E.A. (2)
45	U	McCammon and Golder (30)
844,846 848,851	V	Kirby and Roussel (19)
856	M	Confidential
325	FV	Hutchinson and Jensen (18)
67	U	Peck (35)
43	U	McCammon and Golder (30)
443,449	U	Peck (36)
368,369	U	Peck (35)
435,436 437,438	U	Peck (35)
70	U	Togrol (52)
998	U	Endley, Ulrich and Gray (11)
106	M	Stermac, Selby and Devatta (51)
547,549	U	U.S.A.C.E. (53)
31,32	U	Woodward, Lungren and Boitano (54)
829,830	Q	O'Neill, Hawkins and Mahar (34)
495,497	M	Heerema (17)
*	M	Rigden et al (46)
23,24	U	Fox, Sutton and Oksuzler (13)

Note: U = Unconfined compression; Q = Quick triaxial; V = Laboratory miniature vane; FV = Field vane; M = Other test

Source documents for the data are identified on Table 2. Inspection of the documents indicates that a few of the soil profiles had a relatively thin, surficial sand stratum that would have made a modest contribution to the measured axial capacity. Unit skin friction in these surficial sands was taken as $0.4\bar{\sigma}_v$, and the pile capacity attributed to cohesive resistance was adjusted accordingly. Only in one case did the calculated adjustment approach 10 percent of capacity, adjustments for the few remaining cases being less than 5 percent.

The pile lengths in Table 1 are those embedded in clay beneath surficial sand, excavation or casing over the upper part of the pile where these occurred.

Soil Shear Strength

Undrained shear strengths are given in Table 1. Sampling quality was generally good with thin-walled pushed samplers being used at almost all sites.

Table 2 indicates how the reference soil strength was measured. The most common test type was unconfined compression so this value was usually selected from the API listing even if other measurements were available. Exceptions were made for very soft and soft soils ($s_u < 40$ kPa) for which vane strengths were selected, where available, as such soils are too weak for reliable compression testing. Vane strengths were not available for the first three data lines in Table 1, however the soil testing appears to have been performed carefully on 125 mm diameter piston samples. After reviewing each source document, the strength profile given therein was sometimes preferred to the information in the API listing (LTN 106, 487-491, 495). Soil unit weight data for LTN 23 were provided by BP for whom the tests were performed.

Pile Geometry

Pile embedment length, L, has been expressed as a multiple of pile diameter, D, in Table 1. The simple aspect ratio L/D cannot represent all the variables influencing the effects that pile geometry and stiffness characteristics may have on axial capacity. The pile scale or "length effect" (12, 21, 31, 42) is probably related to pile stiffness as it affects either lateral whip during driving or compressibility under axial loading. If pile whip controls, then the ratio of the travelling stress wave length to pile diameter should be considered (38). Another reason for the "length effect" may be progressive failure of strain-softening soil as the pile shaft compresses under axial load. From this viewpoint, it is the relative pile-soil stiffness in axial loading that is important.

Relative axial pile-soil stiffness has been termed π_3 and defined by Murff (32) as the ratio of pile elastic compression, acting as a free standing column, to the local soil displacement required to mobilize peak skin friction. If the displacement corresponding to the limit of elastic soil response is related to the pile diameter, then π_3 for an open pipe pile is a function of the aspect ratio L/D, peak skin friction, and the pile area ratio (steel section as a fraction of gross sectional area). Although pile area ratios are known for most of the data in Table 1, being on average about 10 percent, there is uncertainty over which of the pipe piles were filled with concrete prior to testing thereby altering their axial stiffness. Further, the relevant peak skin friction values are unknown so π_3 values cannot be confidently determined.

The ratio L/D only approximately represents pile geometry and stiffness characteristics. It is an expedient that can be considered as replacing more complex expressions for flexural and axial pile stiffnesses while reflecting more of these factors than does pile length alone.

Conventional α Correlation

The data from Table 1 are plotted as α versus s_u on Fig. 1 in which the L/D value is given for each data point. Criteria recommended by API (1) are represented by the solid curve. The data conform quite well to the API criteria with the exception of the three points for stiff clays having α values of 1 or more. These data represent the sites at Empire and Long Beach (7, 37) where pile sections were installed through casings set sufficiently deeply into normally consolidated clays that the soils tested were of a stiff consistency.

The Empire and Long Beach tests were performed after the API $\alpha - s_u$ criteria were formulated. Their inclusion on Fig. 1 indicates that s_u is not the variable controlling α. Further, this difference in α values for stiff, normally consolidated and stiff, overconsolidated clays underlay the introduction of a positive length-related adjustment factor in the $\alpha - s_u$ formulation of Dennis and Olson (9).

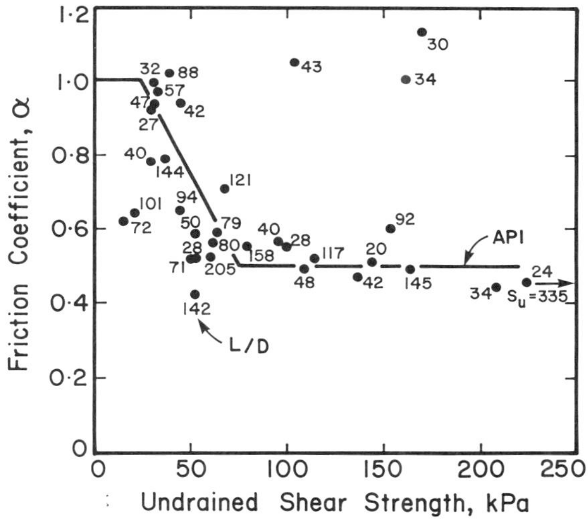

FIG. 1 – Alpha Versus Undrained Shear Strength (1kPa = 20.9 psf)

Peak α Values

The pile test data are presented on Fig. 2 as α versus the strength ratio $s_u/\bar{\sigma}_v$, using logarithmic scales, and again with values of L/D indicated. Inspection discloses a pattern in the L/D values with the relatively longer piles generally having smaller α values for a given strength ratio.

Peak skin friction is measured by stiff piles so the data for the relatively shorter piles, including those tested at Empire and Long Beach, are plotted separately on Fig. 3. The scatter is small, and curves were readily fitted to the data by separate regression analyses for strength ratios greater than unity and in the range 0.35 to 1. For strength ratios below about 0.35, a value of unity was selected for α as these soils are essentially normally consolidated, s_u increases with depth, and the strength ratio tends to be constant over the pile length. In soil profiles having greater average strength ratios, the s_u profile is more uniform, and the strength ratio varies over the pile length with relatively large values occurring near the ground surface.

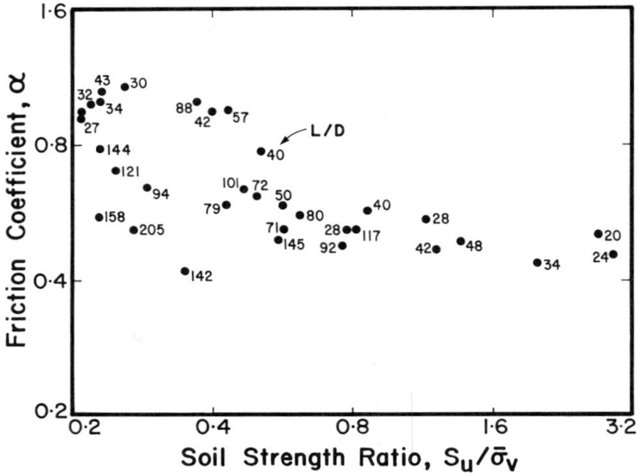

FIG. 2 – Alpha Versus Strength Ratio (All Data)

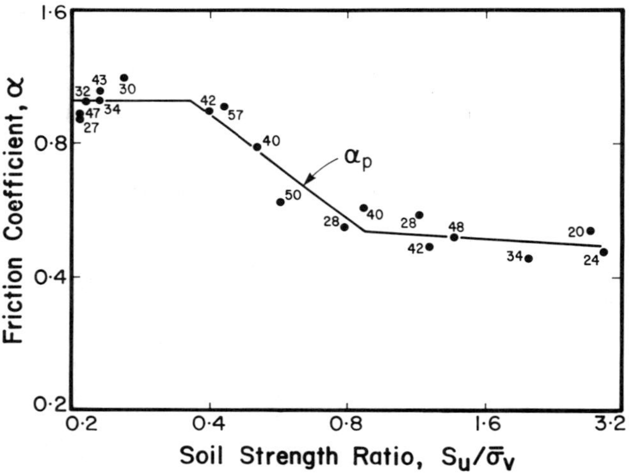

FIG. 3 – Alpha Versus Strength Ratio (L/D < 60)

The curve on Fig. 3 is labelled α_p to suggest that the relationship is for peak skin friction. The curve actually represents piles having L/D = 40 ± 20, and so is likely to be a lower bound to peak α values.

In this formulation, the strength ratio is a measure of soil overconsolidation. No attempt has been made to normalise the strength ratio to account for possible effects of variation in basic soil plasticity characteristics as expressed by plasticity index, PI, or the associated angle of internal friction, ϕ. One possibility is to relate each strength ratio to the value for the same soil in a normally consolidated condition, $(s_u/\bar{\sigma}_v)_{nc}$. There is, however, conflicting information on how this value varies with PI or ϕ (26, 49), while the normally consolidated strength ratio obtained from compression test data can be independent of PI (23). Given this uncertainty, the term $s_u/\bar{\sigma}_v$ has been taken to represent soil overconsolidation without further generalisation.

The correlation between α_p and strength ratio on Fig. 3 bears some resemblance in form to the existing API α - s_u criteria. It reflects conventional wisdom that α is unity for normally consolidated clays and 0.5 for heavily overconsolidated clays.

Pile Length Effect

Having established a correlation for peak skin friction from the relatively shorter piles, the effect of greater pile embedments has been evaluated borrowing from the technique used by Dennis and Olson (9). For each strength ratio in Table 1, a value of α_p was obtained from the correlation on Fig. 3, and this was used to normalise the corresponding measured value α_m. The resulting ratios of α_m/α_p are shown on Fig. 4 as a function of the pile aspect ratio, L/D. Approximate bounds to the data are indicated together with an average line obtained by regression analyses.

The average value of unity for α_m/α_p for the relatively shorter piles is of course conditioned by the fact that these piles (average L/D = 40) were used to define α_p. Values of α_m/α_p for relatively longer piles fall progressively

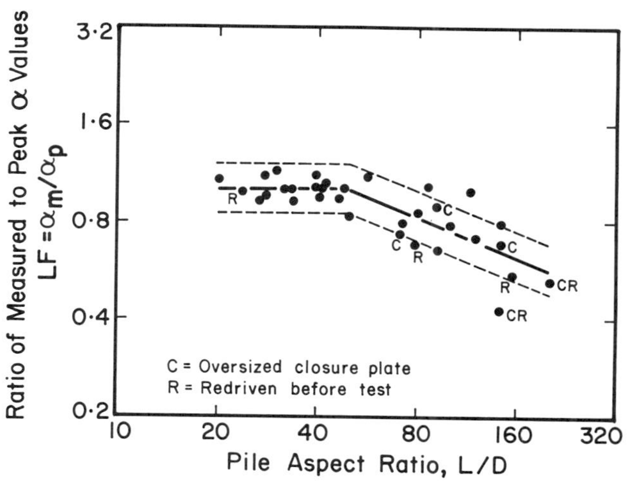

FIG. 4 – Length Factor Versus Aspect Ratio

below unity. The data scatter is due partly to the limitations of L/D in representing pile stiffness, and to other factors that indicate a "length effect" for individual test piles. These factors are discussed subsequently.

Adopting the term LF, or length factor, for α_m/α_p, the solid line between the data bounds on Fig. 4 provides a pile embedment factor to be applied to the α_p values from Fig. 3. The average unit skin friction for a pile is then LF $\alpha_p s_u$ where s_u is the average value over the pile length.

Evaluation

Computed values of pile shaft capacity, Q_{sc}, based on LF $\alpha_p s_u$ are compared with the measured values, Q_{sm}, on Fig. 5. Agreement is good, the data falling within a fairly tight scatter band. The average ratio of measured to computed pile capacity is 1.01 with a standard deviation of 0.15.

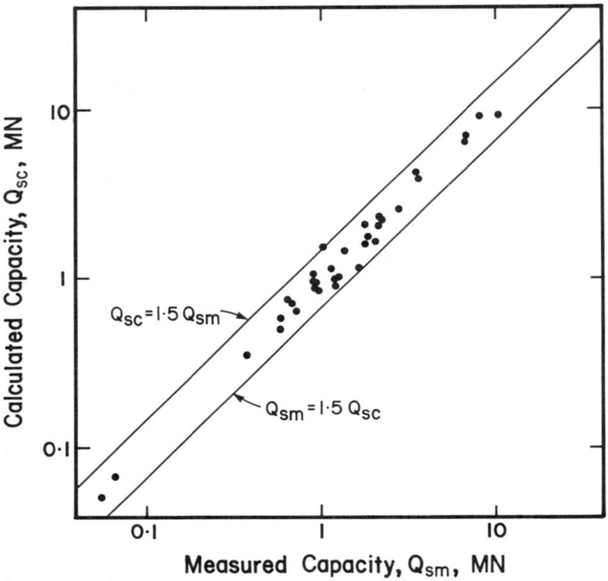

FIG. 5 – Comparison of Measured and Calculated Capacities (1 MN = 225 kips)

The results on Fig. 5 indicate that the $\alpha - s_u/\bar{\sigma}_v$ - L/D correlations provide a good fit to the pile test data which supports an approach based on peak skin friction modified, where appropriate, to account for pile scale. Examination of the pile load test documents does, however, indicate certain deficiencies in execution of some tests that will have biased the data towards low α values. These effects were discussed by Kraft et al (21) who adjusted their data interpretation to minimise this bias.

Length Factor Adjustment

Olson and Dennis (33) note that piles redriven and reloaded to failure exhibited capacities lower than on initial loading. They also recognised the harmful effect on skin friction of an oversized closure plate at the pile tip. The data to which these considerations apply are identified on Fig. 4. With one exception they are all relatively long piles so the bias is incorporated into the pile length factor, LF. Of the six LF values less than 0.7, all but one of the piles were either redriven or had an oversized closure plate, or both as in the case of the two smallest values. As evidence of LF less than 0.7 derives from particularly adverse circumstances, we consider an LF value of 0.7 to be a reasonable lower limit. It should also be remembered that the α_p correlation on Fig. 3 was derived from piles with L/D = 40 ± 20, thereby incorporating some scale effect and providing a lower bound to peak α values.

If the empirical evidence for the pile length factor is attributed to relative pile-soil stiffness effects in axial loading, then the expected shape of the relationship between LF and L/D can be inferred from theory. As L/D is related to the pile-soil stiffness factor π_3, the characteristic shape of the theoretical LF - π_3 relationship shown on Fig. 6 would suggest a lower limit for LF corresponding to the ratio of residual to peak skin friction. From such considerations, Randolph (43) suggested that the measured capacity of a highly compressible pile could be 70 percent of the capacity based on peak skin friction.

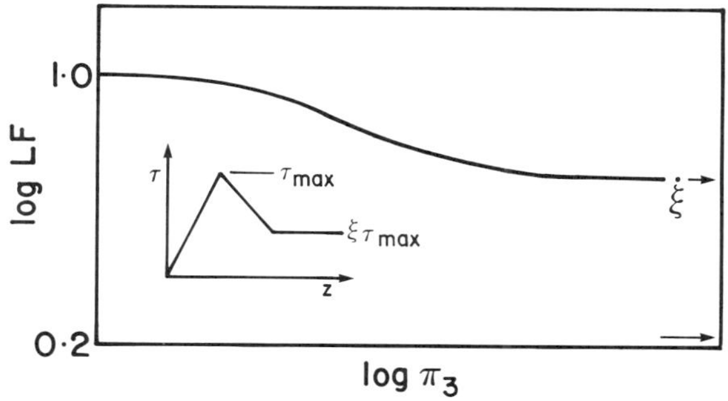

FIG. 6 Theoretical Length Effect

It is not known, however, that the "length effect" is entirely due to progressive soil failure in axial loading. Pile whip during driving may cause irreparable damage to the soil, particularly in overconsolidated clays that cannot flow back to establish a pressure on the pile wall. This effect may partly explain the reduction in α_p with increasing strength ratio. In addition, the nature and magnitude of lateral pile oscillations during driving depend on, among other factors, the lengths of the stress wave and the pile. Some progress has been made in quantifying this behaviour (38). However, an analytical model providing insight into pile whip effects, analogous to that for axial pile-soil stiffness (32), remains to be developed. Pile and stress wave lengths may combine to provide a skin friction reduction effect related to a flexural stiffness ratio for which the pile aspect ratio, L/D, is an approximation.

Modified Relationships

On Fig. 3, the best fit line to the heavily overconsolidated clays has a slight negative slope that is largely conditioned by the two smallest α values in Table 1. In both cases the piles were tested shortly after installation. Further, the result from the most heavily overconsolidated soil in the data base was obtained after several cycles of redriving and testing. Given these unfavourable circumstances, the adoption of a minimum value of 0.5 for α_p appears reasonable. The maximum value of unity for α_p applies to essentially normally consolidated clays which may be defined as those having a strength ratio less than 0.35.

For predicting pile capacity, we believe that the best-fit α-$s_u/\bar{\sigma}_v$ - L/D correlations in Figs. 3 and 4 should be modified to the forms shown on Fig. 7. The main alteration is constraining the length factor, now designated F, to have a minimum value of 0.7. This modification avoids the underestimation of capacity for very long, slender piles that otherwise would result if the α values measured for such piles under unfavourable circumstances were taken at face

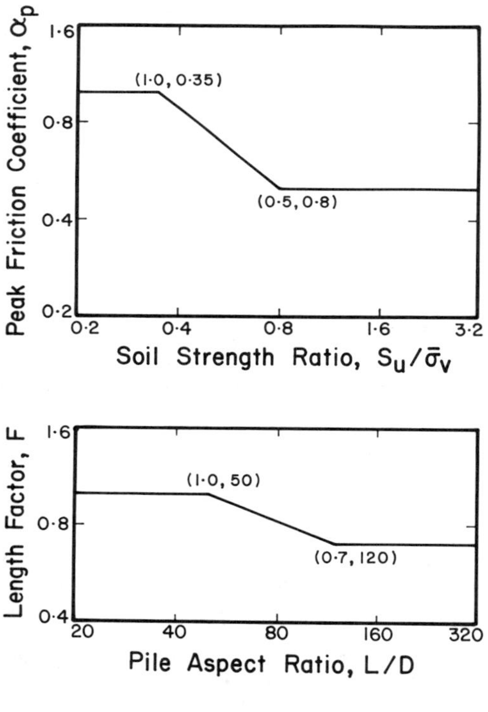

FIG. 7 – Criteria for Capacity Prediction

value. Calculated shaft capacities based on unit skin friction equal to $F \alpha_p s_u$, with F and α_p from Fig. 7, still correlate well with the measured values.

Application to Design

The relationships given on Fig. 7 may be used in predicting pile capacity. Due to the nature of the information available, the skin friction correlation and scale factor were derived using average results for each pile test. An incremental pile capacity method, in which the soil profile is considered layer by layer, is ultimately preferable but its formulation must await the development of an appropriate data base. Layer-by-layer application of the criteria on Fig. 7 would represent a misapplication of the existing test data, and would generally invoke errors on the conservative side.

Standard offshore practice in many areas is to measure s_u by quick triaxial tests on samples recovered by pushing a thin-walled sampler. The criteria developed herein are based on high quality pushed samples, but the reference undrained shear strength was most commonly measured by the unconfined compression test. However, this test is not appropriate to stiff, structured clays for which quick triaxial tests are necessary. In other clays, unconfined and quick triaxial compression tests give comparable results where sampling disturbance is minimised. Accordingly, we recommend that the reference shear strength be taken as that of a pushed sample tested in triaxial compression at natural water content. Where other sampling and testing techniques are applied, equivalent soil strengths can be obtained using adjustment factors where these are available from correlation studies of similar type soils (e.g. 40, 56)

If sand is present in a soil profile, the most rational approach is to apply the criteria on Fig. 7 using average values of s_u and $\bar{\sigma}_v$ for the clays, and to select the scale factor using the total embedded pile length. This gives the average skin friction in clay from which the shaft capacity in clay can be determined. The shaft capacity in sand can be separately assessed in the usual way and added to give the total shaft capacity. At present, pile length effects in sand are broadly accounted for in offshore practice by the use of limiting skin friction values.

The prospects for improving pile design by appropriate site-specific testing are increased by finding that an empirical approach based on peak skin friction adjusted for pile scale effects accurately represents existing pile capacity measurements. Peak skin friction can be measured either by testing a pile segment or, more economically, by the use of precision in-situ instruments (4) calibrated against pile test measurements. If site-specific data are available then analyses incorporating pile stiffness characteristics can be used in an independent assessment of likely scale effects.

Comparison with Current Practice

The criteria on Fig. 7 have been applied to three representative offshore sites for comparison with capacity predictions obtained from the current API methods. The API methods used are those given in Paragraph 2.6.4b of the 1982 edition of RP2A (1). Method 1, commonly termed "f = c" (or skin friction equals undrained shear strength), is restricted to highly plastic clays such as are found in the Gulf of Mexico. Method 2, for other clays, corresponds to the line designated API on Fig. 1.

Normally Consolidated Clay.— Site investigation practice in the Gulf of Mexico has involved percussive sampling with miniature vane or unconfined compression testing to determine shear strength of clays. The criteria on Fig. 7 are intended for use with quick triaxial strengths from pushed samples. Quiros et al (40) have presented a comprehensive set of soil strength data to about

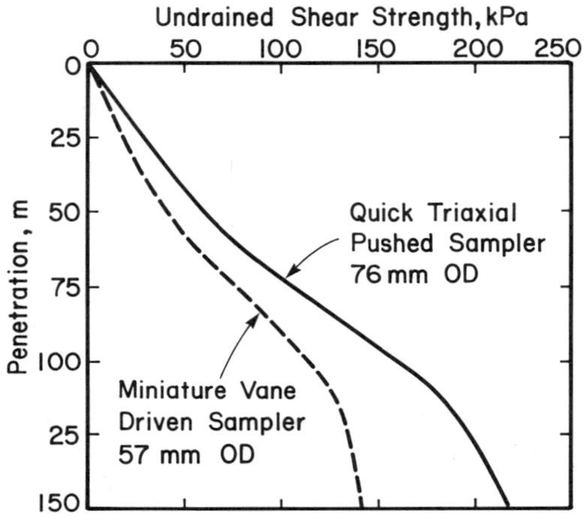

FIG. 8 – Strength Profiles for Normally Consolidated Clays Offshore Louisiana (1kPa = 20.9 psf, 1 m = 3.28 ft)

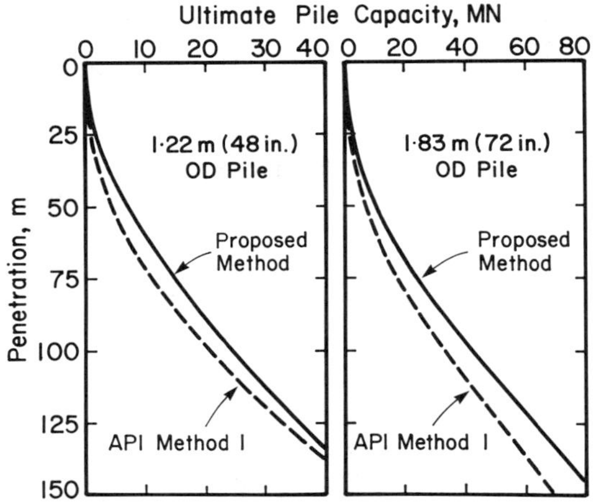

FIG. 9 – Compressive Pile Capacities for Normally Consolidated Clays Offshore Louisiana (1 MN = 225 kips, 1 m = 3.28 ft)

150 m (500 ft) penetration in a normally consolidated clay at a location offshore Louisiana which they designate "Site 2". Undrained shear strength profiles appropriate to the API method and to the criteria on Fig. 7 are shown on Fig. 8. The quick triaxial strength profile on Fig. 8 is essentially the same as the interpreted shear strength profile for the site given in the source publication.

Compressive capacity curves developed using API Method 1, the criteria on Fig. 7, and the appropriate shear strength profiles on Fig. 8 are presented for two different pile diameters on Fig. 9. Unit end bearing was taken as 9 x s_u at the pile tip. The curve for the 1.83 m (72 in.) OD pile attributed to the API method agrees with the corresponding pile capacity curve given by Quiros et al (40).

The α - $s_u/\bar{\sigma}_v$ - L/D criteria produce the greater capacities on Fig. 9. In percentage terms, the differences are greater for the larger diameter pile which is less affected by the length factor in the new procedure. At penetrations of interest for the 1.22 m (48 in.) OD pile, the proposed method gives predicted capacities about 10 percent greater than the API method, producing a small reduction in required pile length of the order of 5 percent. The corresponding increase in predicted capacity for the 1.83 m OD pile is about 25 percent which reduces the required pile penetration by about 10 percent.

Heavily Overconsolidated Clay. Fig. 10 gives a shear strength profile for the Heather Field located about 100 km (60 miles) east of the Shetland Islands in the North Sea. Quick triaxial shear strength and vertical effective stress profiles were obtained from Durning and Rennie (10). The other curve shown on Fig. 10 represents the average value of $s_u/\bar{\sigma}_v$ to a given depth. These silty clays are heavily overconsolidated and particulary strong, the average $s_u/\bar{\sigma}_v$ and s_u values for the profile being about 2 and 500 kPa (10 ksf), respectively.

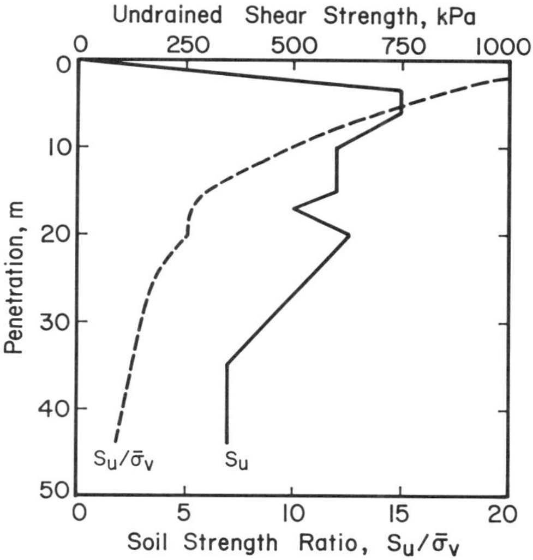

FIG. 10 - Soil Information for Heather Field, North Sea (1kPa = 20.9 psf, 1 m = 3.28 ft)

For the 1.83 m (60 in.) OD piles installed at Heather Field, the criteria on Fig. 7 yield an α value of 0.5 which is also the value given by API Method 2 commonly used for North Sea installations. In hard, heavily overconsolidated silty clays the $\alpha - s_u/\bar{\sigma}_v$ - L/D criteria proposed herein will generally give the same pile capacity as the existing API method.

Overconsolidated Clay.—North Sea cohesive soil profiles are commonly less heavily overconsolidated than at Heather Field. Fig. 11 gives quick triaxial shear strength profiles for two pushed sample borings at Magnus Field (47), which is located about 200 km (125 miles) northeast of the Shetland Islands. A thin sand stratum occurred in the upper part of the profile at this site. Also indicated on Fig. 11 are curves of the average $s_u/\bar{\sigma}_v$ values that reduce to about 0.5 to 0.6 at the depths of interest. The Magnus piles were 2.13 m (84 in.) OD and the target penetration was about 80 m (47).

As shown on Fig. 12, pile capacity curves for the profiles derived using API Method 2 and Fig. 7 are the same to about 60 m penetration, both methods giving $\alpha = 0.5$ to this depth. At 80 m penetration, the capacity predicted by the $\alpha - s_u/\bar{\sigma}_v$ - L/D method is about 30 percent greater then the API value which again corresponds to $\alpha = 0.5$. The associated reduction in required pile penetration is about 10 m, or 12 percent of the embedment length required by the API method.

Although similar shear strength profiles were obtained from the two borings, there are differences which are reflected in the pile capacity predictions for each method. However, the range in pile capacity is reduced using the new criteria. Of the two profiles, Boring 11 gave the greater average shear strength. This is compensated by a lower α value, as the $s_u/\bar{\sigma}_v$ ratio is also greater for Boring 11. The Magnus example demonstrates the reduced

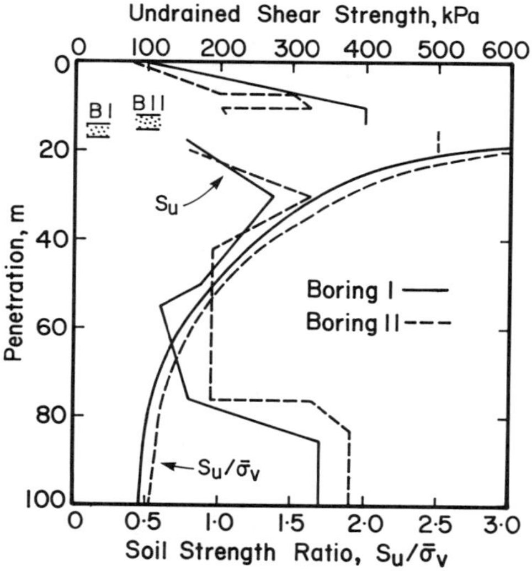

FIG. 11 – Soil Information for Magnus Field, North Sea, (1kPa = 20.9 psf, 1 m = 3.28 ft)

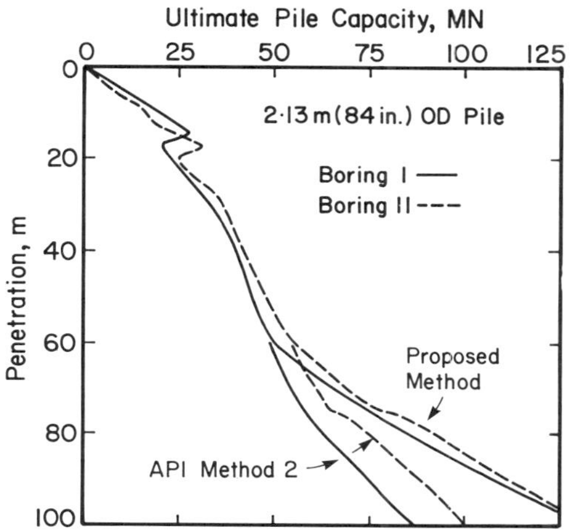

FIG. 12 – Compressive Pile Capacities for Magnus Field, North Sea (1 MN = 225 kips, 1 m = 3.28 ft)

sensitivity to undrained shear strength, and hence to measurement errors, in stiff overconsolidated clay. This can be deduced from inspection of the $\alpha - s_u/\bar{\sigma}_v$ relationship on Fig. 7.

Conclusions

Measured shaft capacities of driven pipe piles in clay represented in the 1982 API data base can be reproduced quite accurately by correlations based on the concept of a peak skin friction reduced to account for pile scale effects. Peak skin friction is obtained from a correlation between α and the strength ratio $s_u/\bar{\sigma}_v$ which indicates the degree of soil overconsolidation. The reduction factor depends on pile aspect ratio, L/D. It represents the influence of pile stiffness in controlling the degree to which either progressive failure can develop in static loading or the soil is damaged by pile whip during driving, or both.

Interpreting pile test data within a framework of soil overconsolidation ratio and pile stiffness is believed to reflect sound physical principles that increases confidence in extrapolating the pile test information to large offshore foundations. Criteria suggested for use in pile design are given on Fig. 7.

When compared with current practice, as embodied in the 1982 edition of API RP2A, the $\alpha - s_u/\bar{\sigma}_v$ - L/D criteria generally produce more optimistic results. For typical piles and normally consolidated highly plastic clays in the Gulf of Mexico the required pile lengths are 5 to 10 percent less. At the other extreme of hard, very heavily overconsolidated silty clays there is no difference in the predicted capacity of offshore piles. Pile lengths are reduced by 10 to 15 percent in clays that are intermediate in terms of strength and degree of overconsolidation.

Consideration should also be given to applying the underlying concepts on a site-specific basis as a means of improving pile design for offshore structures. Site-specifically, peak skin friction can be economically measured and the scale adjustment assessed analytically.

Acknowledgements

The study reported herein was initially part of a state-of-practice review performed in connection with BP's planning of a pile load test programme in the UK. We are grateful to BP International Limited for permission to publish this paper.

The work was further stimulated by publication of the API data base compiled by R.E. Olson and his co-workers at the University of Texas at Austin, and by deliberations of the API RP2A Geotechnical Work Group on possible new methods of predicting skin friction.

We particularly wish to thank J.A. Focht of McClelland Engineers, Inc., and M. F. Randolph of Cambridge University who have through their own work, stimulating discussion, and constructive criticism, directly contributed to the ideas given in this paper. The contents are, of course, entirely our responsibility.

Appendix 1. - References

1. API RP 2A, "Recommended Practice for Planning, Designing, and Constructing Fixed Offshore Platforms," American Petroleum Institute, Dallas, Texas, 13th Edition, 1982, p. 30.
2. A.R.E.A., "Steel and Timber Pile Tests, West Atchafalaya Floodway-New Orleans, Texas and Mexico Railway," American Railway Engineering Association, Bulletin 489, 1950, pp. 149-202.
3. Baligh, M.M., "The Strain Path Method in Geotechnical Engineering", Publication No. R84-01, Department of Civil Engineering, Massachusetts Institute of Technology, Jan., 1984.
4. Boggess, R.L., Bogard, J.D., and Hamilton, T.K., "Advanced In-Situ Instruments for Studying the Behaviour of Cyclically Loaded Friction Piles," ASCE National Convention, Houston, Texas, Oct., 1983.
5. Briaud, J.L., Garland, E. and Felio, G.Y., "Rate of Loading Parameters for Vertically Loaded Piles in Clay", Proceedings of the Sixteenth Offshore Technology Conference, Houston, Vol. 1, 1984, pp. 407-412.
6. Burland, J.B., "Shaft Friction of Piles in Clay - A Simple Fundamental Approach," Ground Engineering, Vol. 6, No. 3, 1973, pp. 1-15.
7. Cox, W.R., Kraft, L.M., and Verner, E.A., "Axial Load Tests on 14-inch Pipe Piles in Clay", Proceedings of the Eleventh Offshore Technology Conference, Houston, Vol. 2, 1979, pp. 1147-1158.
8. Darragh, R.D., and Bell, R.A., "Load Tests on Long Bearing Piles", Performance of Deep Foundations, ASTM, STP 444, 1969, pp. 41-67.
9. Dennis, N.D., and Olson, R.E., "Axial Capacity of Steel Pipe Piles in Clay," Proceedings of the Conference on Geotechnical Practice in Offshore Engineering, ASCE, Apr., 1983, pp. 370-388.
10. Durning, P.J. and Rennie, I.A., "Determining Pile Capacity and Pile Driveability in Hard, Overconsolidated North Sea Clay", Proceedings of the European Offshore Petroleum Conference, London, Vol. 1, 1978, pp. 383-392.
11. Endley, S.N., Ulrich, E.J., and Gray, J.B., "A study of Axial Pile Load Tests," Preprint 3766, ASCE National Convention, Atlanta, Oct., 1979.
12. Focht, J.A., and Kraft, L.M., "Prediction of Capacity of Long Piles in Clay, A

Status Report," presented at the December, 1981, Symposium on Geotechnical Aspects of Offshore and Nearshore Structures, held at Bangkok, Thailand, 33 pp.
13. Fox, D.A., Sutton, V.J.R., and Oksuzler, Y., "North Sea Platform Piling Development of the Forties Field Piles from West Sole and Nigg Bay Experience and Tests," Proceedings of the Conference on Design and Construction of Offshore Structures, Institution of Civil Engineers, London, 1976, pp. 47-59.
14. Francescon, M., "Some Preliminary Results from Instrumented Model Pile Tests in Clay," Proceedings of the Second International Conference on Numerical Methods in Offshore Piling, Austin, Texas, 1982, pp. 367-378.
15. Gallagher, K.A., and St. John, H.D., "Field Scale Model Studies of Piles as Anchorages for Bouyant Structures," Proceedings of the Second European Offshore Petroleum Conference, London, 1980.
16. Gallagher, K.A., Report on "Recent Developments in Understanding the Axial Capacity of Piles in Clay," Ground Engineering, Vol. 16, No. 2, Mar., 1983. pp. 6-7.
17. Heerema, E.P., "Pile Driving and Static Load Tests on Piles in Stiff Clay," Proceedings of the Eleventh Offshore Technology Conference, Houston, Paper No. 3490, 1979, pp. 1135-1146.
18. Hutchinson, J.N., and Jensen, E.V., "Loading Tests on Piles Driven into Estuarine Clays at Port-of-Khorramshahr, Iran, and Observations on the Effect of Bitumen Coatings on Shaft Bearing Capacity," Publication No. 78, Norwegian Geotechnical Institute, 1968, pp. 1-12.
19. Kirby, R.C., and Roussel, G., ESACC Project Field Model Pile Load Test, Hamilton Air Force Base Test Site Norato, California, Amoco Production Company, 1979.
20. Kirby, R.C., Esrig, M.I., and Murphy, B.S., "General Effective Stress Method for Piles in Clay," Proceedings of the Conference on Geotechnical Practice in Offshore Engineering, ASCE, Apr., 1983, pp. 457-498.
21. Kraft, L.M., Focht, J.A., and Amerasinghe, S.F., "Friction Capacity of Piles Driven into Clay," Journal of the Geotechnical Engineering Division, ASCE, Vol. 107, No. GT11, Nov., 1981, pp. 1521-1541.
22. Kraft, L.M., "Effective Stress Capacity Model for Piles in Clay," Journal of the Geotechnical Engineering Division, ASCE, Vol. 108, No. GT11, Nov., 1982, pp. 1387-1404.
23. Larsson, R., "Undrained Shear Strength in Stability Calculation of Embankments and Foundations on Soft Clays," Canadian Geotechnical Journal, Vol. 17, 1980, pp. 591-602.
24. Mansur, C.I., and Focht, J.A., "Pile Loading Tests, Morganza Floodway Control Structure," Transactions, ASCE, Vol. 121, 1956, pp. 555-576.
25. Matlock, H., Lam, I., and Cheang, L., "Analytical Interpretation of Pile Installation and Axial Performance," Proceedings of the Second International Conference on Numerical Methods in Offshore Piling, Austin, Texas, 1982, pp. 133-162.
26. Mayne, P.W., "Cam-Clay Predictions for Undrained Strength," Journal of the Geotechnical Engineering Division, ASCE, Vol. 106, No. GT11, Nov., 1980, pp. 1219-1242.
27. McAnoy, R.P.L., Cashman, A.C., and Purvis, P., "Cyclic Tensile Testing of a Pile in Glacial Till," Proceedings of the Second International Conference on Numerical Methods in Offshore Piling, Austin, Texas, 1982, pp. 257-292.
28. McClelland, B., "Design of Deep Penetration Piles for Ocean Structures," Journal of the Geotechnical Engineering Division, ASCE, Vol. 100, No. GT7, July, 1974, pp. 709-747.
29. McClelland, B., and Cox, W.R., "Performance of Pile Foundations for

Offshore Structures," Proceedings of the First International Conference on the Behaviour of Offshore Structures, Trondheim, Norway, Vol. 1, 1976, pp. 528-544.

30. McCammon, N.R., and Golder, H.Q., "Some Loading Tests on Long Pipe Piles," Geotechnique, Vol. 20, No. 2, 1970, pp. 171-184.

31. Meyerhof, G.G., "Bearing Capacity and Settlement of Pile Foundations," Journal of the Geotechnical Engineering Division, ASCE, Vol. 102, No. GT3, Mar., 1976, pp. 197-228.

32. Murff, J.D., "Pile Capacity in a Softening Soil," Journal, Numerical and Analytical Methods in Geomechanics, Vol. 4, No. 2, 1980, pp. 185-189.

33. Olson, R.E., and Dennis, N.D., "Review and Compilation of Pile Test Results, Axial Pile Capacity," PRAC Project 81-29, American Petroleum Institute, Dallas, Texas, Dec., 1982, 297 pp.

34. O'Neill, M.W., Hawkins, R.A., Mahar, L.J., "Load Transfer Mechanisms in Piles and Pile Groups," Journal of the Geotechnical Engineering Division, ASCE, Vol. 108, No. GT12, Dec., 1982, pp. 1605-1623.

35. Peck, R.B., "A Study of the Comparative Behaviour of Friction Piles," Special Report No. 36, National Research Council, Highway Research Board, 1958, 78 pp.

36. Peck, R.B., "Records of Load Tests on Friction Piles," Special Report No. 67, National Research Council, Highway Research Board, 1961, 418 pp.

37. Pelletier, J.H., and Doyle, E.H., "Tension Capacity in Silty Clays - Beta Pile Test," Proceedings of the Second International Conference on Numerical Methods in Offshore Piling, Austin, Texas, 1982, pp. 163-182.

38. Poskitt, T.J., and Ward, G., "The Effect of Stick-Up on Pile Drivability," Journal of the Society for Underwater Technology, Vol. 10, No. 1, 1984, pp. 23-31.

39. Puech, A.A., "Base Data for the Design of Tension Piles in Silty Soils," Proceedings of the Third International Conference on Behaviour of Offshore Structures, Massachusetts, Vol. 1, Aug., 1982, pp. 141-160.

40. Quiros, G.W., Young, A.G., and Pelletier, J.H., "Shear Strength Interpretation for Gulf of Mexico Clays", Proceedings of the Conference on Geotechnical Practice in Offshore Engineering, ASCE, Apr., 1983, pp. 144-165.

41. Randolph, M.F., and Wroth, C.P., "Recent Developments in Understanding the Axial Capacity of Piles in Clay," Ground Engineering, Vol. 15, No. 7, 1982, pp. 17-25.

42. Randolph, M.F., "Design Considerations for Offshore Piles," Proceedings of the Conference on Geotechnical Practice in Offshore Engineering, ASCE, Apr., 1983, pp. 422-439.

43. Randolph, M.F., "Development of API Axial Pile Database Correlations," Amoco Production Company, Sept., 1983.

44. Randolph, M.F., and Murphy, B.S., "Shaft Capacity of Driven Piles in Clay," Proceedings of the Seventeenth Offshore Technology Conference, Houston, 1985. (accepted for publication)

45. Raymond Step Taper Piles, Raymond International, Inc., Brochure ST-82, 1982, p. 25.

46. Rigden, W.J., Petit, J.J., St. John, H.D., and Poskitt, T.J., "Developments in Piling for Offshore Structures," Proceedings of the Second International Conference on Behaviour of Offshore Structures, London, Vol. 2, 1979, pp. 279-296.

47. Rigden, W.J. and Semple, R.M., "Design and Installation of the Magnus Foundation: I-Prediction of Pile Behaviour", Proceedings of the Conference on Developments in the Design and Construction of Offshore Structures", Institution of Civil Engineers, London, 1983, pp. 29-43.

48. Semple, R.M., Discussion of Session 8, Proceedings of the Conference on

Recent Developments in the Design and Construction of Piles, Institution of Civil Engineers, London, 1980, pp. 397-399.

49. Skempton, A.W., "A Study of the Geotechnical Properties of some Post-Glacial Clays," Geotechnique, Vol. 1, 1948, pp. 7-22.

50. Steenfelt, J.S., Randolph, M.F., and Wroth, C.P., "Instrumented Model Piles Jacked into Clay," Proceedings of the Tenth International Conference on Soil Mechanics and Foundation Engineering, Stockholm, Sweden, Vol. 2, 1981, pp. 857-864.

51. Stermac, A.G., Selby, K.G., and Devata, M., "Behaviour of Various Types of Piles in Stiff Clay," Proceedings of the Seventh International Conference on Soil Mechanics and Foundation Engineering, Mexico City, Vol. 2, 1969, pp. 239-245.

52. Togrol, E., "Bearing Capacity By Load Test," Proceedings of the Eighth International Conference on Soil Mechanics and Foundation Engineering, Moscow, Vol. 2, 1973, pp. 109-11.

53. U.S.A.C.E., "Test Pile Report for Lock D Project," Tennessee -Tombigbee Waterway, United States Army Corps of Engineers, Mobile District, 1979.

54. Woodward, R.J., Lundgren, R., and Boitano, J.D., "Pile Loading Tests in Stiff Clay," Proceedings of the Fifth International Conference on Soil Mechanics and Foundation Engineering, Paris, Vol. 2, 1961, pp. 177-184.

55. Wroth, C.P., "Discussion on the Design and Performance of Deep Foundations," Proceedings of the Conference on the Performance of Earth and Earth-Supported Structures, ASCE, Vol. 3, 1972, pp. 231-234.

56. Young, A.G., Quiros, G.W., and Ehlers, C.J., "Effects of Offshore Sampling and Testing an Undrained Shear Strength", Proceedings of the Thirteenth Offshore Technology Conference, Houston, Vol. 1, 1983, pp. 193-204.

Appendix II. - Notation

The following symbols are used in this paper.

D	=	pile outside diameter
F	=	modified length factor
K_s	=	lateral earth pressure coefficient
k	=	constant of proportionality
L	=	pile embedded length
LF	=	length factor
OCR	=	overconsolidation ratio
PI	=	plasticity index
Q_{sc}	=	calculated shaft capacity
Q_{sm}	=	measured shaft capacity
s_u	=	undrained shear strength
t	=	unit skin friction
z	=	relative soil-pile displacement
α	=	skin friction coefficient = t/s_u
β	=	skin friction coefficient = $t/\bar{\sigma}_v$
π_3	=	pile-soil stiffness ratio
$\bar{\sigma}_v$	=	vertical effective stress
ϕ	=	angle of internal friction

LIMITING TIP AND SIDE RESISTANCE: FACT OR FALLACY?

Fred H. Kulhawy, F. ASCE*

The concept of a critical depth for deep foundations, below which the tip and side resistances are constant, is examined. It is argued that this concept is incorrect. A more fundamental examination of the problem provides a rational explanation of observed phenomena and should be used for design.

INTRODUCTION

It has become commonplace in many recent textbooks and papers to state, as a matter-of-fact, that side and tip resistances of deep foundations achieve a maximum value at a critical depth of some 10 to 20 foundation diameters. Below this critical depth, the resistances are assumed to be constant with depth. This concept is commonly attributed to Vesić (7). However, this concept actually was a preliminary working hypothesis developed from laboratory model tests. More recent and definitive work by Vesić (10) contains no mention of this concept and recommends more rational procedures. Unfortunately it was an attractive approach for design and rapidly was included in texts as an explanation of how deep foundations actually behave. This is not the case. In this paper a general approach is presented which demonstrates rationally how tip and side resistances develop.

EQUILIBRIUM CONSIDERATIONS AND FAILURE MODE

As illustrated in Fig. 1, the applied compressive load on a deep foundation is resisted by a combination of tip resistance and side resistance. Considering equilibrium, we have:

$$Q_c + W = Q_{tc} + Q_{sc} \tag{1}$$

in which Q_c = applied load at the butt, W = weight of foundation and enclosed soil within the volume B × L × D, Q_{tc} = tip resistance, and Q_{sc} = side resistance. Both the tip and side resistances develop as a function of foundation displacement, and the peak values of each are unlikely to occur at the same displacement.

On loading, the foundation moves downward with respect to the surrounding soil, mobilizing side resistance. Fig. 1b illustrates the typical load transfer, and Fig. 1c shows the corresponding unit side resistance, f. For small loads, small displacements occur, and most of the

* Professor, School of Civil and Environmental Engineering, Cornell University, Ithaca, NY 14853

load is supported by side resistance. As the loads increase, larger displacements occur and the full side resistance is reached. From this point on, further applied loads are resisted at the tip. Fig. 1b illustrates this point, from the dotted line at small loads to the solid line at large loads. If no strain-softening occurs, the unit side resistance will remain approximately constant.

The full side resistance is mobilized at very small displacements, typically in the range of 0.2 to 0.4 in. (5 to 10 mm), and is relatively independent of the foundation or soil type, or the foundation dimensions. On the other hand, relatively large displacements are needed to mobilize the tip resistance and these are apparently a function of foundation type and size. This displacement is about 8% of the diameter for driven piles and up to 30% of the diameter for drilled shafts (10). However, most of the tip resistance for most drilled shafts is mobilized at a displacement of about 10% of the diameter.

As described above, side resistance evaluation is a problem in determining the shearing resistance of the foundation-soil interface over the surface of the foundation. Depending on foundation shape, this is a cylindrical to rectangular shear model. Tip resistance is a bearing capacity problem.

In the following, for the sake of simplicity, it will be assumed that the foundations are vertical, with vertical concentric loads, and that the ground surface is horizontal.

SIDE RESISTANCE IN DRAINED LOADING

The side resistance is expressed as:

$$Q_{sc} = \int_{surface} \tau(z)\, dz \tag{2}$$

in which $\tau(z)$ = shearing resistance along the foundation-soil interface.

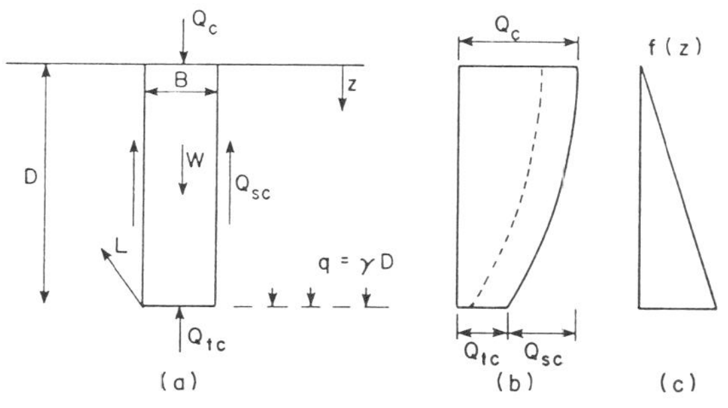

Figure 1. General Description of Deep Foundation

For drained conditions, c = 0 and Eq. 2 then can be given as:

$$Q_{sc} = \int_0^D P(z)\ \bar{\sigma}_h(z)\ \tan\ \delta(z)\ dz \qquad (3)$$

in which P = foundation perimeter, $\bar{\sigma}_h$ = horizontal effective stress, which acts as a normal stress on the soil-foundation interface, and δ = effective stress interface friction angle. Expressing the horizontal stress in terms of the vertical stress ($\bar{\sigma}_v = \bar{\gamma}z$) results in:

$$Q_{sc} = \int_0^D P(z)\ \bar{\gamma}(z)\ z\ K(z)\ \tan\ \delta(z)\ dz \qquad (4)$$

in which $\bar{\gamma}$ = effective soil unit weight and K = coefficient of horizontal soil stress ($\bar{\sigma}_h/\bar{\sigma}_v$). From Eq. 4, terms are often grouped to define $\beta = K \tan \delta$ and $f = \beta\bar{\gamma}z$ = unit side resistance. The critical terms to evaluate in these equations are K and δ.

Interface Angle of Friction. The effective stress angle of friction for the soil-concrete interface is important for determining the frictional strength of the interface. Details of interface behavior studies have been summarized by Kulhawy, et al. (4), and the results have shown that δ can be expressed in terms of $\bar{\phi}$, the effective stress angle of friction for the soil. These results are presented in Table 1. These values are also approximately correct for loading of clays.

Coefficient of Horizontal Soil Stress. The most complex term to evaluate is K, the coefficient of horizontal soil stress. This coefficient is a function of the original in-situ horizontal stresses, and the stress changes caused in response to construction, loading, and time. Taking these factors into account, there may be either an increase or decrease from the original in-situ K_o, the coefficient of horizontal soil stress at-rest. Values of K reported in the literature range from

Table 1

INTERFACE FRICTION ANGLES

Interface Materials	Ratio of Interface Angle of Friction to Soil Angle of Friction, $\delta/\bar{\phi}$	Typical Field Analogy
Sand/rough concrete	1.0	Cast-in-place
Sand/smooth concrete	0.8 to 1.0	Precast
Sand/rough steel	0.7 to 0.9	Corrugated
Sand/smooth steel	0.5 to 0.7	Coated
Sand/timber	0.8 to 0.9	Pressure-treated

a low of about 0.1 to a high of around 5. These bounds roughly correspond to the range from minimum active to maximum passive stress states.

To illustrate some of the principles which lead to these variations, consider the example in Fig. 2. For simplicity, assume 100 ft. (30.5 m) of uniform sand, water table at the ground surface, and a submerged unit weight of 50 pcf (7.9 kN/m^3). The soil angle of friction varies as shown, from 40° at ground surface to 37° at the stress level present at a depth of 100 ft. Also, δ is equal to $\bar{\phi}$. If the sand is truly normally consolidated (NC), then K_0 is nearly constant with depth, β is nearly constant with depth, and f increases almost linearly with depth. However, if the sand is overconsolidated (OC) to some degree, the situation changes markedly. Some overconsolidation is the rule, rather than the exception, in most soil deposits. In the example given, the lower half of the sand is normally consolidated, and overconsolidation increases with elevation. Therefore, K_0 is largest near the ground surface and it decreases with depth until it becomes constant in the normally consolidated range. The values for β follow the same pattern. The values of unit side resistance, f, increase rapidly at shallow depths, appear to be constant at intermediate depths, and then increase again at greater depths. The shape of these curves is the reason for much of the variation in these parameters reported in the literature, because average values are most commonly computed and reported. For example, a 20 ft. (6.1 m) deep foundation would suggest an average $\beta \approx 1.5$, but a 100 ft (30.5 m) deep foundation would suggest

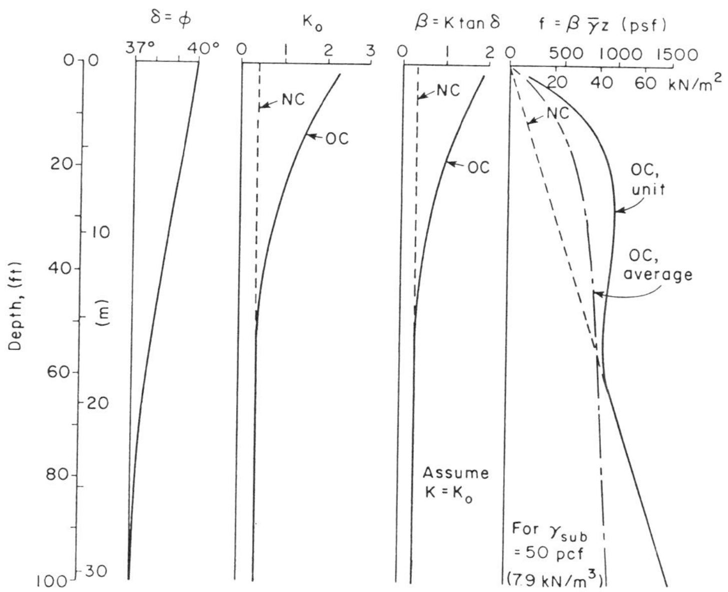

Figure 2. Example Illustrating Principles of Side Resistance in Sand

an average $\beta \approx 0.6$. Furthermore, the shape of the unit or average side resistance curves might suggest that there is a limiting value of f, particularly if the foundation was shorter than 60 to 70 ft. (18 to 21 m). In fact, this suggestion has been made by a number of authors and, as stated previously, has been accepted too casually. However, this example shows that the apparent limiting value of f is only a coincidence which occurs because of decreasing β and increasing z nullifying each other over a limited range in depth. The actual values of f continue to change with depth.

The example further illustrates that ultimately the foundation capacity is controlled by the in-situ stresses. These are determined by the original in-situ value of K_o at rest, and the stress changes caused in response to construction, loading, and time. At the present time, there are three approaches to evaluate K_o. The first is field measurement using in-situ devices such as the pressuremeter. Although some problems exist in the conduct and interpretation of the test, it will suffice to say that the pressuremeter can be used to provide a field estimate of the in-situ K_o. The second approach is to estimate K_o based on a reconstruction of the stress history of the soil (5). The third approach is empirical correlations with field or lab index parameters (e.g., 6).

Once the variation of the in-situ K_o with depth is known, the effects of construction and loading can be introduced to determine the value of K for computing the side resistance. Unfortunately much of the available data is not detailed sufficiently to evaluate these effects precisely. This is because the importance of stress history phenomena which lead to significant variation of K_o with depth has not been fully appreciated until relatively recently, so these data have not been obtained in conjunction with most load tests. However, recent studies by Kulhawy, et al. (4) have led to the recommendations given in Table 2. These have been evaluated using an extensive load test data base (6) and the agreement has been good.

Table 2

HORIZONTAL SOIL STRESS COEFFICIENTS

Foundation Type and Method of Installation	Ratio of Horizontal Soil Stress Coefficient to In-Situ Value, K/K_o
Jetted pile	1/2 to 2/3
Drilled shaft, cast-in-place	2/3 to 1
Driven pile, small displacement	3/4 to 5/4
Driven pile, large displacement	1 to 2

SIDE RESISTANCE IN UNDRAINED LOADING

The side resistance in undrained loading can be expressed in a manner similar to that for drained loading, except that all parameters must be expressed in terms of the effective stresses representative of those during initial loading. Although this problem is not fully resolved, Stas and Kulhawy (6) have shown that the side resistance is essentially similar to that shown in Fig. 2 and that no major differences in behavior occur between drained and undrained conditions.

TIP RESISTANCE

The evaluation of tip resistance is a bearing capacity problem, given by:

$$Q_{tc} = q_{ult} A_{tip} \tag{5}$$

in which q_{ult} = ultimate bearing capacity at the tip and A_{tip} = area of the foundation tip. As shown in Fig. 3, bearing capacity failure occurs as a shear failure in the soil supporting the foundation. Three principal failure modes can occur: general shear, local shear, and punching shear. Fig. 4 illustrates these modes. The boundaries between these three failure modes are not distinct, primarily because the changes in soil behavior modes are gradual. For deep foundations, the typical failure mode is punching shear (10).

Different methods of analysis often are used for the different failure modes. For the general shear mode, a rational approach based on limiting states of equilibrium is employed. The approach is based on theory and has been confirmed by laboratory and field testing. For the local and punching shear modes, a variety of approaches have been suggested, including simplified general shear models, reduced strength parameters in the general shear model, and the onset of plastic yield in an elastic medium. None of these approaches is strictly correct, although most yield computed capacities which are at least somewhat comparable to the field results. A more rational approach, used herein, has been suggested by Vesić (9), which incorporates the stress-deformation characteristics of the soil and is therefore applicable over a wide range of soil behavior.

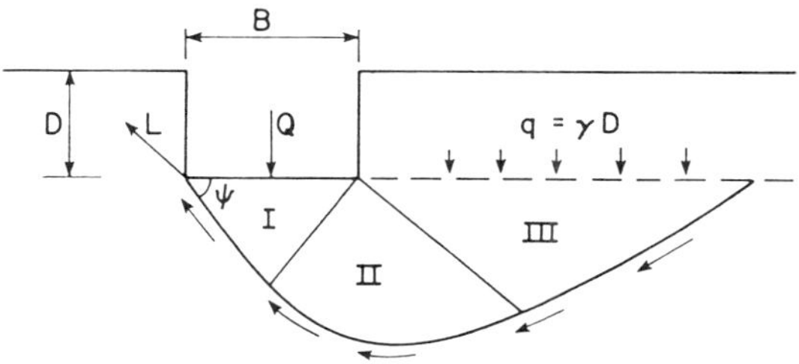

Figure 3. General Description of Bearing Capacity

GENERAL SHEAR FAILURE

The ultimate bearing capacity for the general shear mode is a problem in plastic equilibrium, utilizing the shear surface illustrated in Fig. 3 for $L = \infty$ (i.e., a plane strain solution). The soil within the shear surface is assumed to behave as a rigid-plastic medium and is idealized as three zones: an active Rankine zone (I), a radial Prandtl zone (II), and a passive Rankine zone (III). The soil above the foundation base is modeled as an equivalent surcharge, shown in Fig. 3.

The general solution is the Buisman-Terzaghi equation given below:

$$q_{ult} = c\,N_c + \frac{1}{2} B\gamma\,N_\gamma + q\,N_q \qquad (6)$$

in which: c = soil cohesion, q = surcharge (= γD), B = foundation width, D = foundation depth, γ = soil unit weight, and N_c, N_γ, and N_q = dimensionless bearing capacity factors. Eq. 6 is not a rigorous solution; it is a superposition of solutions which leads to a conservative result, ranging from 17 to 20% for $\phi = 30°$ to $40°$ to a minimum of 0 for $\phi = 0°$ (9).

Eq. 6 includes the Prandtl and Reissner solutions for a foundation on a weightless medium (γ = 0 below depth D), resulting in the first and third terms in the equation and the following values for the bearing

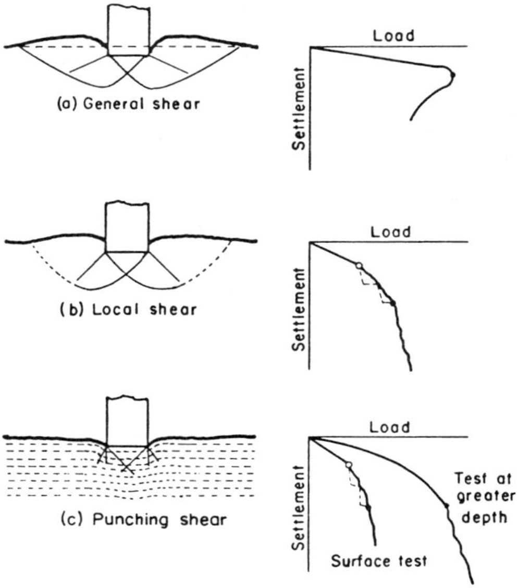

Figure 4. Modes of Bearing Capacity Failure

capacity factors:

$$N_q = e^{\pi \tan\phi} \tan^2 (45° + \phi/2) \tag{7}$$

$$N_c = (N_q - 1) \cot\phi \qquad \text{(note: as } \phi \to 0, N_c \to 5.14) \tag{8}$$

in which ϕ = angle of friction for the soil (alternately known as the angle of shearing resistance). Values of N_c and N_q are given in Fig. 5.

The second term in Eq. 6 results from a solution for cohesionless soil without overburden (c = 0, D = 0). The resulting values for N_γ vary widely, depending on the assumed shear surface and the value of ψ in Fig. 3. The numerical solution by Caquot and Kerisel (1) for $\psi = 45° + \phi/2$ is perhaps the most widely accepted, and is given as a dotted line in Fig. 5. Vesić (9) showed that the Caquot-Kerisel solution can be conveniently approximated by:

$$N_\gamma \approx 2(N_q + 1) \tan\phi \tag{9}$$

Values from Eq. 9 are plotted as a solid line in Fig. 5. The approximation is within 5% for ϕ = 20° to 40°.

Eq. 6 is most commonly used in either of two derivative forms, which depend primarily on the soil type and rate of loading. For undrained loading, $\phi = 0$ and therefore $N_c = 5.14$, $N_\gamma = 0$, and $N_q = 1$. The resulting equation is:

$$q_{ult} = 5.14c + q \tag{10}$$

In this case, the cohesion term is actually the undrained shear strength, s_u. For drained loading, c = 0 and Eq. 6 becomes:

$$q_{ult} = \frac{1}{2} B\gamma N_\gamma + q N_q \tag{11}$$

Eqs. 10 and 11 represent the usual cases employed in design.

RIGIDITY INDEX, AND LOCAL AND PUNCHING SHEAR

Fig. 4 illustrates that the mode of failure and the ultimate bearing capacity is primarily a function of the load-settlement (or stress-strain) behavior of the foundation soil. For very stiff or dense soils, the deformation under load is small, and this corresponds well with the assumed rigid-plastic soil behavior for general shear. For softer or looser soils, the deformation under load becomes large. This increased deformability is at variance with an assumption of rigid-plastic soil behavior, and has led to various ad-hoc approaches in the literature.

Vesić (9) suggested a rational solution which incorporates explicitly the soil deformability. He introduced the rigidity index, I_r, which was developed in the theory for an expansion of a cavity in an infinite soil mass (8) and assumed that the ultimate normal stress on the sides of the active Rankine wedge (Zone I in Fig. 3) is equal to the stress needed to expand a cavity in the same soil mass. The rigidity index is defined as:

$$I_r = \frac{G}{c + q \tan\phi} \tag{12}$$

in which G = shear modulus of soil and q is evaluated at a depth of B/2 below the foundation. The shear modulus is computed from the elastic modulus, E, and Poisson's ratio, ν, by the following:

$$G = \frac{E}{2(1 + \nu)} \tag{13}$$

E is normally determined from field or laboratory tests. Poisson's ratio can also be determined by testing but, since the values do not vary greatly, it is normally estimated.

Once I_r has been determined, it is compared to the critical rigidity index, which is defined as:

$$I_{rc} = \frac{1}{2} \exp[(3.30 - 0.45\ B/L) \cot(45° - \phi/2)] \tag{14}$$

Values of the critical rigidity index are given in Fig. 6. When $I_r >$

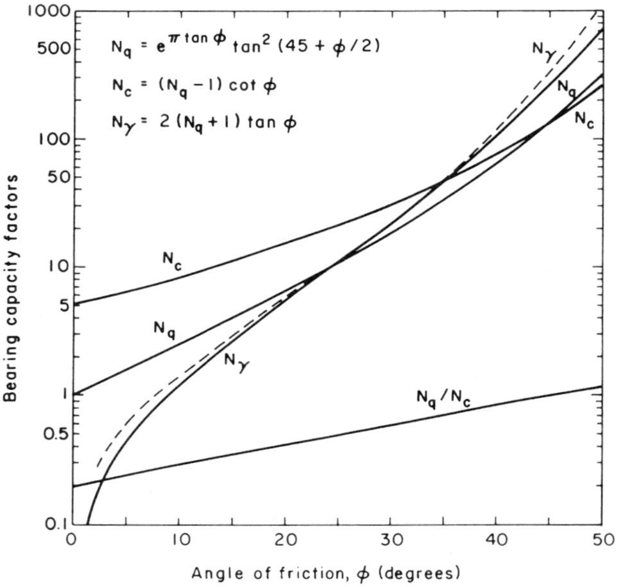

Figure 5. Bearing Capacity Factors for Shallow Foundations

I_{rc}, the soil behaves, for all practical purposes, as a rigid-plastic material and therefore the general shear failure equation would be used. When $I_r < I_{rc}$, local or punching shear would occur and Eq. 6 would be modified as below to account for the decreased rigidity of the soil:

$$q_{ult} = c\, N_c\, \zeta_{cr} + \frac{1}{2} B\gamma\, N_\gamma\, \zeta_{\gamma r} + q\, N_q\, \zeta_{qr} \tag{15}$$

The ζ terms are the rigidity factors and are equal to:

$$\zeta_{qr} = \exp\{[(-4.4 + 0.6\ B/L)\tan\phi] + [(3.07 \sin\phi)(\log_{10} 2I_r)/(1 + \sin\phi)]\} \tag{16}$$

$$\zeta_{\gamma r} = \zeta_{qr} \tag{17}$$

$$\zeta_{cr} = \zeta_{qr} - (1 - \zeta_{qr})/(N_c \tan\phi) \tag{18}$$

For $\phi = 0$:

$$\zeta_{cr} = 0.32 + 0.12\ B/L + 0.60 \log_{10} I_r \tag{19}$$

Values of the rigidity factors are given in Figs. 7 and 8.

Vesić suggested one further modification, to account for the average volumetric strain, Δ, in the active Rankine wedge. This results in a reduced rigidity index for drained loading, given as:

$$I_{rr} = I_r/(1 + I_r\ \Delta) \tag{20}$$

which would replace I_r in the equations above. The average volumetric strain can be determined reasonably well from the results of drained triaxial tests or field tests such as the pressuremeter. For dense soils ($I_r > I_{rc}$), Δ would be taken as zero. For loose soils, typical values of Δ would be in the range of 0 to 0.025 for a stress range of 1 to 5 tons/ft^2 (100 to 500 kN/m^2), and in the range of 0.025 to 0.05 for a stress range of 5 to 10 tons/ft^2 (500 to 1000 kN/m^2).

The use of the rigidity index concept provides both a generalized, as well as a rational, approach to computing the ultimate bearing capacity. The approach is also convenient because the rigidity factors are not very sensitive to small variations in the rigidity index. Although detailed field confirmation is still needed, this approach is the most rational one currently available, and is recommended.

FOUNDATION SHAPE AND DEPTH

The bearing capacity theory described has been developed for the following assumptions: (1) horizontal ground surface, (2) horizontal, infinitely long strip foundation at ground surface, and (3) vertical loading, concentrically applied. Comparable theory exists for a circular foundation. All of these assumptions are not met in practice.

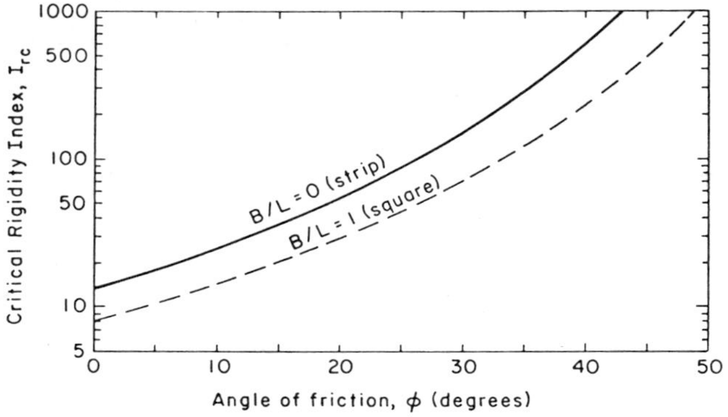

Figure 6. Critical Rigidity Index

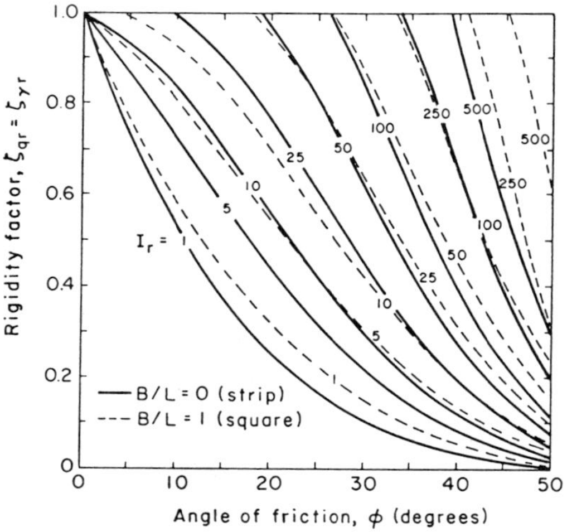

Figure 7. Rigidity Factors, $\zeta_{qr} = \zeta_{\gamma r}$

Various recommendations have been made to adjust the basic bearing capacity equation to take into account these variations. Some have a theoretical basis, but most are empirical correlations with laboratory and field tests. Accordingly the adjustments cited in the literature vary with the data base employed and the judgment of the interpreter of the data. Those described below are based upon the consistent interpretation of Vesić (9).

To take into account the variations which occur in practice, the basic bearing capacity equation is modified by a number of factors which separately treat each variation. To account for foundation shape and depth, the bearing capacity equation is given as:

$$q_{ult} = c\ N_c\ \zeta_{cr}\ \zeta_{cs}\ \zeta_{cd} + \frac{1}{2} B\gamma\ N_\gamma\ \zeta_{\gamma r}\ \zeta_{\gamma s}\ \zeta_{\gamma d} + q\ N_q\ \zeta_{qr}\ \zeta_{qs}\ \zeta_{qd} \qquad (21)$$

The ζ factors are doubly subscripted to indicate which term it applies to (c, γ, or q) and which phenomenon it describes (r for rigidity of the soil, s for shape of the foundation, and d for depth of the foundation). The r factors have already been described.

Shape factors have been developed from the limiting solutions for infinite strip foundations and circular foundations, coupled with experimental observations, and are given below:

$$\zeta_{cs} = 1 + (B/L)(N_q/N_c) \qquad (22)$$

$$\zeta_{qs} = 1 + (B/L)\ \tan\phi \qquad (23)$$

$$\zeta_{\gamma s} = 1 - 0.4\ (B/L) \qquad (24)$$

These factors equal 1 for an infinite strip ($L \rightarrow \infty$), and B/L equals 1

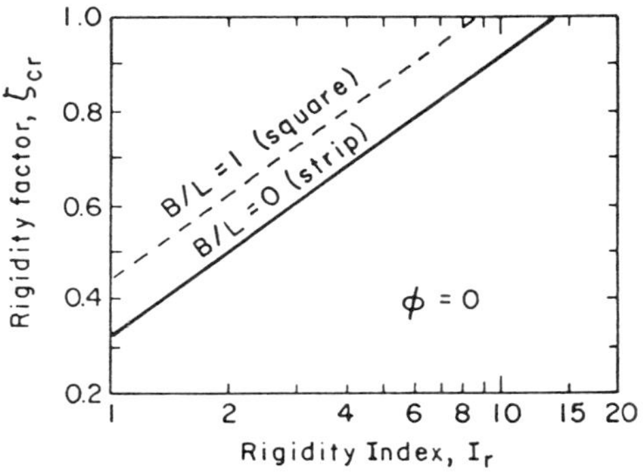

Figure 8. Rigidity Factor, ζ_{cr}

for a square or circle.

Foundations normally are imbedded to a depth, D, below the ground surface. If the foundation is installed so that the surrounding soil above the foundation level remains undisturbed, then the shear surface in Fig. 3 would continue above the foundation to the ground surface. The result would be an increased bearing capacity. This situation is common with properly installed drilled shaft foundations, where soil disturbance is minimal, and with driven piles, which tend to densify the surrounding soil.

For these cases, depth factors can be introduced. These are based largely upon solutions for strip foundations, but should be applicable to the more general problem. The commonly used approximate formulas proposed by Hansen (3) contain an inconsistency because the formulas for $D/B < 1$ do not agree with those for $D/B > 1$ when a comparison is made at $D/B = 1$. The slightly more conservative formulas are recommended and are given below:

$$\zeta_{qd} = 1 + 2 \tan\phi \, (1 - \sin\phi)^2 \tan^{-1} (D/B) \tag{25}$$

$$\zeta_{\gamma d} = 1 \tag{26}$$

$$\zeta_{cd} = \zeta_{qd} - (1 - \zeta_{qd})/(N_c \tan \phi) \tag{27}$$

For $\phi = 0$, the following is recommended:

$$\zeta_{cd} = 1 + 0.33 \tan^{-1} (D/B) \tag{28}$$

in which the $\tan^{-1}$ terms are expressed in radians. The original work leads to a factor of 0.40 in place of the 0.33 in Eq. 28; however, the 0.33 yields results more consistent with the data for shallow and intermediate depths and it gives the well-known result of $q_{ult} = 9.4\ c + q$ for very deep circular or square foundations.

TIP RESISTANCE IN UNDRAINED LOADING

For undrained conditions, $\phi = 0$ and therefore $N_c = 5.14$, $N_\gamma = 0$, $N_q = 1$ and $\zeta_{qr} = \zeta_{qs} = \zeta_{qd} = 1$. Eq. 21 then simplifies to:

$$q_{ult} = 5.14\ c\ \zeta_{cr}\ \zeta_{cs}\ \zeta_{cd} + q \tag{29}$$

Substituting for the ζ terms, assuming a circular or square foundation ($B = L$), the first term in Eq. 29 yields the results shown in Fig. 9. For a rigidity index of 8 or more, the values are constant. Many clays fall into this category. Also the bearing capacity factor changes little by the time D/B reaches 5. Increasing D/B from 5 to ∞ only increases the bearing capacity factor by about 4.5%. From these results it can be seen that the common assumption of 9 for the deep bearing capacity factor is a reasonable one.

Eq. 29 is often presented in the literature using the net ultimate bearing capacity, which subtracts out the weight of the foundation and enclosed soil, W. By this definition, the equation becomes:

$$q_{net} = 5.14\ c\ \zeta_{cr}\ \zeta_{cs}\ \zeta_{cd} + q - W/A_{tip} \tag{30}$$

In some situations, the q and W/A_{tip} terms are comparable in magnitude and are disregarded, resulting in the simplified expression:

$$q_{net} \approx 9\ c \tag{31}$$

Although it is tempting to use this simple approximation, the more correct Eq. 29 should be used, particularly since the weight term has already been included in Eq. 1.

Eq. 29 shows clearly that q_{ult} increases with increasing depth principally because q increases. In addition, c increases for normally consolidated clays because the consolidation stress increases with depth.

TIP RESISTANCE IN DRAINED LOADING

For drained conditions, c = 0 and Eq. 21 simplifies to:

$$q_{ult} = \frac{1}{2}\ B\gamma\ N_\gamma\ \zeta_{\gamma r}\ \zeta_{\gamma s}\ \zeta_{\gamma d} + q\ N_q\ \zeta_{qr}\ \zeta_{qs}\ \zeta_{qd} \tag{32}$$

For a circular or square foundation (B = L), $\zeta_{\gamma s} = 0.6$ and $\zeta_{\gamma d} = 1$, resulting in:

$$q_{ult} = 0.3\ B\gamma\ N_\gamma\ \zeta_{\gamma r} + q\ N_q\ \zeta_{qr}\ \zeta_{qs}\ \zeta_{qd} \tag{33}$$

The dimensionless portion of the first term is given in Fig. 10. This first term rapidly becomes small, compared to the second term, as D/B

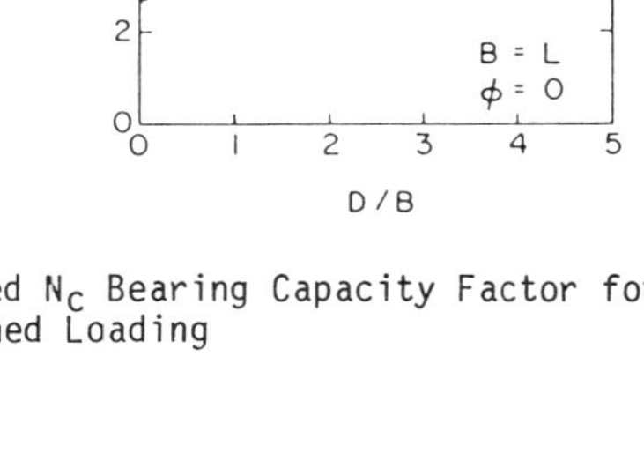

Figure 9. Modified N_c Bearing Capacity Factor for Deep Foundations in Undrained Loading

increases. In fact, by the time D/B equals 4 to 5, the first term typically is less than 10% of the second term. Therefore, for large D/B ratios, the first term can be safely disregarded.

The dimensionless portion of the second term is presented in Fig. 11 for a circular or square foundation, with D/B ratios of 0 and 5. Increasing D/B from 5 to ∞ only increases the results by 2 to 4.5%, so the value for 5 can be used conservatively for D/B > 5. Caution must be exercised with the use of this figure because the bearing capacity factor is not constant with depth. (The same is true for Fig. 10.) It is well established in soil mechanics that, as the stress magnitude increases, both the angle of friction and the rigidity index for the soil decrease. This means that, as the foundation depth increases, both ϕ and I_r decrease, resulting in decreasing values of the bearing capacity factors with depth.

As an illustration of this principle, consider the following

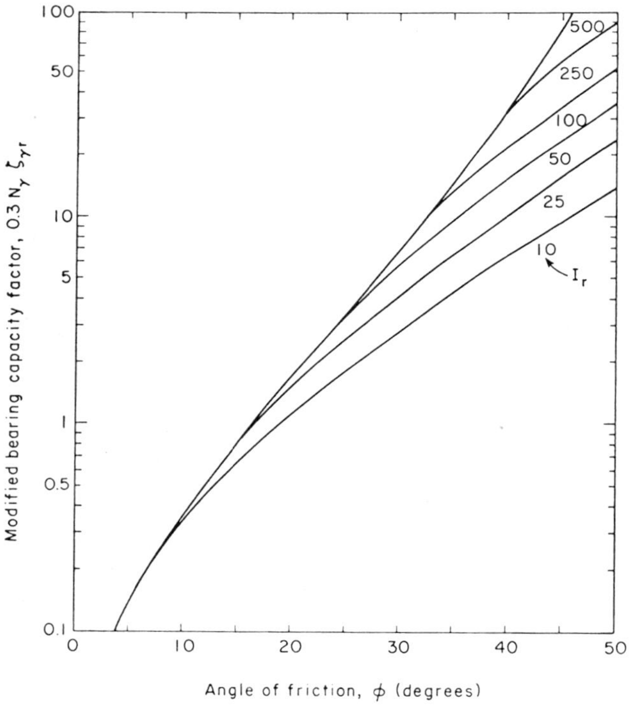

Figure 10. Modified N_γ Bearing Capacity Factor for Deep Foundations in Drained Loading

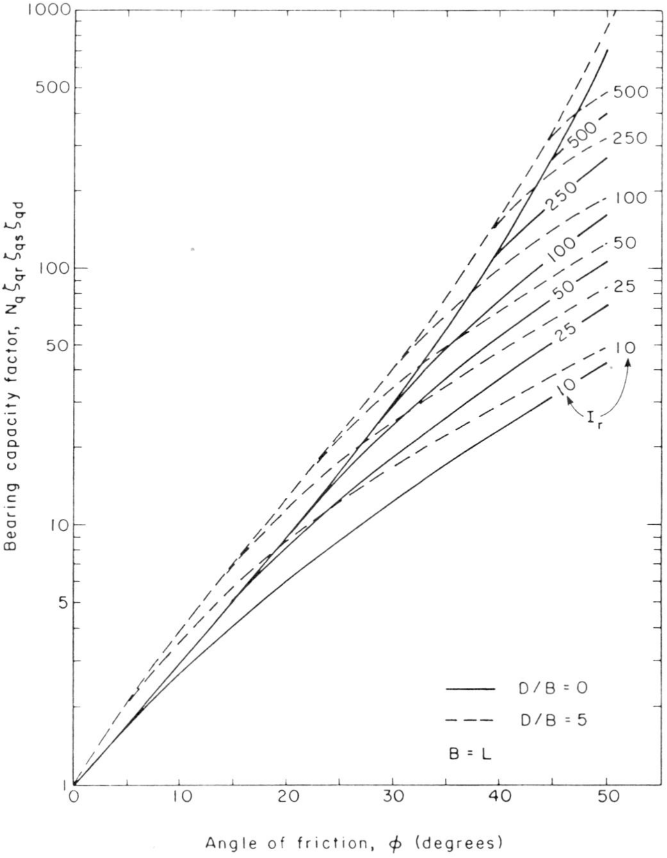

Figure 11. Modified N_q Bearing Capacity Factor for Deep Foundations in Drained Loading

simplified example. For drained conditions, the rigidity index is given by:

$$I_r = \frac{G}{q \tan\phi} = \frac{E}{2(1 + \nu)\ q \tan\phi} \tag{34}$$

Assuming a hyperbolic stress-strain curve (2), the modulus can be approximated by:

$$E \approx \kappa\, p_a \left(\frac{q}{p_a}\right)^n (1 - S)^2 \tag{35}$$

in which κ = modulus number, n = modulus exponent, S = stress level (percentage of failure stress), and p_a = atmospheric pressure in the desired units. Typical values for κ range from 300 for loose sand to 1200 for dense sand, while n is about 0.5 for sand. We will assume that the modulus at 50% stress level is representative, so S = 0.5. Typical values for ν are 0.35 for dense sand and 0.20 for loose sand. Representative limiting effective stress friction angles range from 40° for dense sand to 28° for loose sand, both at low stress magnitudes. If we consider the stresses from ground surface to a depth of about 100 ft. (30.5 m) in dry sand, the friction angles will decrease some 2 to 3° for dense sand and 1 to 2° for loose sand. Substituting these values, the rigidity index is approximated by:

$$I_r \approx \frac{30}{q^{1/2} \tan\,(28° \text{ to } 26°)} \quad \text{in loose sand, for q in ton/ft}^2 \qquad (36a)$$

$$I_r \approx \frac{110}{q^{1/2} \tan\,(40° \text{ to } 37°)} \quad \text{in dense sand, for q in ton/ft}^2 \qquad (36b)$$

With I_r given by Eqs. 36, dry unit weights of 115 and 125 pcf (18.1 and 19.7 kN/m^3) for loose and dense sand, and friction angles varying as above, the approximate tip ultimate bearing capacity can be computed, as given in Fig. 12. Because of the decreasing bearing capacity factor with depth, the tip capacity increases at a decreasing rate. It is clear that no limiting values develop.

Fig. 12 may be used as a rough first approximation for tip capacity in uniform simple sand deposits, realizing that simplifying assumptions and generalized values are included. Linear interpolation may be used for intermediate densities. If the water table is at ground surface, the dotted curves should be used.

Furthermore, the type of deep foundation and construction procedure will influence the tip capacity. For driven displacement piles, a small increase is likely because of densification and stress increase caused by driving. For bulb piles, a large increase is likely because of greater densification and further stress increases caused by creating the bulb. For drilled shafts, a small decrease is likely because of loosening at the tip during drilling, incomplete clean-out before concreting and, for slurry construction, softening because of the slurry.

SUMMARY AND CONCLUSIONS

The tip and side resistances of deep foundations do not reach a limit at a critical depth; instead, they increase with depth as shown in this paper. For the tip, the rate of increase decreases with depth, primarily because of decreasing rigidity with depth. For the side, the rate is a function of increasing overburden and decreasing K_o with depth. The approach outlined is a more rational explanation of the actual behavior of deep foundations and should be used in design.

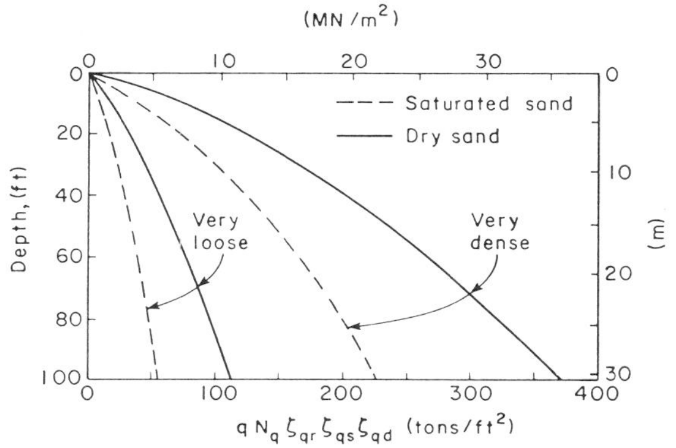

Figure 12. Approximate Tip Ultimate Bearing Capacity for Deep Foundations in Sand

ACKNOWLEDGMENTS

The research described in this paper was supported by the Electric Power Research Institute under Research Project 1493-1, and is reported in more detail in Reference 4. P. G. Landers was the EPRI Project Manager for this study. L. Crouse typed the text.

REFERENCES

1. Caquot, A. and Kerisel, J., "Sur le Terme de Surface Dans le Calcul des Fondations en Milieu Pulvérulent", Proceedings, 3rd International Conference on Soil Mechanics and Foundation Engineering, Vol. 1, Zurich, 1953, pp. 336-337.

2. Duncan, J. M. and Chang, C.-Y., "Nonlinear Analysis of Stress and Strain in Soils", Journal of the Soil Mechanics and Foundations Division, ASCE, Vol. 96, No. SM5, Sept. 1970, pp. 1629-1653.

3. Hansen, J. B., "A Revised and Extended Formula for Bearing Capacity", Bulletin No. 28, Danish Geotechnical Institute, Copenhagen, 1970, pp. 5-11.

4. Kulhawy, F. H., Trautmann, C. H., Beech, J. F., O'Rourke, T. D., McGuire, W., Wood, W. A. and Capano, C., "Transmission Line Structure Foundations for Uplift-Compression Loading", Report EL-2870, Electric Power Research Institute, Palo Alto, CA, Feb. 1983, 412 p.

5. Mayne, P. W. and Kulhawy, F. H., "K_0-OCR Relationships in Soil", Journal of the Geotechnical Engineering Division, ASCE, Vol. 108, No. GT6, June 1982, pp. 851-872.

6. Stas, C. V. and Kulhawy, F. H., "Critical Evaluation of Design Methods for Foundations Under Axial Uplift and Compression Loading", Report on Project 1493, Electric Power Research Institute, Palo Alto, CA, 1984, (in press).

7. Vesić, A. S., "A Study of Bearing Capacity of Deep Foundations", Final Report, Project B-189, Georgia Institute of Technology, Atlanta, GA, Mar. 1967, 279 p.

8. Vesić, A. S., "Expansion of Cavities in Infinite Soil Mass", Journal of the Soil Mechanics and Foundations Division, ASCE, Vol. 98, No. SM3, Mar. 1972, pp. 265-290.

9. Vesić, A. S., "Bearing Capacity of Shallow Foundations", Chapter 3 in Foundation Engineering Handbook, Ed. by H. Winterkorn and H. Y. Fang, Van Nostrand Reinhold Company, New York, NY, 1975, pp. 121-147.

10. Vesić, A. S., "Design of Pile Foundations", Synthesis of Highway Practice No. 42, Transportation Research Board, Washington, DC, 1977, 68 p.

CYCLIC DEGRADATION OF PILE PERFORMANCE IN CALCAREOUS SOILS

by

H.G. POULOS[1], M.ASCE.

ABSTRACT

In this paper, laboratory data on the skin friction characteristics of jacked or driven model piles in calcareous sand is presented. It is demonstrated that the ultimate skin friction is generally reduced following cyclic loading and that this reduction depends primarily on the amplitude of cyclic displacement. Significant cyclic degradation of skin friction commences when the cyclic displacement amplitude reaches the static displacement required to mobilise full skin friction. Other factors which influence cyclic degradation are the method of installation of the pile, the relative density of the soil and the overconsolidation ratio. However there appears to be relatively little effect of cyclic loading on the soil modulus or the ultimate base bearing capacity.

Using a previously-developed theoretical analysis, parametric solutions are obtained for the loss of ultimate shaft load capacity and stiffness of driven offshore piles subjected to cyclic loading. It is demonstrated that shorter stiffer piles may suffer a greater percentage loss in shaft resistance than larger, more flexible piles. Tension piles tend to be less affected by cyclic loading than compression piles, because of the smaller cyclic displacements that are developed under normal working conditions.

[1] Professor and Head, School of Civil and Mining Engineering, University of Sydney, Australia, 2006.

INTRODUCTION

Calcareous and carbonate soils are widely found in offshore deposits in the more temperate regions of the earth. Many of these are of biogenic origin and are composed primarily of the skeletons of marine organisms living in the upper waters of the ocean. One of the features of such soils is that the particles have a very low strength compared to quartz-based silica sands.

A review of some available performance data and current design practice for driven piles in carbonate sands has been made by Datta et al (2). Driven piles have been found to encounter very low driving resistance and to have a low static load capacity when compared to corresponding piles in silica sand. It is postulated that, as a pile is driven into a carbonate sand stratum, crushing of the particles, and hence a volume reduction, occurs due to crushed fines moving into existing voids. Consequently, the lateral stress developed between the pile and the soil is low, resulting in reduced skin friction. Recent model tests by Nauroy and Le Tirant (7) confirm that the pile driving process can actually lead to a reduction in lateral stress near the pile-soil interface, in contrast to the marked increase which occurs near piles driven into silica sands. Moreover, the relatively high compressibility of the soil in the vicinity of the pile tip results in a lower end bearing capacity than for silica sands.

Despite the increasing use of piles in carbonate sands, there appears to have been relatively little systematic research carried out into the factors which influence pile skin friction in such soils, and in particular, the effects of cyclic loading. Such considerations are essential for the design of piles subjected to wave forces. There are indications from small-scale field tests (4) that cyclic loading may cause major reductions in skin friction.

This paper has the following objectives:

i) to present laboratory data on the skin friction characteristics of jacked or driven model piles in calcareous sand under static and cyclic loading conditions

ii) to summarise a theoretical approach to the analysis of cyclic axial response of piles

iii) to present parametric solutions which enable an estimate to be made of the reduction in ultimate load capacity and pile head stiffness of driven offshore piles when subjected to cyclic axial tensile or compression loading.

LABORATORY TESTS ON MODEL PILES

Apparatus and Procedure

A detailed description of the apparatus and results obtained is given by Poulos and Chan (12) and hence only a brief summary will be presented herein. The apparatus developed for the model tests is illustrated in Fig. 1. The vessel containing the soil and pile base is 180 mm (7 in) internal diameter and is 256 mm (10 in) deep. The soil can be consolidated under one-dimensional conditions by an equal pressure applied at the flexible top and base, and is restrained laterally by the sides of the test vessel. In this way, both the consolidation (overburden) pressure and the overconsolidation ratio (OCR) may be controlled. The model piles tested were 20 mm (0.79 in) diameter, 256 mm (10 in) long solid aluminium sections which were jacked into the sand prior to final consolidation of the sand. Some tests were also performed on "buried" piles, in which case the pile was initially placed in position and the sand rained around it. Because of the absence of restraint at the pile tip, only the load-settlement characteristics of the pile shaft are measured. Moreover, because the soil top and base are stress-controlled, a meaningful interpretation can be made of the load settlement curve to obtain both the soil modulus and ultimate skin friction.

The tests generally consisted of the following stages:

i) initial static loading of the pile shaft to failure
ii) cyclic loading of the pile between pre-determined displacement limits for either 10 or 100 cycles
iii) final static loading of the pile to failure.

All tests were carried out at a sufficiently slow rate of loading to ensure that drained conditions prevailed. Attention was concentrated on the following aspects of behaviour:

a) ultimate skin friction and Young's modulus of soil under static loading
b) changes in ultimate skin friction and Young's modulus following cyclic loading.

The latter changes can be conveniently expressed in terms of degradation factors, D_τ for skin friction, and D_E for soil modulus. In each case, the degradation factor is defined as:

$$D = \frac{\text{property after cyclic loading}}{\text{property for static loading}} \tag{1}$$

Soil Characteristics

The soil used in these tests was obtained from Bass Strait, Australia, and consisted of uncemented grey biogenic particles. The soil was well graded with maximum and minimum particle sizes of about 3 mm and 0.08 mm respectively, and a value of D_{50} of 0.3 mm. The minimum and maximum dry densities were 1.28 t/m^3 (79.9 lb/ft^3) and 1.77 t/m^3 (110.4 lb/ft^3), and the average carbonate content was 88%. Drained triaxial compression tests indicated that the drained friction angle ϕ' decreased with increasing effective confining pressure as follows:

$$\phi' = 57.0 - 5.0 \log_{10}\sigma_c' \tag{2}$$

where σ_c' = effective confining pressure, in kPa (1 kPa = 0.17 lb/sq. in).

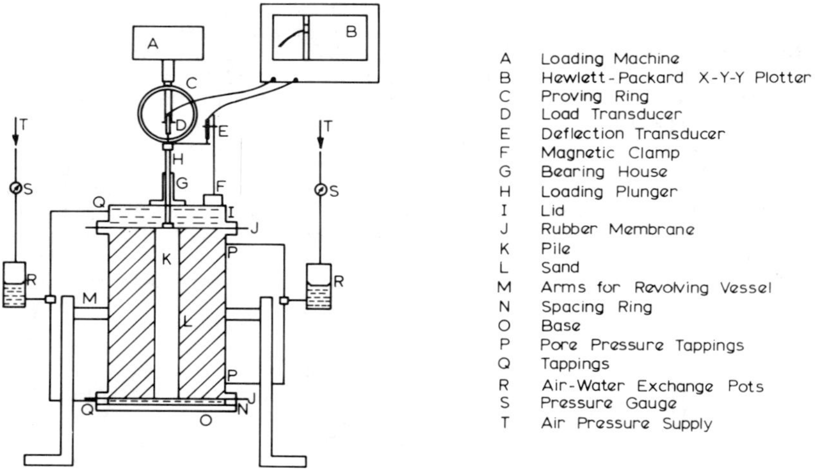

Figure 1 Experimental Apparatus for Model Tests

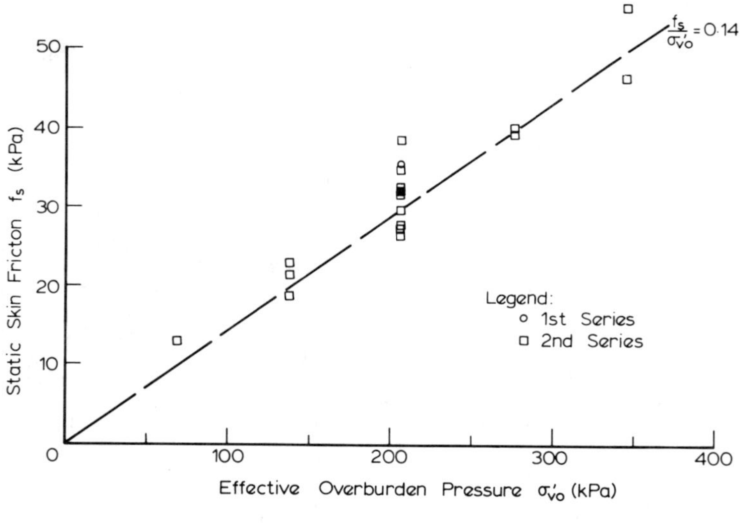

Figure 2 Static Skin friction for Jacked Piles in Calcareous Sand

For a deviator stress level of 50% of the ultimate, the drained secant Young's modulus E'_{50} was found to increase approximately linearly with σ_c';

$$E'_{50} = 8.8 + .046\ \sigma_c' \qquad (3)$$

where E'_{50} is in MPa, and σ_c' is in kPa.

The drained Poisson's ratio ν_s' is not very sensitive to the deviator stress level, but decreases with increasing σ_c', from about 0.44 at σ_c' = 50 kPa (7 lb/sq in) to about 0.23 at σ_c' = 500 kPa (72 lb/sq in).

One dimensional compression tests showed that the compression index C_c increased significantly with increasing initial void ratio e_o. For e_o = 1, under normally-consolidated conditions, C_c was about 0.1.

Behaviour Under Static Conditions

For an initial relative density of 67%, Fig. 2 plots the variation of static ultimate skin friction f_s with effective overburden pressure σ_{vo}' for jacked piles in normally consolidated soil, and demonstrates that the relationship is linear (f_s/σ_{vo}' = 0.14). Tests in overconsolidated soil show that the ratio f_s/σ_{vo}' increases substantially as OCR increases; for example, for OCR = 6, f_s/σ_{vo}' = 0.4. Increasing the relative density also leads to an increase in the normalised skin friction, with f_s/σ_{vo}' being about 0.25 for a relative density of 90%.

Similar tests in silica sand of comparable relative density reveal that the ultimate skin friction is generally about 50% greater than in the calcareous sand.

Although data for the pile tip ultimate capacity f_b have not been obtained directly, laboratory tests on model footings on calcareous sand (1) show that the ultimate bearing capacity reaches a limit of about 3 MPa (435 psi) and that this value is substantially less than for a silica sand, primarily because of the higher compressibility of the calcareous sand. Similar values have been reported from field tests (3).

Values of drained Young's modulus E_{50}' of the soil, for a load level of 50% of the ultimate, were backfigured from the load-settlement curves, using the elastic analysis of Randolph and Wroth (14). Fig. 3 plots E_{50}' versus σ_{vo}', and shows that E_{50}' increases with increasing σ_{vo}'. The values of E_{50}' in Fig. 3 are considerably larger than the values derived from conventional triaxial tests (Eq. 3), due to the effects of pile installation, and are also somewhat larger than values obtained in an earlier series of tests (13) and from field tests (4); these tests indicate that E_{50}'/σ_{vo}' is about 125, which would constitute a conservative lower bound to the results in Fig. 3.

Behaviour Under Cyclic Loading

The ultimate skin friction following cyclic loading, f_c, was generally found to be less than the static skin friction f_s. The ratio f_c/f_s (i.e. the skin friction degradation factor D_τ) was found to depend on the amplitude of cyclic displacement $\mp\rho_c$. For N = 10 cycles, Fig. 4 plots the variation of f_c/f_s with normalised cyclic displacement, ρ_c/ρ_{fs}, where ρ_{fs} is the displacement required to develop the ultimate static skin friction (typically, ρ_{fs} in these tests was 0.002d). It may be deduced from Fig. 4 that significant cyclic degradation of skin friction will not occur unless the cyclic axial displacement exceeds the value of ρ_{fs}. Indeed, for small cyclic displacements, a slight increase of skin friction can occur. This finding is consistent with other laboratory test data (5, 11, 13)and with the postulate of Randolph (15). Thus, ρ_{fs} may be considered to be a critical cyclic displacement amplitude, and is referred to as such below.

For large cyclic displacements (e.g. 10 or more times ρ_{fs}), the skin friction is reduced to about half the initial static value. Comparative tests on piles in silica sand reveal that, in this case, the loss of skin friction is much less severe than in calcareous sands, and typically, the maximum loss of skin friction does not exceed 25%.

In contrast to earlier test results the current tests indicated that the majority of cyclic degradation occurred within the first few cycles, and that beyond about 10 cycles, very little further degradation occurred. Hence, the results in Fig. 4 apply for $N > 10$ cycles of loading.

The overconsolidation ratio OCR and relative density R_D have some influence on cyclic degradation of skin friction. D_τ tends to increase as OCR increases or as R_D increases.

A few tests carried out on driven model piles (13) indicated that the cyclic degradation of skin friction was similar to that of jacked piles, and thus, the results presented herein have been presumed to be applicable also to driven piles at least approximately.

Contrary to expectations, the soil modulus E_s' for the jacked model piles does not appear to be significantly reduced by cyclic loading. Tests by Chua (1) also show that the ultimate bearing capacity of a footing is not seriously affected by cyclic loading, and therefore it may be inferred that the ultimate base bearing capacity of piles is similarly not affected.

In summary, cyclic loading may lead to a reduction of skin friction, this reduction being a function of the cyclic displacement relative to the static displacement required to cause full shaft slip. Neither the soil modulus nor the ultimate base bearing capacity appear to be seriously influenced by cyclic loading.

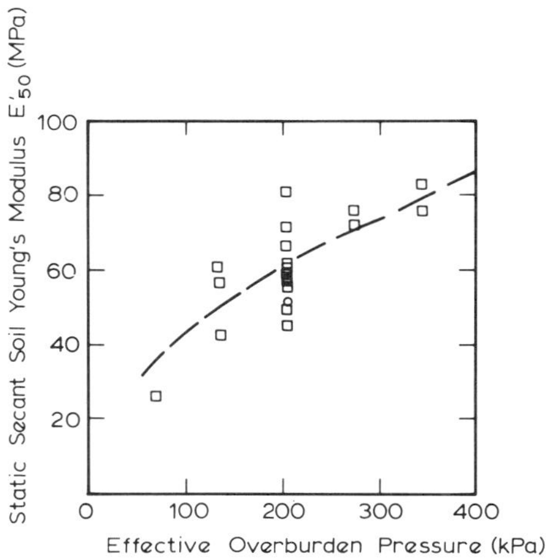

Figure 3· Backfigured Soil Young's Modulus for Jacked Piles in Calcareous Sand

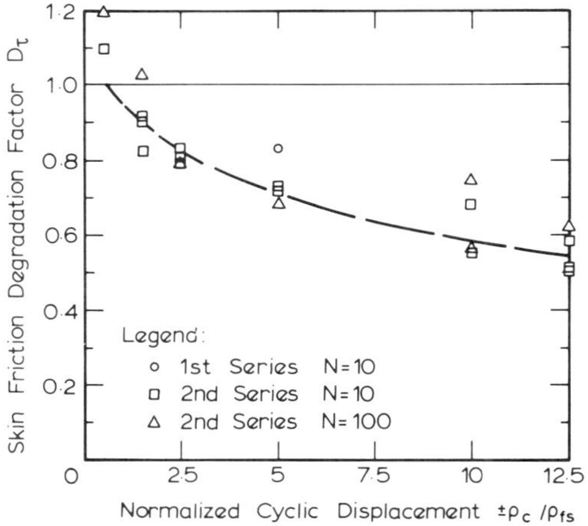

Figure 4 Skin Friction Degradation Factors for Jacked Piles in Calcareous Sand N = 10 & 100

ANALYSIS OF CYCLIC AXIAL RESPONSE

Simplified analyses of cyclic axial response have been described by Poulos (11). These utilise a boundary element approach in which the soil mass is modelled as an elastic continuum, but slip can occur at the pile-soil interface. This type of analysis, although approximate, has been shown to give very similar results to finite element analyses for static load-settlement behaviour, and has been successfully applied to field problems (8). The following characteristics of behaviour are considered in the cyclic response analysis:

i) degradation of skin friction, described by a skin friction degradation factor D_τ
ii) degradation of ultimate base resistance, described by a base resistance degradation factor D_b
iii) degradation of soil modulus, described by a modulus degradation factor D_E
iv) the effect of loading rate, whereby the ultimate resistance and stiffness of a pile increases as the rate of load application increases. A loading rate factor D_R, which increases linearly with the logarithm of loading rate, can be used to quantify this effect.

Two analysis methods may be used, both relying on essentially the same theoretical basis, but differing in their treatment of cyclic loading. One is a single-step analysis, in which a single iterative computation is performed to determine the behaviour of the pile after N cycles. The other is a cycle-by-cycle analysis in which the application of each of the N cycles is modelled in sequence. After each cycle, the degradation factors are determined and used to adjust the ultimate skin friction, ultimate base resistance and soil modulus for the next cycle of load. The detailed computational procedure for this latter approach is as follows:

1) Starting with the static values of E_s, f_s and ultimate base pressure f_b, modified by the rate factor D_R, the pile is analysed for the maximum load P_{max}.

2) The load is then decreased to the minimum value P_{min} and the analysis repeated. Finally, the load is returned to zero, thus completing the first cycle of loading.

3) The cyclic displacement $\pm\rho_c$ at each element is calculated and the degradation factors D_τ, D_b and D_E determined, these being dependent on the ratio ρ_c/ρ_{fs}.

4) The values of E_s, f_s and f_b for the next cycle are determined by multiplying the static values by the appropriate degradation factor and the rate factor D_R; the next cycle of loading is then applied, as in steps 1) and 2), and the procedure repeated until the required number N of cycles have been simulated.

A program, TAPCYC, has been written to carry out this analysis. In the latest version of this program, provision is also made for the specification of a simple "strain-softening" behaviour of the interface, of the type described by Murff (6). It can be implemented on a microcomputer with 128 k memory.

It has been found that both the "single-step" and "cycle-by-cycle" methods give virtually identical cyclic response (11), and although the cycle-by-cycle method involves a greater amount of computation, it is more versatile in that a wider variety of degradation models can be employed and "parcels" of cyclic loading of different amplitude can be readily analysed. For these reasons, this method has been preferred and has been used in obtaining the results described in this paper.

Parameters Required for Analysis

To perform the above analysis, the following parameters are required:

a) for static loading, the ultimate static skin friction f_s along the pile shaft, the ultimate base resistance of the pile f_b, the distribution of Young's modulus E_s along and below the pile and the Poisson's ratio ν_s of the soil (this latter parameter is relatively unimportant)

b) for cyclic loading, the degradation factors D_τ, D_b and D_E for ultimate skin friction, base resistance and soil modulus respectively, and the loading rate factor D_R. Also, because the degradation factors depend on the normalised cyclic displacement, the static displacement required for full slip, ρ_{fs}, is required.

Estimation of ρ_{fs} requires further discussion. The simplest approach appears to be to use elastic theory for pile response to determine the relationship between shear stress and displacement at various points along the pile and then, on the assumption of linear response until slip occurs, to determine the displacement required to cause the shear stress to reach the ultimate skin friction. In effect, this procedure assumes bilinear "t-z" curves (5). A particularly convenient and simple expression for ρ_{fs} is provided by Randolph (15);

$$\rho_{fs} = \zeta \cdot \frac{f_s}{G_s} \cdot \frac{d}{2} \qquad (4)$$

in which f_s = ultimate static skin friction

G_s = soil shear modulus

= $E_s/2(1 + \nu_s)$

ζ = parameter depending on the length to diameter ratio of the pile, ν_s, and the distribution of soil modulus with depth; typically ζ ranges between 3.5 and 5.5.

Comparisons with boundary element solutions show that Eq. 4 provides a good estimate of ρ_{fs}. An attractive feature of Eq. 4 is that it involves the parameters f_s and E_s, which are already used for the static loading response. However, it should also be noted that it is only applicable to single isolated piles; hence, if pile groups are to be considered, ρ_{fs} should be estimated from a boundary element analysis.

Parametric Study of Pile Response

Solutions were obtained for driven steel tube piles of 25, 50 and 100 m (82, 164 and 328 ft) length and 1.0 and 2.0 m (3.3 and 6.6 ft) diameter in order to study the effects of cyclic loading on the axial load capacity and stiffness. The wall thicknesses for the 1 and 2 m piles were 37.5 mm (1.5 in) and 60 mm (2.4 in) respectively. Fig. 5 illustrates the problem considered. The pile was divided into ten shaft elements and one base element. The following assumptions were made:

i) the calcareous soil profile was uniform and extended to a depth of 200 m (656 ft)

ii) an isolated single pile is considered, unless otherwise stated

iii) the ultimate skin friction f_s increased linearly with depth, over a distance of 10 pile diameters, and then remained constant at a maximum value of 20 kPa (2.9 psi) thereafter. This assumption is not consistent with the laboratory data for static skin friction but reflects the limited available field experience with driven piles in calcareous sands and makes some allowance (albeit extremely approximate) for the poorly understood phenomenon of "arching" around piles in granular media

iv) the ultimate base bearing capacity f_b is 3 MPa (435 psi) in compression

v) the soil modulus E_s increases linearly with depth as follows:

$$E_s = 0.75z \tag{5}$$

This expression represents a conservative interpretation of available laboratory and field evidence.

vi) Poisson's ratio ν_s is constant with depth, and equal to 0.35

vii) the skin friction degradation characteristics under cyclic loading are described by Fig. 4, in which the value of ρ_{fs} is determined from Randolph's expression (Eq. 4); however, the effects of other assumptions regarding ρ_{fs} are also examined

viii) there is no cyclic degradation of the ultimate base resistance f_b or the soil modulus E_s

ix) there are no loading rate effects i.e. the factor D_R is unity. No experimental evidence on this factor has yet been gathered, but it would appear unlikely that there would be a substantial rate effect for sands

x) no excess pore pressures are developed during cyclic loading, so that no allowance is made for the possibility of skin friction reduction due to a decrease in effective normal stress arising from a gradual increase in pore pressure adjacent to the pile shaft. Such a possibility may need to be considered in cases involving large diameter piles in fine sand deposits.

Attention below is focussed on the loss of ultimate shaft capacity and pile head stiffness due to cyclic loading. Earlier studies (10, 11) have demonstrated that both depend primarily on the amplitude of cyclic loading and not on the ambient or mean load. Consequently, the results presented will apply to both compression and tension piles, provided of course that the extreme load (mean plus cyclic) does not exceed the ultimate load capacity.

Loss of Shaft Capacity

For 1 m diameter piles, Fig. 6 shows the percentage loss of ultimate shaft capacity as a function of the normalised cyclic load amplitude, $\pm P_c/P_u$ where $\pm P_c$ is the amplitude of cyclic load (about any mean load P_o) and P_u is the ultimate static compressive load capacity (shaft load plus base load) of the pile (note that $P_o + P_c$ cannot exceed P_u, nor can $P_o - P_c$ exceed the ultimate tensile load capacity). These figures show that, as would be expected, the greater the normalised cyclic load, the greater is the reduction in shaft capacity because of the greater cyclic displacements generated. Also, the reduction is more severe for shorter piles, and for the larger-diameter piles.

An attempt was made to determine whether the results could be further normalised in terms of a dimensionless axial stiffness such as that proposed by Murff (6) or Poulos (9), but it was not found possible to obtain comparable results between 1 m and 2 m diameter piles of equivalent dimensionless axial stiffness.

Because the cyclic degradation does not depend on the mean load P_o, Fig. 6 provides data on cyclic degradation for both tension and compression piles, and reveal an interesting difference. For tension piles, the ratio P_c/P_u that can be applied is relatively small, because of the relatively small uplift load capacity. Hence, only a small percentage loss in ultimate shaft capacity will occur. On the other hand, compression piles can be subjected to a greater cyclic load and thus can suffer a greater percentage loss of ultimate shaft capacity. This point can be demonstrated by the example of a 1 m diameter 50 m long pile. From Fig. 6, for a cyclic load of 50% of the uplift load capacity (i.e. about 27% of the compressive load capacity), a tension pile suffers a loss of only about 2% of the ultimate static load capacity. However, for a cyclic load of 50% of the compressive load capacity, Fig. 6 shows the compression pile to suffer a loss of about 9% of the ultimate shaft capacity. Therefore, the perhaps unexpected conclusion is reached that compression piles may be more susceptible to cyclic degradation of shaft load capacity than are tension piles (at least for the case considered herein).

It is interesting to note that Randolph (15) derives the following expression for the "safe" cyclic load P_c which may be applied to a relatively long flexible offshore pile

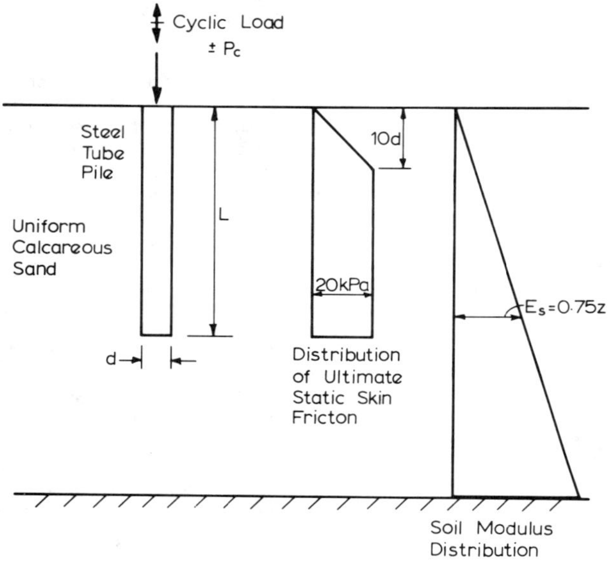

Figure 5 Problem Studied

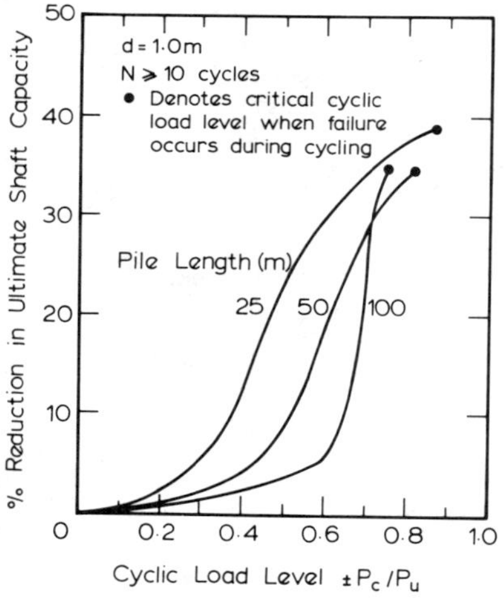

Figure 6 Loss of Total Skin Resistance Due to Load Cycling 1m Diameter Piles

$$\frac{P_c}{P_u} \leqslant \frac{1}{\sqrt{\pi_3}} \tag{6}$$

in which P_c = cyclic load
P_u = ultimate static pile load capacity
π_3 = dimensionless pile stiffness factor,

$$= \frac{\pi\, d\, f_s\, L^2}{AE\rho_{fs}}$$

f_s = static ultimate skin friction
L = pile length
d = pile diameter
A = cross-sectional area of pile
E = Young's modulus of pile material
ρ_{fs} = shaft displacement to cause full slip.

Application of Eq. 6 to the case of the 100 m long piles would give values of the safe cyclic load level, P_c/P_u of 0.37 for d = 1 m and 0.62 for d = 2 m. From the results of the present analyses, the corresponding reductions in ultimate shaft resistance of these cyclic load levels are 2% and 23%. Thus, Randolph's criterion appears to be satisfactory for the 1 m diameter pile, but not for the 2 m diameter pile which is too stiff to be classified as a long flexible pile. Randolph's criterion does not appear to apply to the 25 m or 50 m long piles for the same reasons.

Cyclic Stiffness of Pile Head

Fig. 7 plots, for the 1 m diameter piles, the variation of cyclic pile head stiffness (defined as cyclic load divided by cyclic displacement of pile head) with cyclic load level. The following characteristics are observed:

i) the cyclic stiffness increases with increasing pile length and diameter

ii) the cyclic stiffness generally decreases as the cyclic load level increases. This decrease is more gradual for the shorter or stiffer piles; for longer, more compressible piles, an abrupt change in cyclic stiffness occurs for cyclic load levels in excess of about 50% of the static ultimate compressive load.

Because no cyclic degradation of soil modulus is included in the analysis, degradation of cyclic stiffness arises solely because of the occurrence of slip along the shaft and the consequent reduction in ultimate skin friction. At lower cyclic load levels, a "shakedown" condition of equilibrium is reached after only three or four cycles, but at higher load levels, up to ten cycles may be required for shakedown to

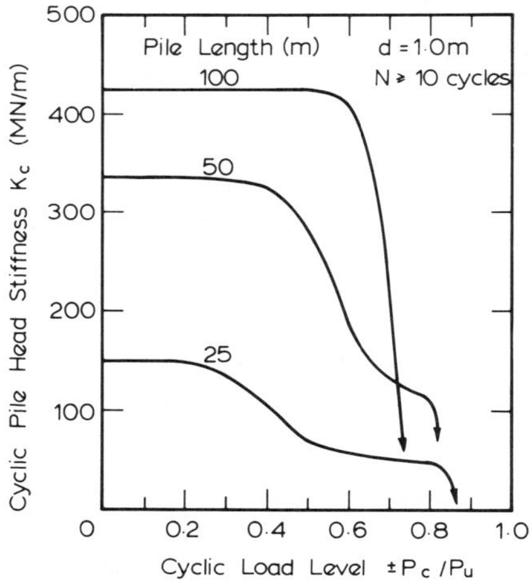

Figure 7 Variation of Cyclic Pile Head Stiffness with Cyclic Load Level 1m Diameter Piles

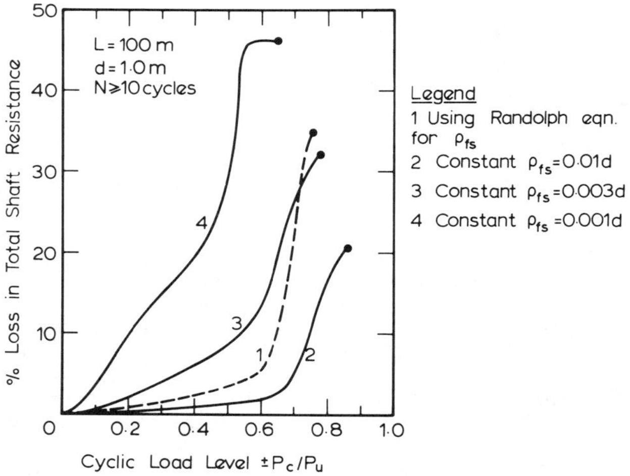

Figure 8 Effect of Assumptions for Critical Cyclic Displacement ρ_{fs}

occur. A range of critical cyclic load levels exists in which failure will develop during cycling; as can be seen from Fig. 7, this cyclic load level generally lies between 0.6 and 0.8 times the ultimate static load for the parameters chosen in this study. If some degradation of soil modulus had also been included, a more severe degradation of cyclic stiffness would occur, with a correspondingly lower critical cyclic load level.

EFFECTS OF ASSUMPTIONS

In this Section, the effects of several assumptions made in obtaining the foregoing parametric solutions will be examined. The case of a 100 m long 1 m diameter pile will be considered, and attention will be concentrated on the loss of shaft load capacity due to cyclic loading.

Critical Cyclic Displacement

The pile displacement for full slip, ρ_{fs}, may be considered as a critical cyclic displacement as it appears to define the onset of significant cyclic degradation of skin friction. In the solutions presented above, ρ_{fs} has been determined from the Randolph expression (Eq. 4), but alternative estimates could be made. For example, it is frequently assumed that shaft slip occurs at a displacement equal to some small percentage of the shaft diameter (eg 1%); in contrast, the laboratory model tests indicate a much smaller value of ρ_{fs} of about 0.1% to 0.2% of diameter. The solutions for loss of shaft resistance, making various assumptions about ρ_{fs}, are shown in Fig. 8. The use of a constant value of ρ_{fs} along the pile in the analysis, rather than the variable value given by Eq. 4, leads to a different pattern of cyclic degradation, with the maximum degradation occurring near the top of the pile and gradually progressing downward as the cyclic load level increases. This contrasts with the upward progression of cyclic degradation from the pile tip, when the Randolph expression (Eq. 4) is used. In this latter case, although the cyclic displacement ρ_c is relatively small near the tip, the value of ρ_{fs} is also small, and the ratio ρ_c/ρ_{fs} is larger than near the top of the pile; consequently degradation commences near the tip and works upwards. However, the solution for loss of shaft resistance in this case lies between the solutions for a constant ρ_{fs} of about 0.3% and 1.0% of the shaft diameter, and in this respect, the results are not over-sensitive to the assumed distribution of ρ_{fs}.

Distribution of Soil Modulus

As has been previously demonstrated (8), the assumed distribution of soil modulus E_s will influence the computed shear stress distributions and hence the pattern of pile-soil slip along the shaft. It will also therefore influence the pattern of cyclic degradation of skin friction. Solutions have been obtained for the cases of a linearly increasing soil modulus with depth, and a constant soil modulus equal to the value at mid-length of the pile.

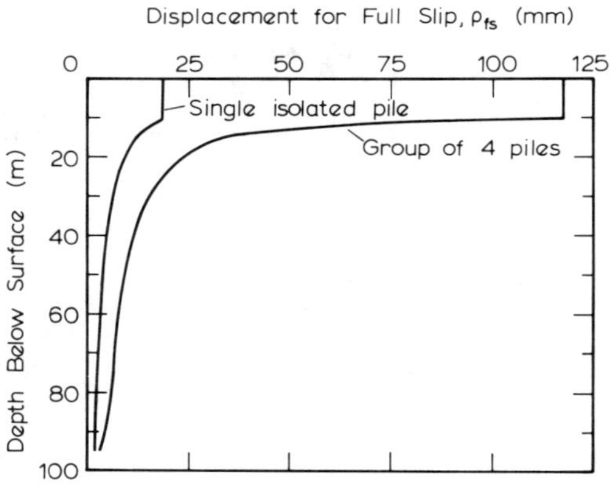

Figure 9 Distribution of Displacement for Full Slip

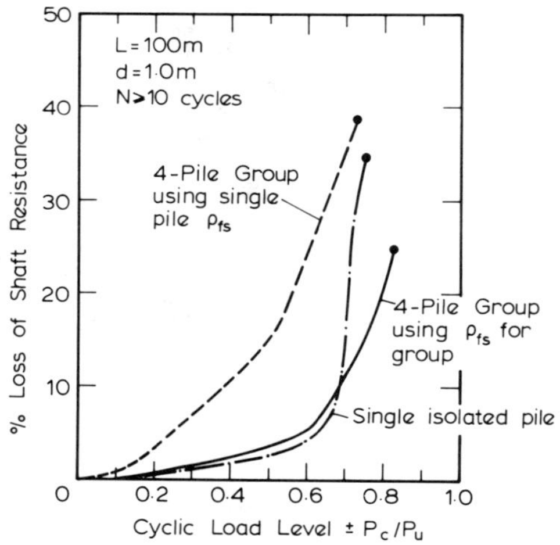

Figure 10 Solutions for 4-Pile Group

At any cyclic load level, there is a greater loss of shaft resistance for the case of a constant E_s, and the critical cyclic load level is slightly less than for a linearly increasing E_s. The major reason for this difference is that a greater amount of cyclic degradation occurs in the upper half of the pile when E_s is constant, both because the shear stresses are greater and also because the values of ρ_{fs} from Eq. 4 are smaller near the pile head.

Consequently, it may be concluded that the assumed soil modulus distribution may significantly influence the results of a cyclic degradation analysis.

Pile Groups

The grouping of piles will have at least two counterbalancing effects in relation to cyclic loading:

i) the cyclic displacements will be increased by interaction between the piles, thus tending to increase cyclic degradation of skin friction

ii) the static displacement for full slip, ρ_{fs}, will increase, thus tending to reduce the amount of cyclic degradation.

Earlier analyses (10, 11) took only the first effect into account, and concluded that a pile group could suffer a much more serious reduction in ultimate load capacity than a single isolated pile subjected to the same average cyclic load.

Analyses have been carried out for a symmetrical four-pile group, located on a pitch circle of 6d diameter, using two alternative assumptions for ρ_{fs}:

i) the values for a single isolated pile, determined from Eq. 4.

ii) the values for a group of four piles, estimated from an elastic boundary element analysis.

These two distributions of ρ_{fs} are shown in Fig. 9, and the resulting solutions for loss of shaft resistance are shown in Fig. 10. There is a much greater loss of shaft resistance when the single pile ρ_{fs} value is used. When the ρ_{fs} values for the four-pile group are used, the loss of skin friction is generally similar to that for a single isolated pile. Consequently, it is now concluded that group effects should not have a major effect on cyclic degradation of pile shaft load capacity. This conclusion remains to be verified by laboratory and field tests.

CONCLUSIONS

Laboratory data is presented on the static and cyclic behaviour of model pile sections in calcareous sand. It is demonstrated that cyclic degradation of ultimate skin friction is controlled by the amplitude of cyclic displacement, relative to a critical cyclic displacement which is equal to the static displacement required for full slip, ρ_{fs}. For large relative cyclic displacements, a reduction in skin friction of about 50% can occur. However, there appears to be little degradation of ultimate base resistance or soil modulus with cycling.

Using a theoretical analysis based on a simplified form of boundary element procedure, and parameters based largely on the laboratory tests, solutions have been obtained for a range of offshore pile lengths and diameters. Attention has been focussed on the loss of ultimate shaft resistance due to cyclic loading and to the variation of cyclic pile head stiffness with cyclic load amplitude. The following conclusions have been drawn:

i) for a given cyclic load level relative to the ultimate compressive load capacity, shorter stiffer piles suffer a greater loss in shaft resistance than longer flexible piles

ii) because of the limited uplift capacity of tension piles, only relatively small cyclic loads can be sustained, and only relatively small cyclic displacements occur. They therefore suffer only a minor loss of shaft resistance due to cyclic loading.

iii) cyclic pile head stiffness decreases with increasing cyclic load level, more gradually for stiffer piles than for longer compressible piles.

iv) the assumptions made in the analysis can materially alter the solution obtained. The assumed magnitude and distribution of ρ_{fs} can have a marked effect on the computed cyclic behaviour, and the distribution of soil modulus and the inclusion of group effects may also be important.

It is felt that the results presented in this paper may give a useful guide to the possible extent of reduction of axial shaft load capacity of driven offshore piles dues to cyclic loading. However, it should be emphasised that there is no reliable published field data presently available to assess the reliability of these results. Furthermore, the important question of cumulative displacement of piles during cyclic loading about a nonzero mean load is not addressed. It would appear likely that this displacement depends not only on the cyclic response of the pile-soil system, but also on the creep properties of the soil. There is clearly a need to investigate further this aspect of behaviour.

ACKNOWLEDGEMENTS

The work described in this paper forms part of a project into the Behaviour of Offshore Foundations, being carried out at the University of Sydney with the support of a grant from the Australian Research Grants Scheme. Support for work on the properties of calcareous sands has also been provided by the Marine Sciences and Technology Council. The author gratefully acknowledges the contributions of K. Chan, E.W. Chua and T.S. Hull, who have participated in the laboratory work.

APPENDIX I - REFERENCES

1. Chua, E.W. "Bearing Capacity of Shallow Foundations in Calcareous Sand". M.Eng.Sc. thesis, Univ. of Sydney, 1983.

2. Datta, M., Gulhati, S.K. and Rao, G.V. "An Appraisal of the Existing Practice of Determining the Axial Load Capacity of Deep Penetration Piles in Calcareous Soils". Proc. 12th Annual OTC Conf, Houston, 1980, Vol.4, pp.119-130.

3. Hagenaar, J. "The Use and Interpretation of SPT Results for the Determination of Axial Bearing Capacities of Piles Driven into Carbonate Sands and Coral". Proc. 2nd Eur. Pen. Testing, Amsterdam, A.A. Balkema, 1982, Vol.1, pp.51-55.

4. King, R.W., Van Hooydonk, W.R., Kolk, H.F. and Windle, D. "Geotechnical Investigations of Calcareous Soils on the North West Shelf, Australia". Proc. 12th Annual OTC Conf, Houston, 1980, Paper OTC 3772.

5. Matlock, H. and Foo, S.C. "Axial Analysis of a Pile Using a Hysteretic and Degrading Soil Model". Proc. Conf. Num. Meth. in Offshore Piling, I.C.E. London, 1979, pp.165-185.

6. Murff, J.D. "Pile Capacity in a Softening Soil". Int. Jnl. Num. Methods in Geomechs, Vol.4, No.2, 1980, pp.185-189.

7. Nauroy, J.F. and Le Tirant, P. "Model Tests of Piles in Calcareous Sands". ASCE Spec. Conf. on Geot. Practice in Offshore Eng, Austin, 1983, pp.356-369.

8. Poulos, H.G. "Settlement of Single Piles in Non-Homogeneous Soils". Jnl. Geot. Eng. Divn, ASCE, Vol.105, No.GT5, 1979, pp.627-641.

9. Poulos, H.G. "The Influence of Shaft Length on Pile Load Capacity in Clays". Geotechnique, Vol.32, No.2, 1982, pp.145-148.

10. Poulos, H.G. "Influence of Cyclic Loading on Axial Pile Response". Proc. 2nd Conf. Num. Methods in Offshore Piling, Austin, 1982, pp.419-440.

11. Poulos, H.G. "Cyclic Axial Pile Response - Alternative Analyses". ASCE Spec. Conf. on Geot. Practice in Offshore Eng, Austin, 1983, pp.403-421.

12. Poulos, H.G. and Chan, K.E. "Laboratory Study of Pile Skin Friction in Calcareous Sand". Civil Eng. Res. Rept. R486, Univ. of Sydney, 1984.

13. Poulos, H.G., Young, G.S., Hain, S.J. and Lewis, W.M. "Static and Cyclic Skin Friction of Model Piles in Carbonate Sand". Civil Eng. Res. Rept. R485, Univ. of Sydney, 1984.

14. Randolph, M.F. and Wroth, C.P. "Analyses of Deformation of Vertically Loaded Piles". Jnl. Geot. Eng. Divn, ASCE, Vol.104, No.GT12, 1978, pp.1465-1488.

15. Randolph, M.F. "Design Considerations for Offshore Piles". ASCE Spec. Conf. on Geot. Practice in Offshore Eng, Austin, 1983, pp.422-439.

APPENDIX II - NOTATION

The following symbols are used in this paper:

A = cross-sectional area of pile
C_c = compression index
d = pile diameter
D_b = degradation factor for base resistance
D_E = degradation factor for soil modulus
D_τ = degradation factor for skin friction
D_R = rate factor
D_{50} = mean particle size
e_o = initial void ratio
E_s = soil Young's modulus
E_{50}' = drained value of E_s at 50% stress level
f_b = ultimate base pressure
f_c = ultimate skin friction after cyclic loading
f_s = ultimate static skin friction
G_s = soil shear modulus
L = embedded pile length
N = number of cycles
P_c = cyclic load amplitude (half peak-to-peak)
P_u = ultimate static load capacity
P_{max} = maximum load
P_{min} = minimum load
R_D = relative density
ϕ' = effective stress friction of soil
ν_s' = drained Poisson's ratio of soil
ρ_{fs} = static displacement to cause full slip at any point on the pile; also, the critical cyclic displacement
ζ = factor in expression for ρ_{fs}
π_3 = dimensionless pile stiffness factor
σ_c' = effective consolidation pressure
σ_{vo}' = effective overburden pressure

RESIDUAL STRESSES IN PILES AND THE WAVE EQUATION

By Jean-Louis Briaud[1], M. ASCE, and Larry M. Tucker[2], Assoc. M.ASCE

ABSTRACT: A data base of 18 vertical load tests on driven piles in sand is presented. The piles were sufficiently instrumented and the load tests sufficiently detailed to yield residual driving load data. On the average for these 18 piles the residual point load was 56% of the measured point load and the residual friction load 19% of the measured friction load. The residual loads obtained from the load tests are compared with residual loads predicted by the wave equation analysis.

INTRODUCTION

During a hammer blow, a pile will move downward first, then rebound and then oscillate around a final position. At its final position, the pile is in equilibrium under a certain point load and a certain friction load, which cancel out since the load at the pile head is zero. The process repeats itself for each blow and, when the pile reaches final penetration, the residual load distribution in the pile can be as shown on Fig. 1.

During the downward movement of the pile, the pile-soil friction is acting upward on the pile to resist the penetration of the pile; the point soil resistance is also acting upward. During the rebound that follows, the soil under the point pushes the pile back up while the pile decompresses elastically. These two components of the rebound create enough upward movement to reverse the direction of the pile-soil friction which now acts downward at least in the upper portion of the pile. Equilibrium is reached when enough of the friction stresses have reversed themselves in order to keep the bottom of the pile stressed against the soil.

The above explanation shows how the residual stresses phenomenon is governed by the unloading characteristics of the point and friction transfer curves (q-w and f-w curves) on one hand, and by the elastic characteristics of the pile on the other. Residual stresses can also be caused by reconsolidation after driving, if relative movement between the pile and soil occurs. In sands, a significant residual point load can exist because point capacities

[1]Associate Professor, [2]Research Associate, Civil Engineering Department, Texas A&M University, College Station, Texas 77802.

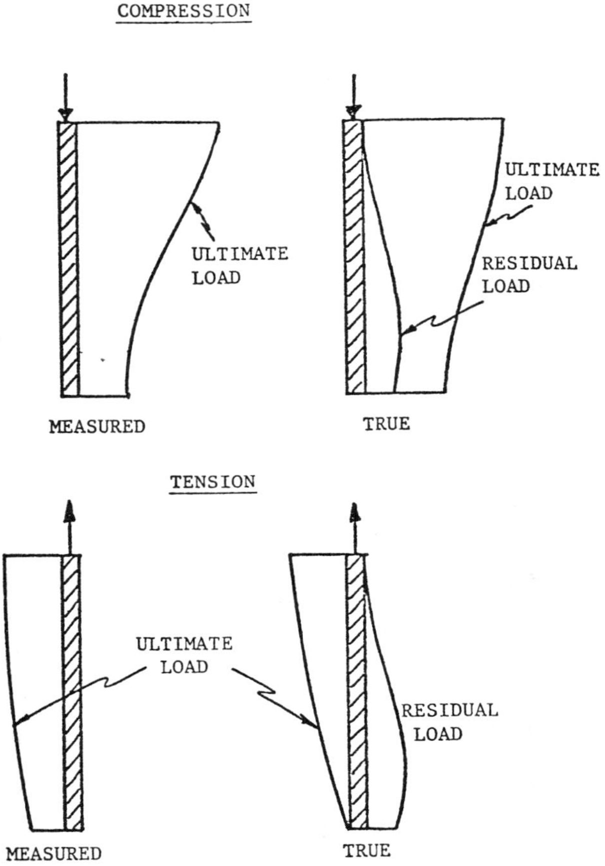

FIG. 1 - Difference between True Loads and Measured Loads in Conventional Load Tests are large and because little movement is needed to unload the friction transfer curve, while much more movement is needed to unload the point transfer curve.

The existence of residual stresses has been known for a long time (7, 8) but has not been routinely included in pile design.

CONVENTIONAL LOAD TESTS

In a conventional load test on an instrumented pile, the following testing sequence is usually observed: first the pile is instrumented, second the pile is driven, third the instrumentation is zeroed, fourth the load test is performed. Zeroing the instrumentation after having driven the pile is equivalent to assuming that zero stresses exist in the pile after driving. Therefore, in a conventional load test residual stresses are not measured.

The difference in load distribution in the pile between the measured loads as described above and the true loads that exist in the pile is shown on Fig. 1 for a compression test and a tension test. As can be seen, the interpretation of the results from a conventional compression test will lead to a point load which is lower than the true point load and to a friction load which is higher than the true friction load.

The interpretation of the results from a conventional tension test on the other hand, will lead to a point load which is larger than the true point load which is zero, and to a friction load which is smaller than the true friction load.

MEASURING RESIDUAL STRESSES

There are several methods that can be used to determine residual stresses. The following four methods are not equivalent. They are presented in order of the authors preference, assuming that all instrumentation results are equally reliable.

Method 1: Read Instrumentation Before and After Driving

The most direct method of measuring residual loads is to hang the pile in driving position under the hammer. While the pile is hanging under its own weight the instrumentation is zeroed. The pile is driven and the instrumentation is read immediately after final penetration is reached. The residual loads are then calculated directly after proper considerations due to the dead weight.

The instrumentation in this case is generally of the strain gage type. In addition, pressure cells are sometimes placed under the pile tip for direct measurements of the residual point pressure. These strain gage and pressure cell type of measurement are difficult to make accurately. All other methods consist of zeroing the instrumentation after the pile is driven and of obtaining the residual load by using a combination of testing sequence and theoretical reasoning.

Method 2: Hunter-Davisson Method

Hunter-Davisson (8) devised a method which involves a given testing sequence during the load test and then a special reasoning sequence during the data analysis.

The testing sequence consists of:

1. Drive the pile.
2. Zero the instrumentation.
3. Load the pile to failure in compression, read the instrumentation.
4. Unload the pile to zero top load, read the instrumentation.
5. Zero the instrumentation.
6. Load the pile to failure in tension, read the instrumentation.
7. Unload the pile to zero top load, read the instrumentation.

The reasoning and analysis sequence consists of (Fig. 2):

1. Curve 1 represents the measured ultimate compression load distribution, assuming no stresses in the pile at start of test.
2. Curve 2 represents the measured compression load distribution after complete release of the applied compressive load, assuming no stresses in the pile at start of test.
3. Curve 3 represents the measured ultimate tension load distribution assuming no stresses in the pile at start of test.
4. Curve 4 represents the measured tension load distribution after complete release of the applied tensile load, assuming no stresses in the pile at start of test.
5. It is assumed that no residual stresses exist at the end of the tension test. Therefore, curve 4 represents the residual compressive loads for the test and should be subtracted from curve 3 in order to give curve 5, the adjusted tension load distribution.
6. Curve 4 includes the original compressive residual loads in the pile due to driving before the compression test, plus the residual loads induced by the compression test. The subtraction of curve 2 from curve 4 gives curve 6, the original compressive residual loads in the pile.
7. Curve 6 is added to curve 1 to give curve 7, the adjusted compression load distribution.

This method is based on the assumption that no residual loads are induced by the tension test. Using linear soil behavior, it can be shown (1, 2, 13) that this assumption is incorrect. For short piles, the error is small however. The error is zero at the top and bottom of the pile and peaks towards the middle of the pile where the residual tension load can be 25% of the ultimate tension load according to theory (1, 2, 13). Therefore, this method will lead to

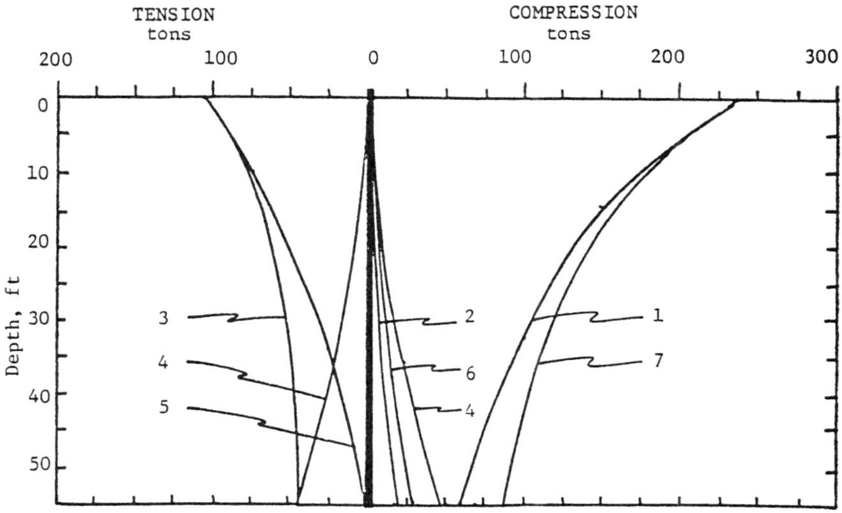

1 - Ultimate Compression
2 - Unload Compression
3 - Ultimate Tension
4 - Unload Tension
5 - Corrected Tension
6 - Residual Load
7 - Corrected Compression.

FIG. 2 - The Hunter-Davisson Method to Obtain Residual Stresses accurate predictions of residual point load but somewhat erroneous residual friction stresses distribution. The accuracy of this method as well as the accuracy of the other three methods are tied to the stability of the instrumentation including no creep and no zero shift.

Method 3: No Unloading Reading Method

Often, the testing sequence used for a load test does not follow rigorously the testing sequence required to apply the Hunter-Davisson method. Method 3 is used when no reading of the instrumentation is taken after unloading the pile, that is to say, when steps 4 and 6 for the testing sequence of Method 2 are not performed. Therefore, compared to Method 2, curves 2 and 4 are not available. In this case the tension load at the point is assumed to be the residual compressive load at the point after driving. This assumes that the change in residual loads created by the compression test is negligible. Curve 4 for Method 3 is assumed to be a line joining the point load of curve 3 and zero load at the top. These added assumptions make Method 3 less reliable than Methods 1 and 2.

Method 4: No Instrumentation Method

This method is the least reliable of the four methods. When the instrumentation of the pile is reduced to the measurement of top load and top movement, the distribution of load in the pile cannot be obtained. In Method 4, a compression test followed by a tension test is necessary. The ultimate tension load is assumed to be the ultimate friction load in compression. This friction load is subtracted from the ultimate compression load to give the ultimate point load. The validity of this assumption can be evaluated from the comparison of the measured ultimate friction load in tension and the measured ultimate friction load in compression. Briaud, et al. (1, 2) presented data which tends to indicate that this assumption may only be valid for long piles.

RESIDUAL STRESSES: LOAD TEST RESULTS

A data base of vertical load tests on instrumented piles driven in sands was collected (2). A review of the literature revealed 10 sites with a total of 33 instrumented piles. The piles varied in length from 10 ft (3.3 m) to 100 ft (33 m), in diameter from 0.66 ft (0.20 m) to 2.05 ft (0.62 m). They included steel pipe piles, H piles, concrete piles, and timber piles. The part of the data base used in this article is presented in Table 1. The complete data base is described in detail elsewhere (2, 13). The four methods of residual stresses determination described earlier were applied to the load test whenever possible in order to quantify the magnitude of residual loads.

The results are summarized in Table 2 for the load tests for which the analyses were possible. In that table Q_{PR} is the

Table 1. - Pile Load Test Information

Site	Pile	Symbol	Pile Type and Material	Dimensions: Diameter ft	Dimensions: Length ft	Dimensions: Area, in.^{2}	Modulus of Elasticity, psi x 10^{-6}	Type Test	Strain Rods	Strain Gages	Soil Data: SPT	Soil Data: CPT	Soil Data: PMT	Other	Reference
(1)	(2)	(3)	(4)	(5)	(6)	(7)	(8)	(9)	(10)	(11)	(12)	(13)	(14)	(15)	(16)
Lock & Dam 4, Arkansas River (1963)	1	1	Steel Pipe	1.20	53.1	17.12	29.0	C,T	X						4, 9
	2	2	Steel Pipe	1.50	52.8	23.86	29.0	C,T	X	X					
	3	3	Steel Pipe	1.70	53.0	27.36	29.0	C,T	X						
	7	4	Steel "H" HP 14 x 73	1.34	52.1	29.33	29.0	C,T	X	X	X				
	10	5	Steel Pipe	1.50	53.1	23.86	29.0	C,T	X	X					
	16	6	Steel Pipe	1.42	52.7	19.62	29.0	C,T		X					
Low Sill Structure, Old River, La. (1956)	2	7	Steel Pipe	1.75	65.1	27.36	29.0	C,T	X						10
	4	8	Steel Pipe	1.42	66.3	22.65	29.0	C,T	X		X				
	5	9	Steel Pipe	1.42	45.1	22.65	29.0	C,T	X						
	6	10	Steel Pipe	1.58	65.2	25.00	29.0	C,T	X						
Ogeechee River (1969)	H-15	11	Steel Pipe	1.50	49.3	27.49	30.0	C		X	X	X		Densities	14, 15
Lock & Dam 26 Replacement Site	3IP-IIIS	12	Steel "H" HP 14 x 73	1.34	54.0	21.40	29.0	C,T		X	X				3
Gregersen (1969)	A	13	Circular Prestressed Concrete	0.92	26.2	95.45	3.15	C		X					5
	D/A	14	Circular Pres. Conc.	0.92	52.5	95.45	3.15	C		X					
	C	15	Tapered Circular Prestressed Concrete	Top 0.92 Bot. 0.66	26.2	Varied	3.15	C		X	X	X			
	B/C	16	Straight Cir. to 26.3 ft Tapered to bottom Pres. Conc.	Top 0.92 Bot. 0.66	52.5	Varied	3.15	C		X					
Sellgren (1981)	A-I	17	Square Pres. Conc.	1.00	35.4	113.00	3.15	C		X			X		11
	A-II	18	Square Pres. Conc.	1.00	35.4	113.00	3.15	C		X					

Note: 1 ft = 0.305 m; 1 psi = 6.89 kN/m^2.

Table 2. - Measured Residual Point Loads, Ultimate Point and Friction Loads for the Piles of the Data Base

		Method 1: Measured			Method 2: Hunter-Davisson			Method 3: No Unload Reading			Method 4: Compression/Tension	
Site (Reference)	Pile	Q_{PR} tons	Q_{PU} tons	Q_S tons	Q_{PR} tons	Q_{PU} tons	Q_S tons	Q_{PR} tons	Q_{PU} tons	Q_S tons	Q_{PU} tons	Q_S tons
Arkansas River (4, 9)	1				37	87	85	32	82	90	77	95
	2				45	120	122	62	137	105	127	115
	3				48	163	109	64	179	93	152	120
	7[a]				25	85	158	25	85	158	175	67
	10				10	83	159				134	108
	16				40	93	82	33	86	89	97	78
Low Sill (10)	2							29	179	241	225	195
	4							37	125	315	215	225
	5							29	54	91	61	84
	6							51	186	244	245	185
Gregersen (5)	A	3.3	6.5	20.5							17	10
	D/A	5.3	12.0	39.0							24	27
	C	1.3	4.5	29.5							18	12
	B/C	5.3	7.0	41.0							25	23
Sellgren (11)	AI	7.0	66.0	11.5								
	AII	13.0	33.5	109.0								
Lock & Dam 26 (3)	3IP-IIIS[a]				24.5							
Ogeechee River (14), (15)	H-15										248	173

[a]H-piles

Note: 1 ton = 8.9 kN

residual point load, Q_{PU} and Q_S are the point load and friction load at failure after consideration of residual stresses. On the average the residual point load (Q_{PR}) was equal to 0.56 times the true ultimate point load (Q_{PU}) and the residual friction load was equal to 0.19 times the true ultimate friction load (Q_S). Also the true ultimate point load was on the average 1.56 times the measured ultimate point load and the true ultimate friction load was 0.81 times the measured ultimate point load. Ultimate or failure load is defined here as the load reached for a pile head settlement equal to 10% of the pile diameter.

A comparison of Methods 2 and 3 was possible for five piles at the Arkansas River site. The residual load given by Method 3 averaged 13 percent higher than that given by Method 2. This discrepancy is due to the fact that Method 3 assumes that no additional residual loads are induced in the piles during the compression test. This is obviously not the case. A comparison of Method 4 with the results of Methods 1, 2 and 3 was possible for 14 piles. Methods 1, 2 and 3 yield a compressive side friction load, whereas Method 4 gives the tensile friction load. A comparison of the results showed that, on the average the friction in tension is only 70 percent of the friction in compression. This percentage varies from 40% to 110% and is lowest for short piles, H piles and tapered piles.

Because of the difference between friction load in tension and friction load in compression, Method 4 was not used in further analyses. The residual loads used in further analyses were those given by the method available for each pile: Method 1 being the most reliable method, Method 3 the least reliable.

WAVE EQUATION ANALYSIS

There has been widespread interest in the use of the wave equation for pile driving analysis since the early 1960's. The main application has been in pile driveability studies and hammer selection. The wave equation has also been used to verify pile capacity during construction. In 1975 Holloway et al. (7) proposed the use of the wave equation for obtaining residual stresses in driven piles.

The pile is modeled as a system of concentrated weights connected by weightless springs. The soil is modeled as a spring and a dashpot (Fig. 3). The displacements and velocities of each of the pile segments and the forces between the segments are then computed using formulas applied by Smith (12).

The soil model used in this analysis is that proposed by Smith (12) and presented by Hirsch et al. (6). The model is a linear elastic-plastic model. The movement necessary to initiate plastic deformation is called the quake and was taken as 0.1 in. (0.25 cm) for both side and point resistances. The values for side and point damping were taken as 0.05 sec/ft (0.15 sec/m) and 0.15 sec/ft

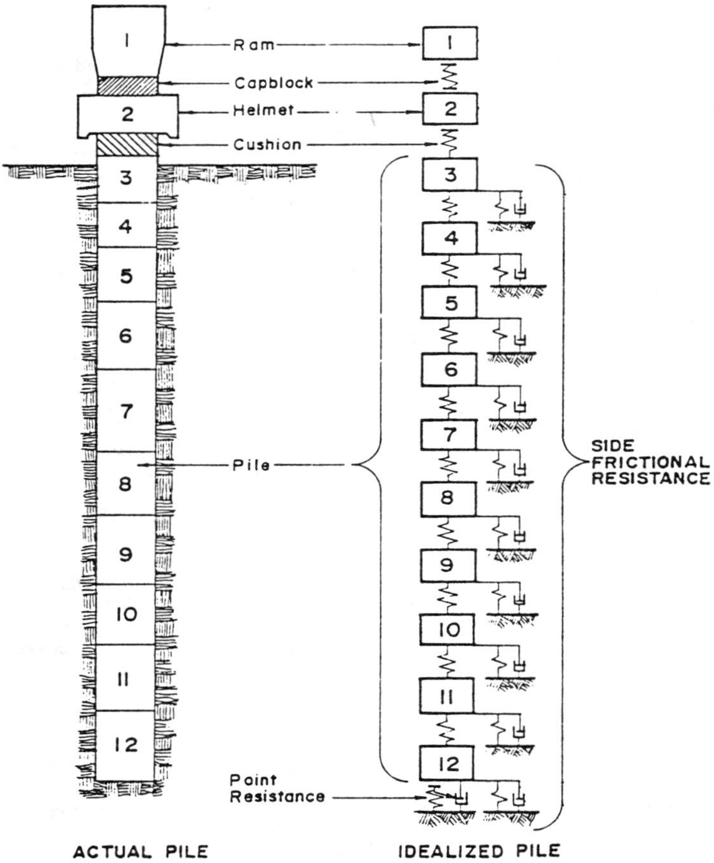

FIG. 3 - Wave Equation: Pile Model

(0.49 sec/m) respectively. The ratio between total soil resistance and point resistance was taken from the pile load test results (Table 2) and the side load was distributed uniformly along the pile.

MULTIPLE BLOW VS. SINGLE BLOW ANALYSIS

Smith's equations are solved incrementally increasing the time by small steps. When the energy and momentum of the system satisfy given criteria the solution is stopped. Any forces left in the pile at this time are residual forces. Locking these forces in the pile and simulating a second blow of the hammer on the pile with the computer program results in new solutions for the residual forces. Fig. 4 shows a plot of the point load versus the time interval. The solution was performed for six blows, each of which ran for 1000 time steps. After the third blow the solution repeated itself for each blow.

One of the conventional output of the wave equation is the static resistance at the time of driving versus blow count curve. Fig. 5 shows such a curve for a single blow analysis and a fifth blow analysis. The single blow analysis does not include residual stresses since in this case the pile is stress free before the hammer blow is simulated. The fifth blow analysis does include the residual stresses created by the four previous blows simulated in the program. As can be seen on Fig. 5, for the same static resistance at the time of driving the pile which is prestressed by the residual stresses is easier to drive at hard driving conditions than the stress free pile.

OBTAINING RESIDUAL STRESSES

As seen in Fig. 4 the point load oscillates about, and converges towards a specific value which is the calculated residual point load. Such a multiple blow wave equation analysis was performed on piles of the data base and in all calculations a series of five blows was always sufficient to obtain constant values of residual stresses. The number of time intervals per blow required to determine the residual point load varied from pile to pile. The number also varied with the level of soil resistance: the lower the soil resistance, the larger the number of iterations required to determine the residual point load. The lower soil resistances also had a tendency to give more erratic results. In all cases studied, 1000 iterations per blow was sufficient to define the residual point load. This number may be decreased for higher soil resistance and may need to be increased for piles longer than 100 ft (30 m).

RESULTS

A wave equation analysis was performed for all piles for which residual point loads were known and hammer data was available. A plot of the results for Arkansas pile 3 is shown in Fig. 6 in terms of the residual point load versus the blowcount on the pile. The residual point loads predicted by this approach are plotted versus

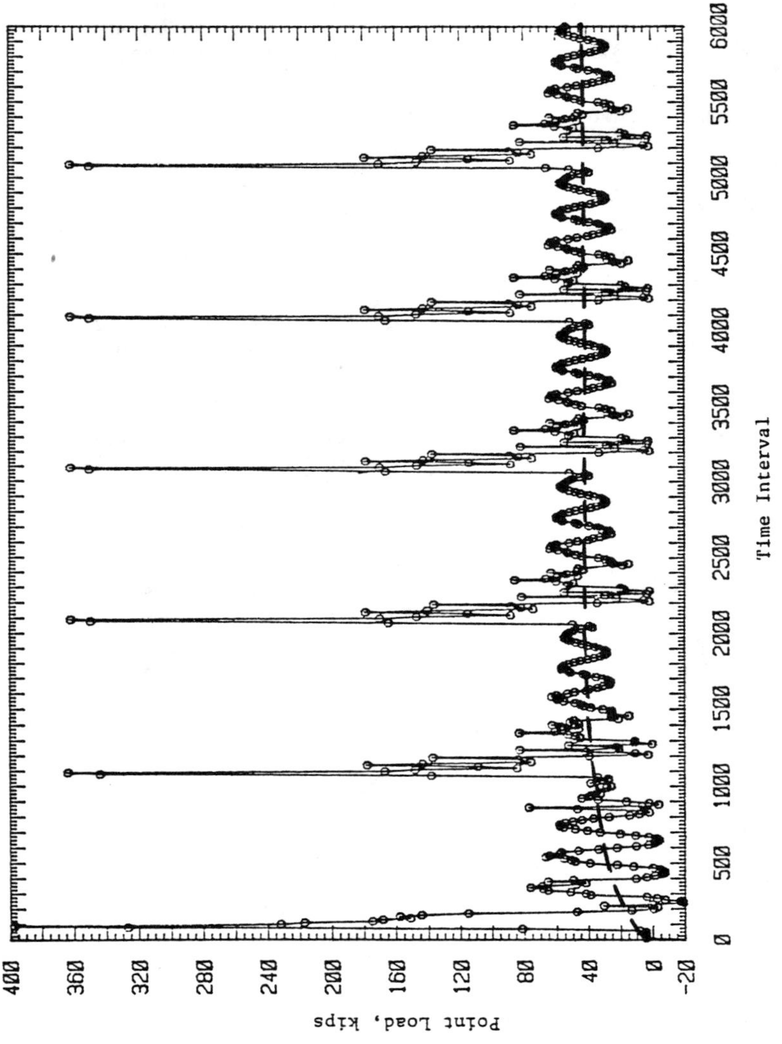

FIG. 4 - Point Load versus Time Interval for Six Hammer Blows

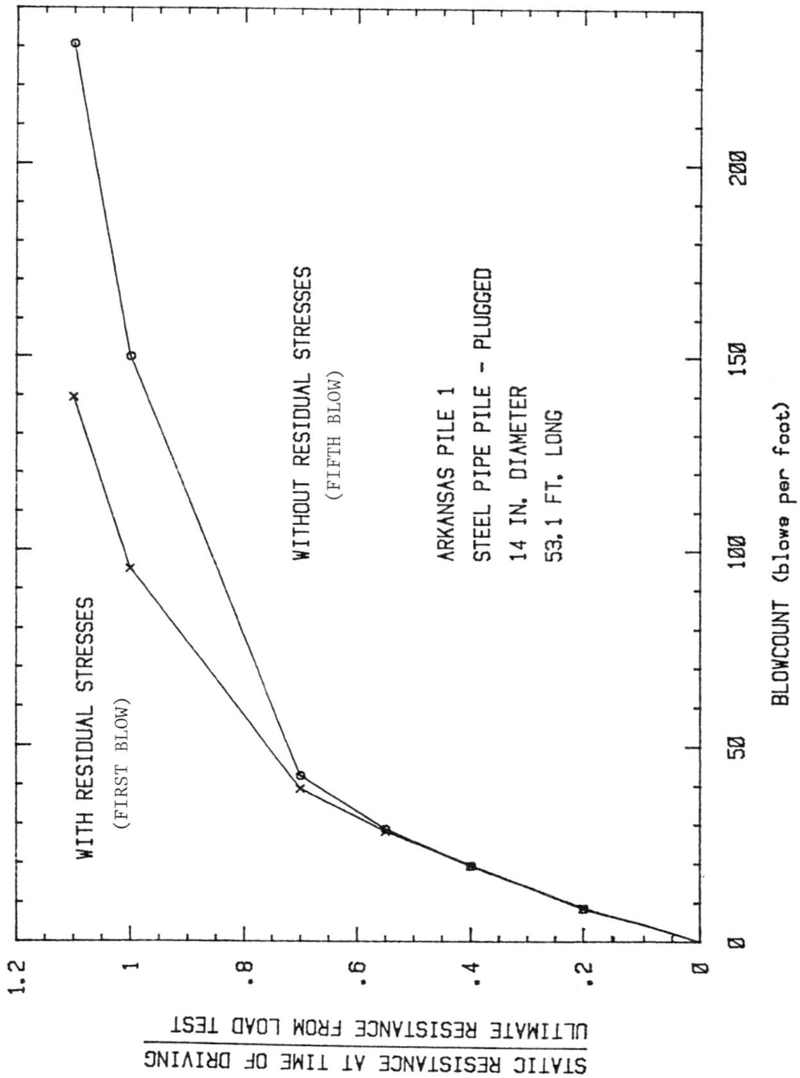

FIG. 5 - Pile Driveability Analysis with and without Residual Stresses

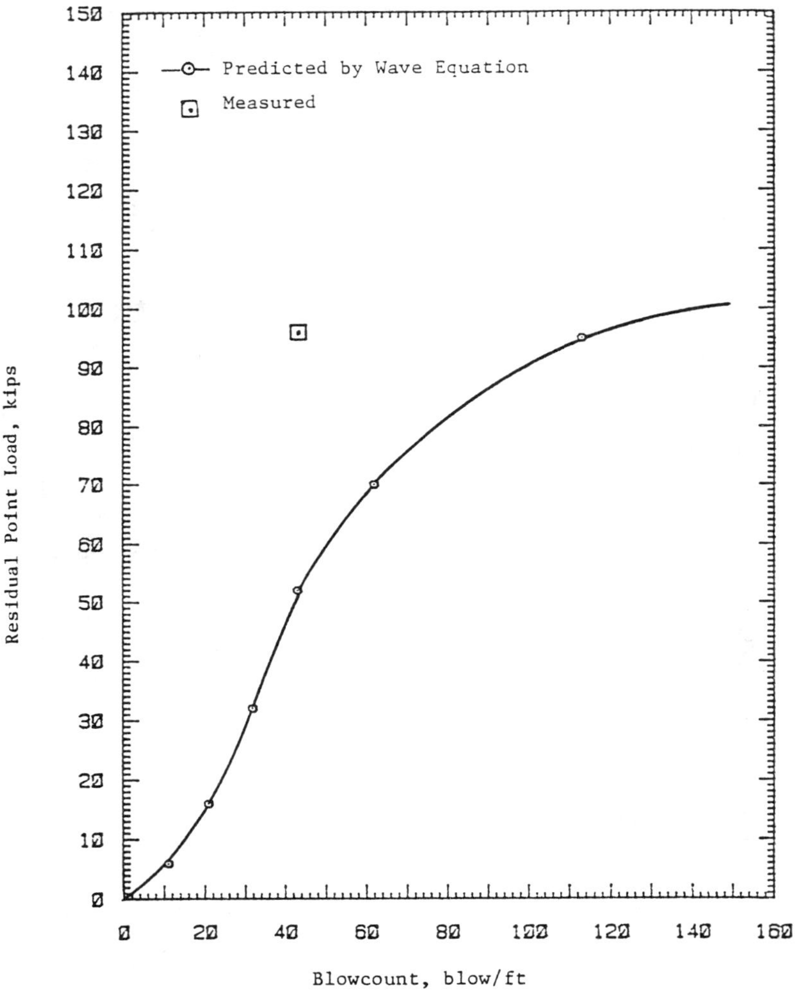

FIG. 6 - Residual Point Load versus Blow Count: Arkansas River Pile 3

the measured residual point loads in Fig. 7 for the piles analyzed.

The results for Arkansas Pile 3 are also plotted in Fig. 8 in terms of residual point load versus total static soil resistance at the time of driving. It was assumed that the static resistance at the time of driving was the same as the ultimate load from the compression load test (0.1 x D failure criterion). A plot of the residual point loads predicted by this approach versus the measured residual point loads is shown in Fig. 9 for the piles analyzed. An analysis of the ratio of predicted residual point load over measured residual point load yields a mean of 1.116 with a coefficient of variation of 0.657. These results provide a better correlation than the previous results. The fact that the residual point load is related to the static total load at failure is consistent with previously proposed theoretical expressions for the residual point load (1, 2, 13).

CONCLUSIONS

Residual stresses are important because of two main reasons.

1. Residual stresses affect the length of pile required. Current methods of pile capacity calculations are based on results from conventional load tests. These data bases of conventional compression tests on instrumented piles do not include residual stresses since the instrumentation is zeroed after driving. As a result the measured ultimate point loads are too low and the measured ultimate friction loads are too high. It was found that on the average the true ultimate point load is equal to 1.56 times the measured ultimate friction load and that the true ultimate friction load is equal to 0.81 times the measured ultimate load.
2. Residual stresses affect wave equation driveability analyses. As shown on Fig. 5 there is a need for simulating several hammer blows with the wave equation program in order to properly include residual driving stresses. Five blows was found to be sufficient and generally piles with residual stresses will drive easier than piles without residual stresses at hard driving conditions.

In conclusion, there is a need to properly measure residual stresses in order to develop a correct data base for driven piles in sands which can then be used to develop a correct design method. Also, pile driveability analyses using the wave equation approach must be used by simulating a minimum of five blows with the program in order to properly include residual stresses.

ACKNOWLEDGMENTS

The results presented in this article are part of a study performed for the United States Geological Survey and the Federal Highway Administration under Contract No. DTFH 61-82-C-0038. The contact persons were Charles Smith and Carl Ealy. The team of researchers included H. M. Coyle, R. A. Hawkins, L. L. Lowery, R. L. Lytton, and M. W. O'Neill.

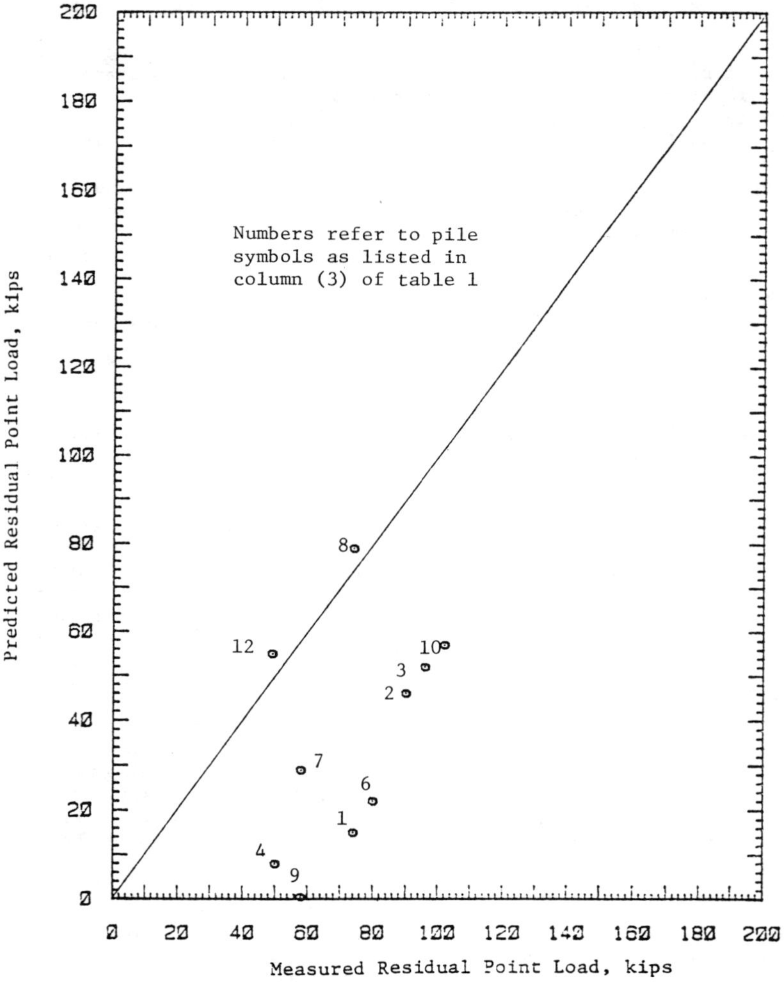

FIG. 7 - Comparison of Predicted versus Measured Residual Point Load Using Blowcount Correlation

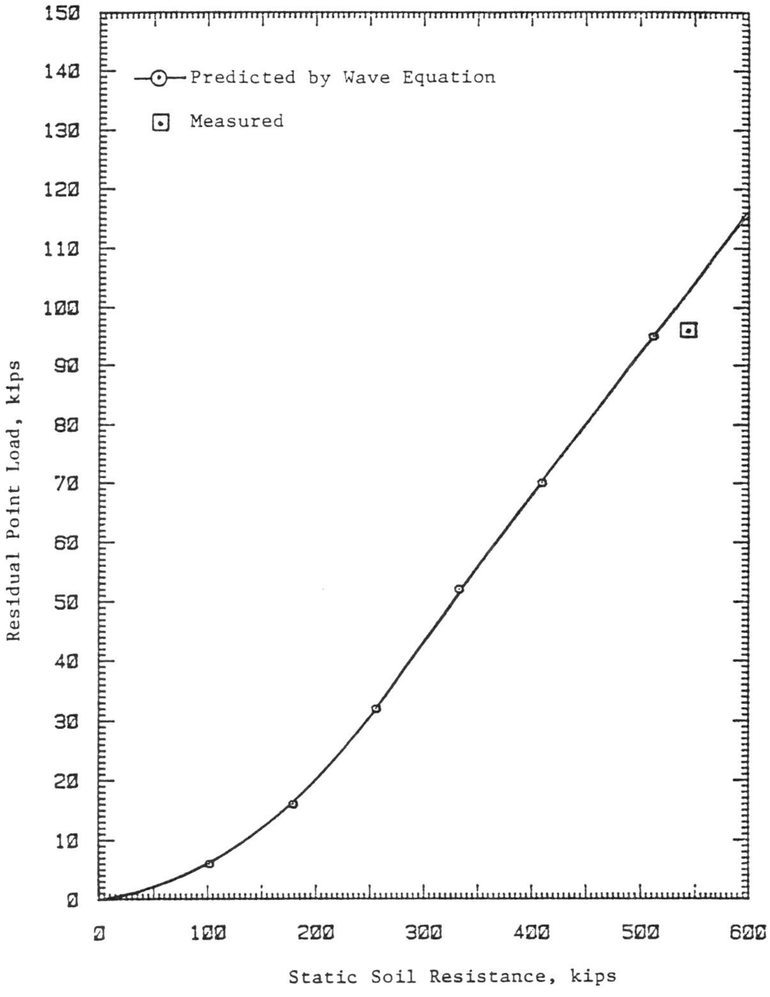

FIG. 8 - Residual Point Load versus Static Soil Resistance: Arkansas River Pile 3

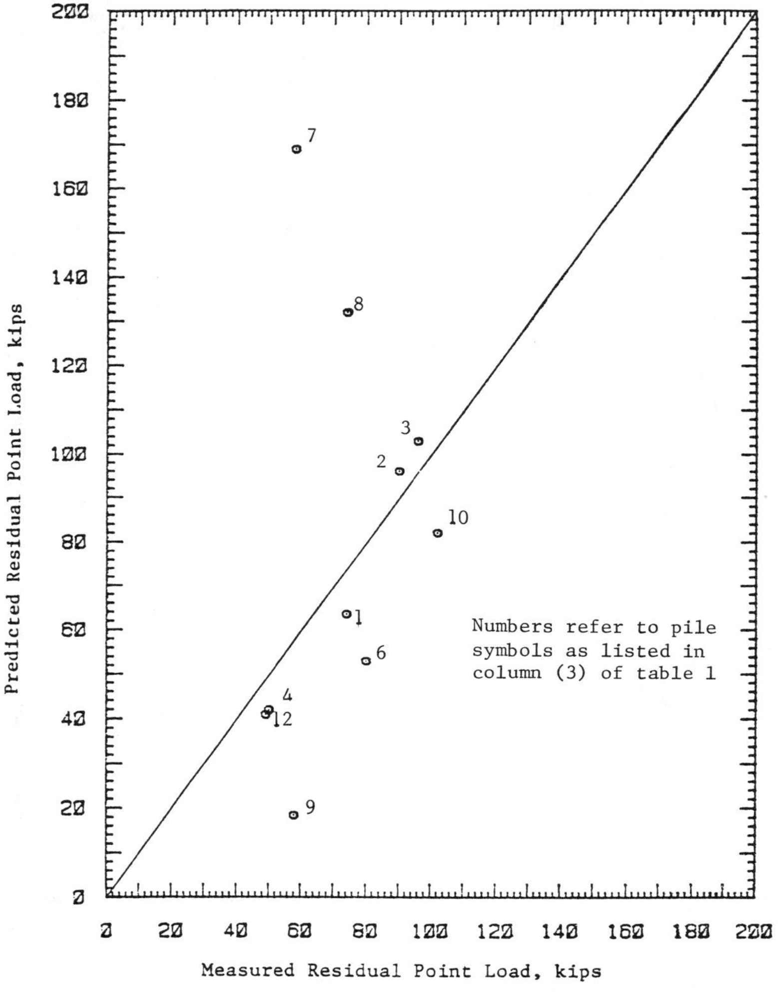

FIG. 9 - Comparison of Predicted versus Measured Residual Point Load Using Ultimate Resistance Correlation

REFERENCES

1. Briaud, J.L., and Tucker, L.M., "Piles in Sand - A Method Including Residual Stresses," accepted for publication in the Journal of the Geotechnical Engineering Division, ASCE, 1984.
2. Briaud, J.L., Tucker, L.M., Lytton, R.L., and Coyle, H.M., "Behavior of Piles and Pile Groups in Cohesionless Soils," Research Report to FHWA, Civil Engineering Department, Texas A&M University, September 1983.
3. Fruco and Associates, "Overwater Steel H-pile Driving and Testing Program," Report for Corps of Engineers, St. Louis, Mo., September 1973.
4. Fruco and Associates, "Pile Driving and Loading Tests," Report for Corps of Enginers, Little Rock, Arkansas, September 1964.
5. Gregersen, O.S., Aas, G., and DiBiaggio, E., "Load Tests on Friction Piles in Loose Sand," Proceedings of the VII International Conference on Soil Mechanics and Foundation Engineering, Moscow, Vol. 2.1, 1973, pp. 19-27.
6. Hirsch, T.J., Carr, L., and Lowery, L.L., Jr., "Pile Driving Analysis-Wave Equation Users Manual," Report No. 76-13.2, U.S. Department of Transportation, Washington, D.C., April 1976.
7. Holloway, D. M., Clough, G. W., and Vesic, A.S., "The Mechanics of Pile-Soil Interaction in Cohesionless Soils," Report for Corps of Engineers, Vicksburg, Miss., December 1975.
8. Hunter, A.H. and Davisson, M.T., "Measurements of Pile Load Transfer," Performance of Deep Foundations, ASTM STP 444, 1969, pp. 106-117.
9. Mansur, C.I. and Hunter, A.H., "Pile Tests - Arkansas River Project," Journal of the Soil Mechanics and Foundation Division, ASCE, Vol. 96, No. SM5, September 1970, pp. 1545-1582.
10. Mansur, C.E. and Kaufman, R.I., "Pile Tests, Low-Sill Structure, Old River, Louisiana," Transactions of ASCE, Vol. 123, 1958, pp. 715-748.
11. Sellgren, E., "Friction Piles in Non-cohesive Soils, Evaluation from Pressuremeter Tests," Thesis, Chalmers University of Technology, Goteborg, Sweden, 1982.
12. Smith, E.A.L., "Pile Driving Analysis by the Wave Equation," Transactions, ASCE, Vol. 127, 1964, Part I, p. 1145.
13. Tucker, L.M., "The Behavior of Piles in Cohesionless Soils," Master's thesis, Civil Engineering Department, Texas A&M University, 1983.
14. Vesic, A.S., "A Study of Bearing Capacity of Deep Foundations," Project B-189 (Final Report), Georgia Institute of Technology, August 1966.
15. Vesic, A.S., "Tests on Instrumented Piles, Ogeechee River Site," Journal of the Soil Mechanics and Foundation Division, ASCE, Vol. 96, No. SM2, March 1970, pp. 561-583.

Compression Load Tests on
Concrete Piles in Alluvium

Thomas M. Gurtowski[1], M.ASCE and
Ming-Jiun Wu[2], M.ASCE

INTRODUCTION

A series of pile load tests was accomplished during the design phase for the West Seattle Freeway Bridge Replacement project. Five different sites, A through E, were selected along the alignment which are shown in Figure 1, Site Plan and Subsurface Profile. At both Sites C and D, which were near the main span bridge piers, 24-inch-diameter (61 cm) open-end steel pipe piles were embedded into the lower glacial soils and load tested in compression and tension. At Sites A, B, and E a 24-inch (61 cm) octagonal prestressed concrete hollow pile with a 60-inch (102 cm) long concrete plug at the tip was load tested in compression. At site E the soils consisted of glacially overconsolidated clay. At Sites A and B the soils consisted of hydraulic fill over alluvium. Details of the bridge project, pile driving, pile dynamic testing and results, and the load testing program are discussed in Yamane and Wu (15). The results of the instrumented concrete pile compression load tests in the alluvial deposits (Sites A and B) will be presented and discussed herein.

Located approximately 50 feet (15 m) southwest of test Site A, a nine-pile group was fully instrumented to monitor and evaluate the long-term performance of an actual bridge pier. Initial monitoring results and a discussion of the instrumentation and installation are presented in Shannon & Wilson, Inc. (11). In the pile group instrumentation program a pile tip load cell was attached to three production piles prior to driving. The production piles were identical to the load test concrete piles. Details about the pile tip load cell are presented in Green, Feldman, and Thordarson (3). One purpose of the load cell was to measure the residual pile tip load after driving. The load test pile instrumentation at Sites A and B were installed into the piles after driving; consequently, the residual tip load after driving was not determined. The results of the pile tip load cell will be included with the pile load test measurements.

The non-linear vertical load-displacement behavior of the test piles at Sites A and B was modeled with pile-base and pile-shaft load-transfer functions. One purpose of the model study was to evaluate the residual pile tip load.

[1] Principal Engineer, Shannon & Wilson, Inc., Seattle, WA.
[2] Associate, Shannon & Wilson, Inc., Seattle, WA.

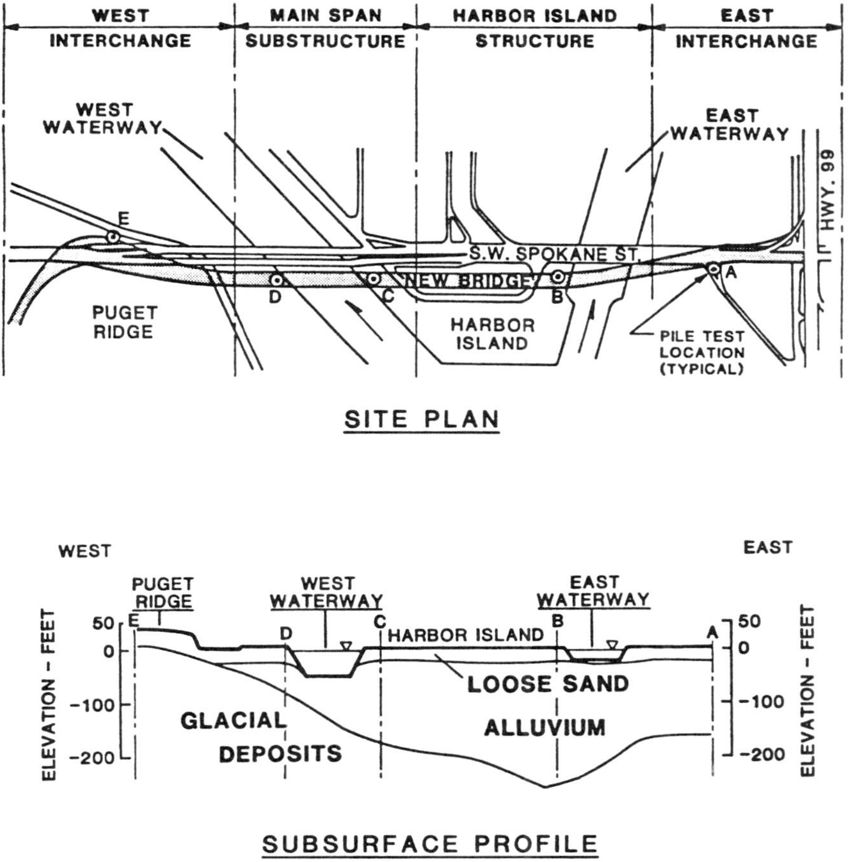

FIG. 1 SITE PLAN AND SUBSURFACE PROFILE
(1ft.=0.305m)

SOIL CONDITIONS

Standard Penetration Resistance N-values and pile driving resistance values as a function of depth at Sites A and B are presented in Figure 2. Generally, the soil conditions at Sites A and B consisted of 7- to 12-foot-thick (2 to 4 m) hydraulic fill consisting of loose to medium dense sand. The fill was underlain by alluvial deposits which consisted of loose to very dense sand with scattered layers of slightly plastic silt. The thicknesses of the alluvium ranged from about 140 feet (43 m) at Site A to 200 feet (61 m) at Site B. Underlying the alluvium were glacially-overridden deposits consisting of hard clay, till, or very dense sand. The groundwater table was near a depth of 10 feet (3 m).

As shown in Figure 2, the Standard Penetration Resistances (N-values) for the alluvial deposit generally increased with depth to about 30 feet (9 m). Below this, the N-values averaged about 40 blows per foot (blows per 30 cm).

PILE DRIVING

The concrete piles were driven with a single-acting Kobe K45 diesel hammer. The average measured hammer energy for the last 5 feet (1.5 m) of penetration was around 74,400 ft.-lbs. (100 kN-m). Generally, the hammer was set at 3/4 throttle to a depth of about 35 feet (11 m) at Site A and 40 feet (12 m) at Site B, and then at full throttle below these depths. A 10.5-inch-thick (267 mm) fir plywood pile cushion was used. Driven pile lengths were 98 feet (29.9 m) and 84 feet (25.6 m) at Sites A and B, respectively.

Generally, the pile driving resistance of the 24-inch (61 cm) concrete piles within the alluvium was about equal to the N-values, as shown in Figure 2. Hard driving conditions were encountered when the pile was driven through dense to very dense sand.

SOIL ENGINEERING PROPERTIES

During the design phase of the bridge project several direct shear tests were performed to measure the frictional resistance between concrete or steel and soil. For the alluvial sand, the direct shear friction angle of sand to sand ranged from 35 to 37 degrees, and 30 to 31 for concrete to sand. Triaxial compression tests performed on the alluvial sand indicated an effective stress friction angle of 34 to 41 degrees. For alluvial silt an effective stress friction angle of 36 to 37 degrees was measured.

PILE LOAD TESTS AND RESULTS

Pile load testing details are presented in Figure 3 which shows the reaction frame, two 700 kip (3,114 kN), calibrated hydraulic ram jacks and jacking arrangement; two Evergreen Weigh, 1,000 kip (4,448 kN) SR-4 calibrated strain gauge load cells; and ball and socket socket swivel joints. The jacking system was equipped with a load maintaining pump, so that constant loads could be held as deformations

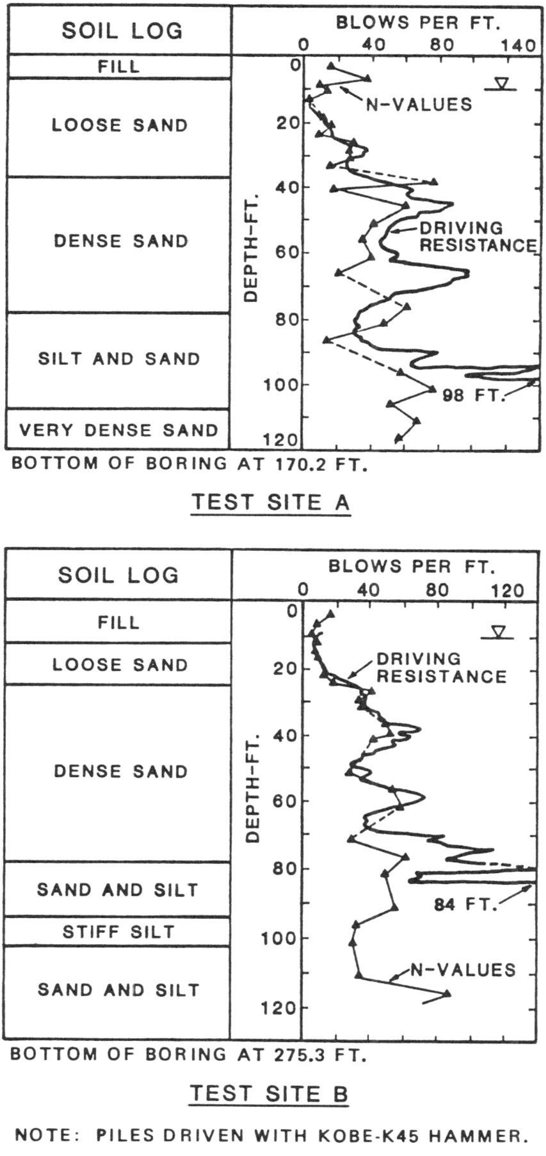

FIG. 2 N-VALUES AND DRIVING RESISTANCES
(1ft.=0.305m)

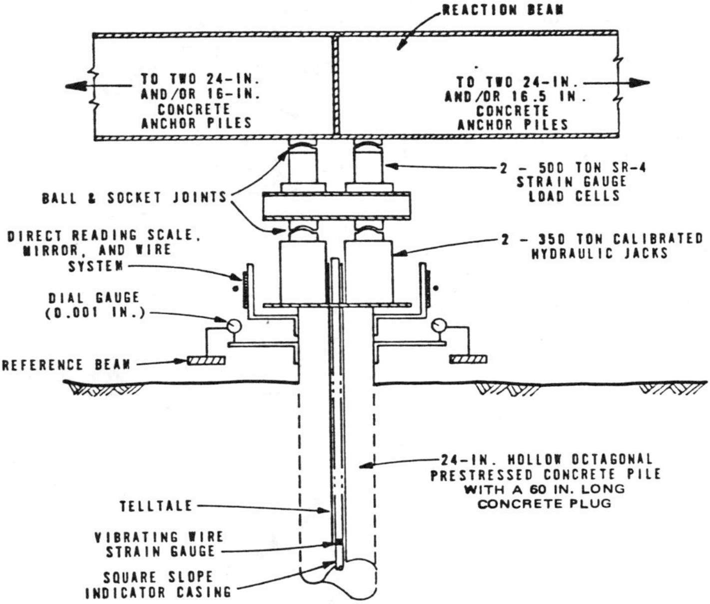

FIG. 3 TYPICAL COMPRESSION TEST SET-UP AT SITES A AND B

(1in.=25.4mm, 1ton=8.9KN)

occurred. Figure 3 also schematically illustrates the various methods used to obtain measurements of pile top settlements. Four dial gauges with an accuracy of 0.001 inches (0.025 mm) were attached to the pile body and spaced equally at 90 degrees. The dial gauges reacted against a frame, supported independently and away from the influence of the pile load test set-up. Two direct reading scale, mirror, and piano wire systems also were attached to the pile body.

Two independent measuring systems were used to measure the strain at various points along the load test piles in order to estimate the load distribution. These were the SINCO vibrating-wire strain gauges and stainless steel telltale rods, 1/4-inch (6.35 mm) in diameter. The strain gauges and the base mounts for the telltale rods were welded to a 2-1/2-inch (63.5 mm) square steel casing with a 1/8-inch (3.2 mm) wall thickness. The instruments were installed in the central hole of the pile after pile driving. Following installation of the instruments, the pile was backfilled with high-early strength sand-cement grout.

One compression load test was performed at each of Sites A and B. The piles were tested to plunging failure in accordance with the Standard Loading Procedures at Site B and the Quick Load Test Method at Site A, as described in ASTM Designation: D 1143-74. Both of these loading procedures were modified to allow time for reading instrumentation.

The results of the vertical compression pile load tests are presented in Figure 4. As shown, the 24-inch (61 cm) concrete piles failed by plunging at butt loads of approximately 1,050 kips (4,760 kN) at Site A and 900 kips (4,003 kN) at Site B. Also included in Figure 4 are the pile tip settlements as a function of the applied pile butt load. These tip settlements were extrapolated from the lowest telltale measurements.

Test data obtained by Bozozuk, et. al., (1) indicate that the ultimate skin friction is fully mobilized when the pile tip settlement is about 0.16 to 0.2 inches (4 to 5 mm). Once the ultimate skin friction is fully mobilized, the frictional resistance may remain constant or may decrease, but would not increase. Any additional load will be transmitted to the pile tip. As shown in Figure 4, at a pile butt settlement of about 0.4 inches (10 mm) the pile tips settled about 0.20 inches (5 mm) under an applied load of 820 kips (3,647 kN) at Site A and an applied load of 740 kips (3,292 kN) at Site B. As the load increased beyond these loads, the pile tip settlement increased at an increased rate.

Load transfer with depth along the test piles for various applied butt loads are presented in Figures 5 and 6 for Sites A and B, respectively. Calculation methods described by Vesic (13) were used to evaluate pile load transfer with depth. It should be noted that the telltale and strain gauge results do not include the effect of residual loads from pile driving. Vesic (13) and Hunter and Davisson (5) found that, after pile driving, residual stresses may remain in the soil below the pile tip, which may indicate a lower end-bearing resistance.

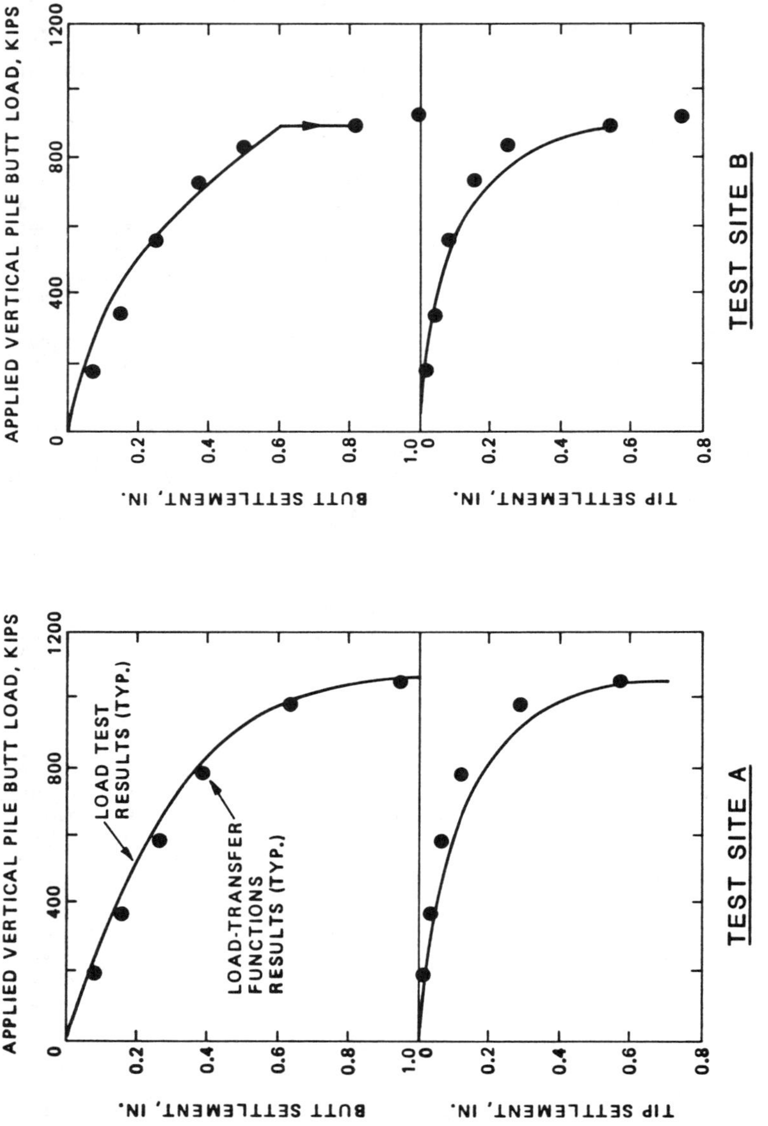

FIG. 4 AXIAL COMPRESSION PILE LOAD VS. SETTLEMENT
(1in.=25.4mm, 1kip=4.45kN)

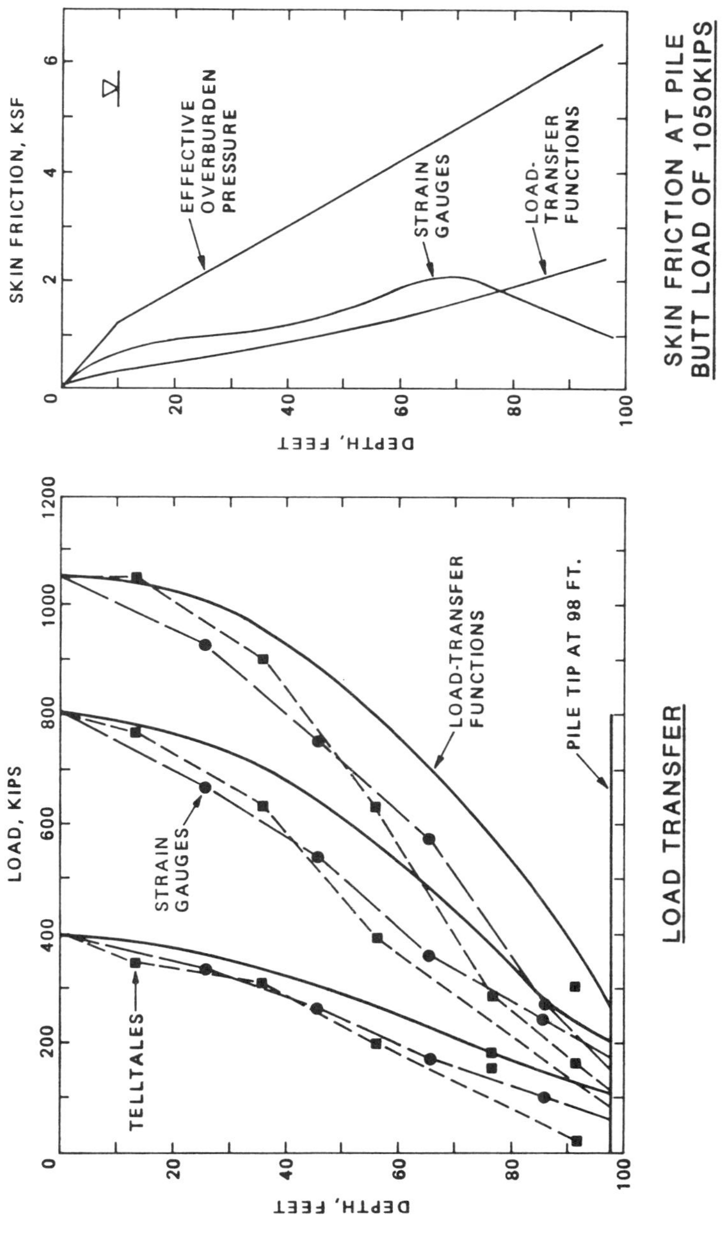

FIG. 5 LOAD TRANSFER AND SKIN FRICTION VS. DEPTH AT SITE A
(1ft.=0.305m, 1kip=4.45kN, 1ksf=47.9kPa)

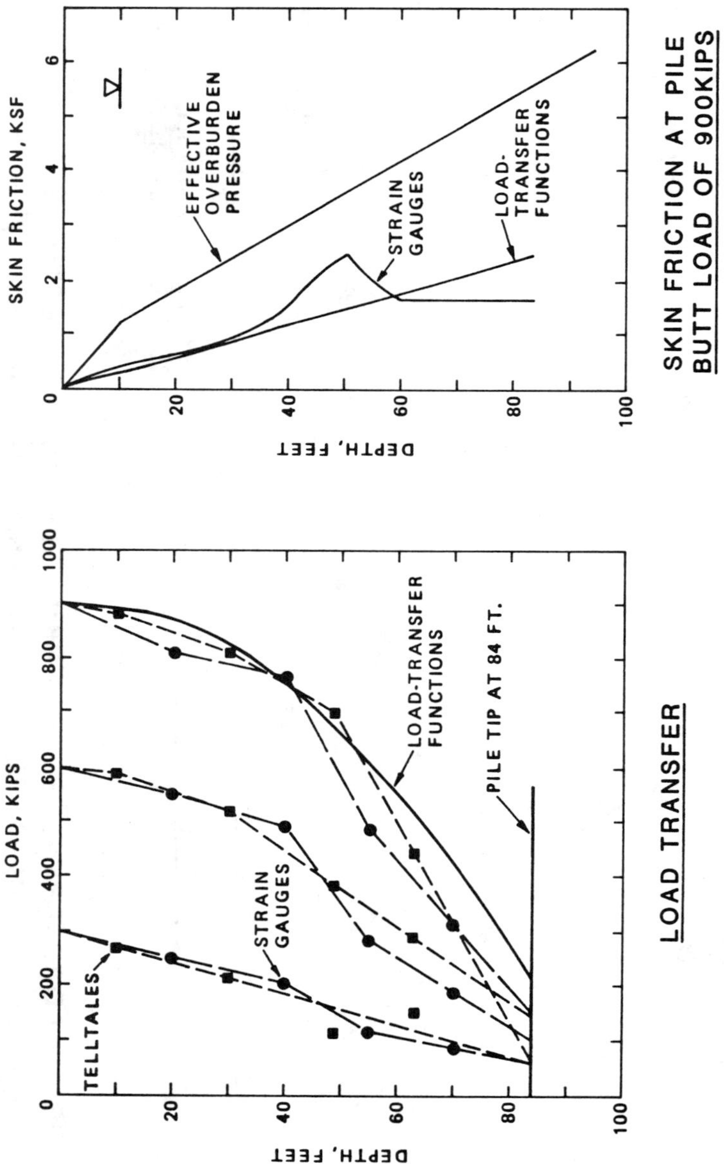

FIG. 6 LOAD TRANSFER AND SKIN FRICTION VS. DEPTH AT SITE B
(1ft.=0.305m, 1kip=4.45kN, 1ksf=47.9kPa)

This phenomenon was observed at the bridge pier adjacent to Site A. The load-transfer functions plots shown in the figures will be discussed in a later section.

Included with the load transfer results in Figures 5 and 6 is the variation of skin friction at the same depths along the test piles under the maximum compression or failure load at each site. The pile skin friction values were estimated from the slope of the appropriate load transfer curve, obtained from strain gauges or load-transfer functions.

RESIDUAL PILE TIP LOAD

As previously mentioned, prior to pile driving load cells were attached to three piles for a bridge pier located about 50 feet (15 m) from Site A. The piles were all driven with the same procedure as the test pile at Site A and to an average tip elevation 6.5 feet (1.8 m) lower than the load test pile. A Kobe KC45 single acting diesel hammer was used to drive the piles. Redrive blow counts after a delay of one to two days were generally greater than 200 blows per foot (30 cm). After the strain gauges and telltales were installed and the piles grouted, the average measured residual pile tip load was 100 kips (444.8 kN). Further details are presented in Shannon & Wilson, Inc. (11). Since the piles are the same, and the soil, groundwater, and pile driving conditions are very similar, a residual pile tip load of approximately 100 kips (444.8 kN) could have existed in the test pile at Site A had it been measured,or a tension pullout test accomplished. This value will be included in the evaluation of the end-bearing capacity and pile skin friction at Site A.

LOAD-TRANSFER FUNCTIONS ANALYSIS

Load-transfer functions relate the amount of loading force transferred to the soil at any particular depth with the magnitude of displacement between the pile and soil. The main assumption of these functions is that the soil behavior at any depth is independent of the soil behavior at all other depths. The analytical method proposed by Seed and Reese (10) was used to determine the axial non-linear movement of the pile under an applied axial compression butt load. The computer program which was adapted and used for this study was developed by Johnson (6). The shaft load-transfer function (t-z) selected was proposed by Kraft, Ray, and Kagawa (7) and the base load-transfer function (q-z) selected was proposed by Vijayvergiya (14). Details about these load-transfer functions are presented in their respective papers.

The load-transfer functions analysis results are presented in Figure 4 as pile butt and pile tip settlements at specific applied butt loads. The pile at Site A was analyzed prior to modeling pile B, because an approximate residual tip load was measured near Site A. The purpose of this analysis was to attempt to model the actual pile load versus settlement measured during the compression load test by including an approximate residual pile tip load.

A constant soil shear modulus of 5,000 psi (34,450 kPa), a curve fitting factor (R_f) of 0.9, and Poisson's ratio of 0.4 were used for the shaft load-transfer function. The input data used to generate the results at Site A other than those mentioned above include an angle of internal friction of 37 degrees, pile-soil angle of friction of 30 degrees, and a lateral earth pressure coefficient (K) of 0.5 for the hydraulic fill and 0.7 for the alluvial deposit. The maximum skin friction resistance was calculated as follows:

f_s = Unit skin friction = $\overline{\sigma}_o$ Ktan δ
$\overline{\sigma}_o$ = Average effective overburden pressure along the pile shaft
K = Lateral earth pressure coefficient
tan δ = Coefficient of friction between pile and soil

Consequently, Ktanδ equals 0.29 for the hydraulic fill and 0.40 for the alluvial deposit.

In order to attempt to model pile tip settlement behavior, a low end bearing capacity factor would be required. Assuming a pile tip residual load of zero the instrumentation results shown in Figure 5 indicate a maximum tip load of approximately 170 kips (756 kN) at Site A, or bearing pressure of 51.4 ksf (2,461 kPa). At an effective overburden pressure of 6.5 ksf (311 kPa) the bearing capacity factor (Nq) would be approximately 8. Assuming a residual pile tip load of 100 kips (444.8 kN) and zero residual stresses during the compression test, Nq would increase to about 12.6. This higher value agrees reasonably well with Terzaghi's (12) bearing capacity factor for a local shear failure at an angle of internal friction of 35 degrees, which is close to the value obtained during laboratory testing. For Terzaghi's general shear failure condition, the required angle of internal friction would be 27 degrees at an Nq of 12.5, which is unreasonable for the relatively dense soils penetrated at Site A. Therefore, it was decided to attempt to model pile tip load settlement with a low Nq-value to calculate the maximum bearing pressure in the base load-transfer function relationship. A "critical displacement" (z_c) of 0.04B or 1-inch (25.4 mm) was assumed.

As shown in Figure 4, the load-transfer functions of Kraft, Ray, and Kagawa (7) and Vijayvergiya (14) using Terzaghi's local bearing capacity factor accurately modeled the measured load pile butt settlement at Site A. The same load-transfer functions and input parameters from Site A, except for a higher K-value of 0.8 for the alluvial deposit (higher N-values), were used to evaluate the pile load test at Site B. As shown in Figure 4, the load-tranfser functions closely modeled the measured results.

It has been established that hammer driven piles always contain residual loads at the pile tip and along the pile shaft. These residual loads can result from the pile attempting to rebound back to its original length after being subjected to a dynamic compressive impulse load. If the compression loads were high enough a residual point load will also exist. These residual loads will mask the real or actual load distribution along the pile if the strain gauges were not zeroed prior to pile driving. However, the total loads measured at the pile butt in compression or tension will not change. In general, adjusting for residual stresses would result in an increase in the total pile tip load and decrease in the total pile side load.

Coyle and Castello (2) have described methods for adjusting or correcting residual loads in hammer driven piles. One of the methods is to follow the procedures outlined by Hunter and Davisson (5). Another method suggested by Holloway, Clough, and Vesic (4) is to assume that the adjusted or corrected total compressive side load should be equal to the tension load obtained in a pullout test. In other words, the adjusted point load is equal to the difference between the total compression test load and the total tension test load. These methods were used to correct for residual loads by Coyle and Castello (2) from four of the Arkansas compression and tension load tests on closed-end steel pipe piles driven into medium dense to dense fine sand with silt and clay. One result of their analysis was that the tension side load measured after the pullout test and the adjusted or corrected compression side load using the Hunter and Davisson (5) method was approximately 0.7 of the unadjusted compression side load measured after the compression load test.

Assuming the load-transfer function load distribution for Site A represents the adjusted compression side load, i.e., includes the correct residual tip and side loads, then the skin friction should be 0.7 of the skin friction measured from the instrumentation. As shown in Figure 5, the load-transfer functions skin friction is about 0.7 of the average strain gauge skin friction under the same applied pile butt load. However, this relationship was not observed at Site B.

REVIEW OF STATIC PILE CAPACITY DESIGN METHODS

Coyle and Castello (2) have presented a noteworthy paper which presents correlations to evaluate skin friction and end-bearing capacities for designing piles in sand. Their paper also reviews in detail the existing theories and generally accepted design criteria for estimating static pile capacity. A summary of relevant pile design parameters suggested by Coyle and Castello (2) are presented in Table 1 along with the results of the pile load tests at Sites A and B. Also included, for comparison are suggested relevant design parameters from Meyerhof (8) and Nordlund (9).

As shown in Table 1 the only design parameter which correlates with the pile load test results at Sites A and B is the unit side resistance (f_s). However, the lateral earth pressure coefficient (K) and combined factor ($K\tan\delta$) suggested by Coyle and Castello (2) and Nordlund (9) are greater than those estimated from the load tests. The estimated unit point resistances (q_o) from the pile load tests assuming a residual tip load of approximately 100 kips (4,788 kN), are 35 to 40 percent of those suggested by Coyle and Castello (2) and Meyerhof (8).

The reason for the relatively low unit point resistance for the piles and soil conditions at load test Sites A and B is not clear. However, one distinct difference is that the 24-inch (61 m) concrete

TABLE 1 SUMMARY OF PILE DESIGN PARAMETERS

DATA BASE	ANGLE OF INTERNAL FRICTION	BEARING RELATIVE DEPTH FACTOR	UNIT POINT RESISTANCE	FRICTION RELATIVE DEPTH FACTOR	PILE-SOIL FRICTION ANGLE	LATERAL EARTH PRESSURE COEFF.	COMBINED FACTOR	UNIT SIDE RESISTANCE AT PILE MIDPOINT
	ϕ	D/B	q_o, ksf	D/B	δ	K	$K\tan\delta$	f_s, ksf
SITE A	37	35	80 (3830kPa)	17	30	0.7	0.40	1.5 (72kPa)
COYLE & CASTELLO (2)	37	35	210 (10055kPa)	17	30	2.4	0.80	1.5 (72kPa)
SITE B	37	28	67 (3208kPa)	14	30	0.8	0.46	1.5 (72kPa)
COYLE & CASTELLO (2)	37	28	190 (9097kPa)	14	30	2.7	0.9	1.4 (67kPa)
MEYERHOF (8)	37	35	200 FIG. 4* (9576kPa)	–	–	–	–	2.0 FIG. 9 (96kPa)
NORDLUND (9)	37	–	–	–	31	1.8	1.1	4.0 (191kPa)

*REFERS TO FIGURES IN MEYERHOF (8)

NOTE: 1 ksf = 47.9 kPa

test piles are about twice the length of most of the piles reported by Coyle and Castello (2). Using the methods proposed by Coyle and Castello (2), an angle of internal friction of 34 degrees would be required to obtain the measured total ultimate static capacity of the test pile at Site A. However, this friction angle does not seem to correlate with the laboratory test results or the results of the load-transfer functions analysis. A friction angle of 34 degrees at 17 pile diameters of penetration into dense sand (at a total depth of about 67 feet (20 m)) indicates an ultimate unit side resistance (f_s) of about 1 ksf (47.9 kPa) which is too low for the approximately 80 to 85 percent relative density sand soils at Site A.

SUMMARY AND CONCLUSIONS

Two instrumented, long, concrete piles were driven into alluvium and load tested to failure in compression. The strain gauges and telltale rods were installed into the load test piles after driving; consequently, residual pile loads were not measured. However, at an adjacent bridge pier a residual tip load was successfully measured with a pile tip load cell installed before pile driving. A load-transfer functions analysis using available shaft and base transfer functions was accomplished in order to evaluate 1) the load-transfer functions, 2) pile load versus settlement, 3) pile load distribution with depth, 4) unit skin friction and end bearing resistances, and 5) the residual pile tip load. In addition, the unit skin friction and end bearing resistances and soil properties were compared to currently available pile design parameters proposed by others.

Using procedures suggested by Seed and Reese (10) and the base and shaft load-transfer functions proposed by Vijayvergiya (14) and Kraft,

Ray, and Kagawa (7), the non-linear vertical load-displacement behavior of the test piles was modeled with relative accuracy for Sites A and B, as shown in Figure 4. The base load-transfer function included Terzaghi's (12) local shear failure bearing capacity factor for calculating the ultimate bearing capacity at a tip displacement of 1-inch. The load-transfer distribution along the Site A test pile obtained from the instrumentation and load-transfer functions are very similar. The load-transfer functions also indicated a residual tip load of 60 to 100 kips (267-445 kN) which was approximately the value measured with the pile tip load cell.

Methods have been described by Coyle and Castello (2) to correct or evaluate residual loads in hammer driven piles. One of the methods, which does not require a tension load test, is to take 0.7 of the unadjusted compressive side load. Assuming the load-transfer functions pile load distribution results have accounted for residual loads and could be considered as adjusted loads, then they would be approximately 0.7 of the strain gauges measured compression side load or unadjusted side load. This trend was observed at Site A, but not at Site B.

The unit side resistance values (f_s) suggested by Coyle and Castello (2) correlate reasonably well with the pile load test results. The combined factors (Ktanδ) suggested by Coyle and Castello (2) shown in Table 1 are twice the values measured at Sites A and B, yet the unit side resistances are almost identical. According to Meyerhof (8) conventional side resistance theory in terms of the lateral earth pressure coefficient (K) cannot be used for piles longer than about 15 to 20 piles diameters because beyond this depth the unit side resistance becomes practically independent of the effective overburden pressure. This indicates that for long piles with deep penetrations, the lateral earth pressure coefficient decreases with depth. The pentrations of Sites A and B test piles exceed the "critical depth ratio" (8) by a factor of about 2. The results summarized in Table 1 confirm Meyerhof's findings. The unit side resistance for long piles should not be evaluated with a lateral earth pressure coefficient. It should be evaluated with a unit side resistance versus relative depth function, similar to that proposed by Coyle and Castello (2).

The suggested unit point resistances (q_o) of Meyerhof (8) and Coyle and Castello (2) are about 250 percent greater than that obtained from the pile load test results including an estimated residual pile tip load. The reason for the relatively low unit point resistance is not clear.

ACKNOWLEDGEMENTS

The authors would like to acknowledge the Seattle Engineering Department for permission to publish this paper. The assistance of many members of the staff of Shannon & Wilson, Inc., especially Mr. H. Ellis, is gratefully appreciated.

REFERENCES

1. Bozozuk, M., Keenan G. H., and Pheeney, P.E., "Analysis of Load Tests on Instrumented Steel Test Piles in Compressible Silty Soil," Behavior of Deep Foundations, STP 670, American Society for Testing and Materials, 1979, pp. 153-180.

2. Coyle, H. M., and Castello, R. R., "New Design Correlations for Piles in Sand," Journal of the Geotechnical Engineering Division, ASCE, Vol. 107, NO. GT7, July 1981, pp. 965-986.

3. Green, G. E., Feldman, A. I., and Thordarson, P., "A Pile Tip Load Cell for a Driven Concrete Pile," presented at the Sept. 5-8, 1983, Field Measurements in Geomechanics International Symposium, held at Zurich, Switzerland.

4. Holloway, D. M., Clough, G. W., and Vesic, A. S., "The Mechanics of Pile-Soil Interaction in Cohesionless Soils," Contract Report S-757-5 (Final Report), United States Army Engineer Waterways Experiment Station, Dec., 1975.

5. Hunter, A. H., and Davisson, M. T., "Measurement of Pile Load Transfer," Performance of Deep Foundations, STP 444, American Society for Testing and Materials, 1969, pp. 106-117.

6. Johnson, L. D., "Methodology for Design and Construction of Drilled Shafts in Cohesive Soil," Technical Report TR-GL-84-5, United States Army Engineer Waterways Experiment Station, Mar., 1984.

7. Kraft, L. M., Ray, R. P., and Kagawa, T., "Theoretical t-z Curves," Journal of the Geotechnical Engineering Division, ASCE, Vol. 107, No. GT11, Nov., 1981, pp. 1543-1561.

8. Meyerhof, G. G. "Bearing Capacity and Settlement of Pile Foundations," Journal of the Geotechnical Engineering Division, ASCE, Vol. 102, No. GT3, Mar., 1976, pp. 195-228.

9. Nordlund, R. L., "Bearing Capacity of Piles in Cohesionless Soils," Journal of the Soil Mechanics and Foundation Division, ASCE, Vol. 89, No. SM3, May, 1963, pp. 1-35.

10. Seed, H. B., and Reese, L. C., "The Action of Soft Clay Along Friction Piles," Transactions, ASCE, Vol. 122, 1957, pp. 731-754.

11. Shannon & Wilson, Inc., "Long-Term Pile Group Study, West Seattle Freeway Bridge Replacement, City of Seattle, Washington; Progress Report No. 1, Instrumentation, Installation and Initial Monitoring, Pier EA-31," prepared for City of Seattle, FHWA and ABKJ, Nov., 1982.

12. Terzaghi, K., Theoretical Soil Mechanics, John Wiley and Sons, Inc., New York, NY, 1943.

13. Vesic, A. S., "Design of Pile Foundations," NCHRP Synthesis of Highway Practice No. 42, Transportation Research Board, 1977.

14. Vijayvergiya, V. N., "Load-Movement Characeristics of Piles," 4th Symposium of Waterway, Port, Coastal and Ocean Division, ASCE, Long Beach, Calif., Vol. 2, 1977, pp. 269-284.

15. Yamane, G., and Wu, M. J., "Pile Foundation West Seattle Freeway Bridge Replacement, Seattle, Washington," Transportation Research Record 884, 1982, pp. 29-37.

Vertical Behavior of Two 16-Year Old Drilled Shafts in Expansive Soil

L. D. Johnson,[1] M. ASCE, and W. R. Stroman,[2] F. SAME

Vertical load tests were performed on drilled shafts subject to lateral confining and heaving pressures from expansive soil over 16 years. Results of load tests indicated mobilized skin friction equal to the maximum soil shear strength. Lateral earth pressures on the shafts were about 3 times the vertical pressure. End bearing capacity was significantly reduced compared to short-term capacity because of long-term wetting and soil softening beneath the shaft base.

Introduction

Seven test shafts were constructed July 1966 in a test site on Lackland Air Force Base to study the performance of drilled shafts in expansive soil. The test site contains a 100- by 100-ft (30- by 30-m) covered area consisting of two impervious neoprene fabric reinforced membranes separated by 8 in. (200 mm) of sand. This cover was constructed in July 1974 to observe trends in long-term heave in expansive soil caused by lack of evaporation beneath lightly loaded areas (9). Two of the shafts, LAFB-1 located 15 ft (5 m) north of the covered area and LAFB-2 located within the southeast corner of the covered area, Figure 1, were selected for additional study.

The test site was instrumented with 10 porous stone, Casagrande type piezometers, 11 freestanding benchmarks, and two deep reference benchmarks 60 (18 m) and 80 ft (24 m) deep as shown in Figure 1 to provide accurate pore water pressure and elevation profiles. Pressure heads recorded in the piezometers indicated a perched hydrostatic water table from 8 ft (2.4 m) below ground surface to about 35 ft (10.7 m). Pore pressures decreased to almost zero at 50 ft (15.2 m). A deep water table was observed 80 ft (24.4 m) below ground surface. The surface of the perched water table is known to vary with time and it was observed only 4 ft (1.2 m) below ground surface in May 1979.

Load tests were originally conducted on most of these shafts within 12 months following construction to evaluate the influence of heaving soil on performance. Soil heave was accelerated by flooding 10-ft (3-m) square diked areas around the test shafts for 1 year beginning November 1966. Results of these tests (17) indicated that

---

[1] Research Civil Engineer, US Army Engineer Waterways Experiment Station, Vicksburg, MS.

[2] Project Engineer, US Army Engineer District, Fort Worth, TX.

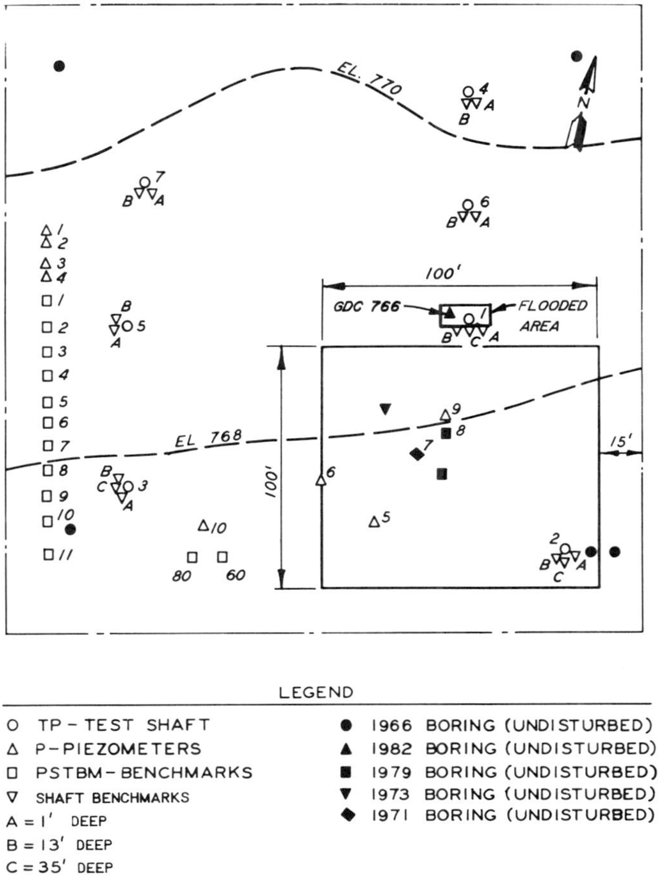

Figure 1. Lackland Air Force Base (LAFB) test site plan

the shafts were only beginning to react to expansion of the swelling soil. Tension forces in the shafts were considerable and appeared to be concentrated in the primary soil formation below 12 ft (3.7 m) of depth.

Vertical load tests were performed in 1982 on shafts LAFB-1 and LAFB-2 to investigate long-term performance in expansive soil. LAFB-1 is 1.5 ft (0.5 m) in diameter by 34 ft (10.4 m) long including a 3-ft (0.9 m) diameter (2:1) bell. LAFB-2 is 2.5 ft (0.8 m) in diameter by 36 ft (11.0 m) long including a 4-ft (1.2 m) diameter (2:1) bell. Both test shafts were instrumented with strain gages secured on the steel reinforcement and distributed at periodic intervals along the full length of the shafts. LAFB-1 was reinforced with 5 No. 9 bars (4 perimeter and 1 center) tied with No. 4 bars at 12-in. (300-mm) centers for 5 $in.^2$ (3226 mm^2) total area of steel reinforcement or 2% steel. Four strain gages were placed on the perimeter reinforcement at each of the following depths: 0.5, 11.5, 19.5, and 32.5 ft (0.2, 3.5, 5.9, and 9.9 m). Single strain gages were also placed on the center bar at 2.5, 5.5, 8.5, 15.5, 23.5, and 27.5 ft (0.8, 1.7, 2.6, 4.7, 7.2, and 8.4 m). LAFB-2 was reinforced with 3 No. 14S bars tied with No. 4 bars at 12-in. (300-mm) centers for 6.75 $in.^2$ (4355 mm^2) total steel reinforcement or 1% steel. Three strain gages were placed at each of the following depths; 0.5, 6, 11, 15, 21, 27, and 32 ft (0.2, 1.8, 3.4, 4.6, 6.4, 8.2, 9.8 m). LAFB-2 was also instrumented with 8 Carlson earth pressure cells. Two cells were mounted in specially prepared holes cut into undisturbed soil on opposite sides of the shaft perimeter at depths of 2.4, 6, 15, and 32 ft (0.7, 1.8, 3.4, and 9.8 m) below ground surface. Compressive strength tests performed on 3-in. (76 mm) diameter concrete core samples taken from LAFB-1 at 3.8 and 11.5 ft (1.2 and 3.5 m) of depth following the load tests indicated a Young's modulus of 216000 tsf (21000 MPa) and compressive strength of about 400 tsf (38 MPa). Tests conducted on cylinders taken from the batch concrete indicated 360000 tsf (34000 MPa) and 312 tsf (30 MPa) for the Young's modulus and the compressive strength, respectively, after being kept 1 year in a fog room.

Description of Soils

The overburden material consists of about 8 ft (2.4 m) of expansive black to gray CH clay and 4 to 5 ft (1.2 to 1.5 m) of GC clayey gravel with caliche. The primary material encountered below the gravel is fissured expansive clay (CH) shale of the Upper Midway group of the Tertiary system. The soils are uniform within the test site. Laboratory tests to classify the soil and to evaluate strength and consolidation parameters were performed on 6-in. (152-mm) diameter undisturbed samples obtained in 1966, 1971, 1973, 1979, and 1982 by Shelby tube or piston samplers. Relatively undisturbed samples could not be obtained from the clayey gravel.

Results of Laboratory Tests. Classification tests, Figure 2, indicate CH soils above and below the clayey gravel. Water contents in 1982 appear in the upper range of observed water contents indicating that the clay shale is wetter than in previous years.

Triaxial unconsolidated undrained (Q) tests under confining pressure equivalent to the in situ total vertical overburden pressure and consolidated undrained tests with pore pressure measurements (R) were performed on 1.4- by 2.8-in. (36- by 71-mm) long undisturbed specimens from samples of boring 6DC-766 obtained in 1982 adjacent LAFB-1. Triaxial Q and drained (S) direct shear (DS) tests had also been performed in 1966. Results of the Q tests, Figure 3, indicate similar strengths in 1966 and 1982. These strengths are about 0.8 tsf (76.6 kPa) from 0 to 8 ft (0 to 2.4 m), 0.9 tsf (86.2 kPa) from 15 to 21 ft (4.6 to 6.4 m), 1.5 tsf (143.7 kPa) from 25 to 35 ft (7.6 to

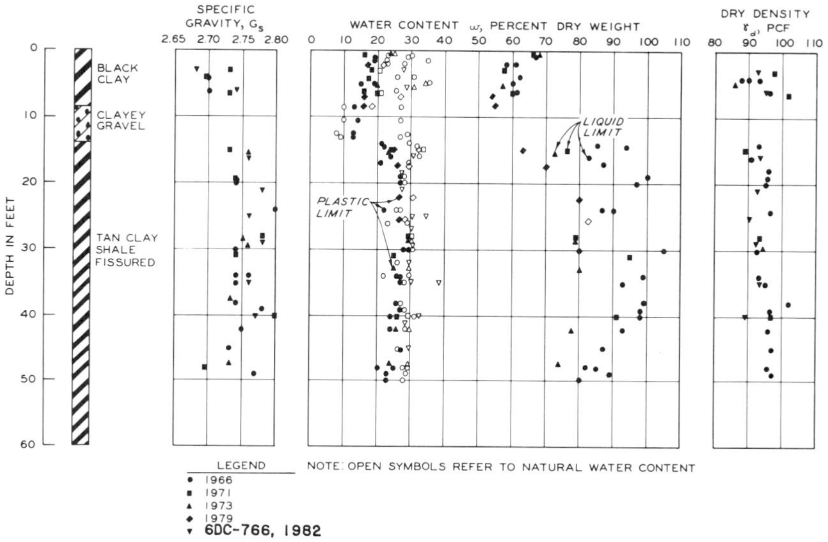

Figure 2. Specific gravity, water content, Atterberg limits, and dry density of Lackland Air Force Base (LAFB) test site soil

10.7 m), and 3.6 tsf (344.9 kPa) at 40 ft (12.2 m). The 1982 R data shown in Figure 3 indicate an effective friction angle of 25 deg at 3 and 6 ft (0.9 and 1.8 m), 21 deg at 16 ft (4.9 m), and 29 to 34 deg below 20 ft (6.1 m). The 1966 direct shear (DS) results of 9 and 12 deg performed on presplit specimens result in a low strength unless the observed 1.0 tsf (95.8 kPa) cohesion for intact samples is included.

Initial elastic moduli found by the hyperbolic model (5) from results of Q and R tests, Figure 3, indicate relatively small values of about 150 tsf (14.4 MPa) above 8 ft (2.4 m) with a tendency to increase linearly below 16 ft (4.9 m) according to E_s = 18.2z tsf (5.72 z MPa) where z is the depth in feet (meters). Some of the initial moduli from the 1966 soil samples exceeded 1000 tsf (95.8 MPa) below 25 ft (7.6 m) of depth.

A perched water table reduces the effective vertical overburden pressure σ'_v down to about 50 ft (15.2 m) as shown in Figure 4. Effective stress analysis is based on this σ_v . The consolidometer data indicate overconsolidation ratios (OCR) from 2 to as much as 18. Coefficients of lateral earth pressure at rest K_o vary from 0.7 above 8 ft to 1.3 below 16 ft (4.9 m) based on tests of samples from 6DC-766 bored in 1982. Closed symbols indicate K_o after Brooker and Ireland (3) and open symbols after Gardner (6). Earlier samples indicate larger coefficients. Compression indices vary from 0.14 to

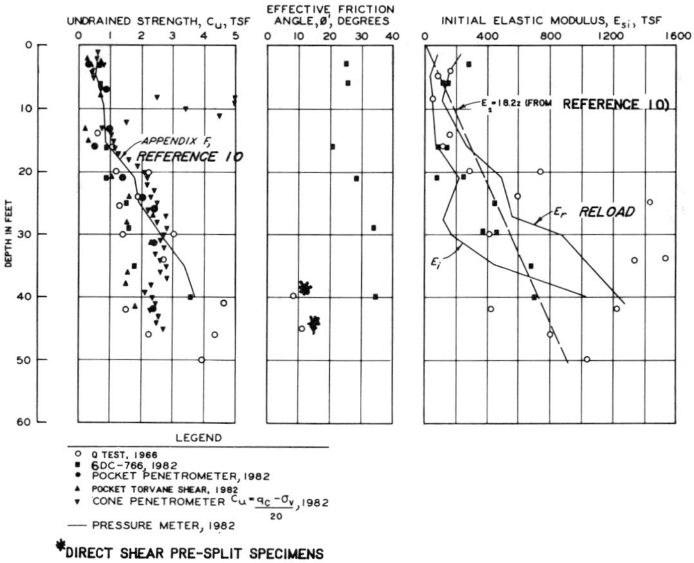

Figure 3. Strength and initial elastic modulus of LAFB test site soil

0.39 and swell indices from 0.04 to 0.14. Swell pressures of the 1982 samples vary from 0.4 tsf (38.3 kPa) above 8 ft (2.4 m) to 4.5 tsf (431.1 kPa) below 16 ft (4.9 m). Swell pressures determined in 1966 were 1 to 1.6 tsf (95.0 to 153.3 kPa) above a depth of 8 ft (2.4 m) and as much as 13 tsf (1245.4 kPa) below a depth of 20 ft (6.1 m) or 2 to 3 times greater in 1966 than in 1982. These data suggest that significant wetting and swell could have occurred in the soil profile during the past 16 years which had reduced the swell pressures. An expected decrease in soil strength from wetting was not measured on these soil samples.

Results of In Situ Tests. Pocket penetrometer and pocket torvane shear tests performed on samples from boring 6DC-766 in 1982 generally indicated undrained strengths similar to those from Q laboratory tests, Figure 3. Cone penetrometer data using a cone bearing capacity factor of 20 indicated an upper limit of undrained shear strength. The cone penetrometer data also indicated an extremely strong stratum from 8 to 12 ft (2.4 to 3.7 m) coincident with the clayey gravel.

A pressuremeter test performed by J. L. Briaud (Appendix F, 10) was used to develop soil modulus, undrained strength, and coefficients of lateral earth pressure at rest. The initial loading modulus E_i, Figure 3, is less than the initial modulus evaluated by the hyperbolic model from results of triaxial strength tests performed on samples from boring 6DC-766. The pressuremeter reloading modulus E_r exceeds

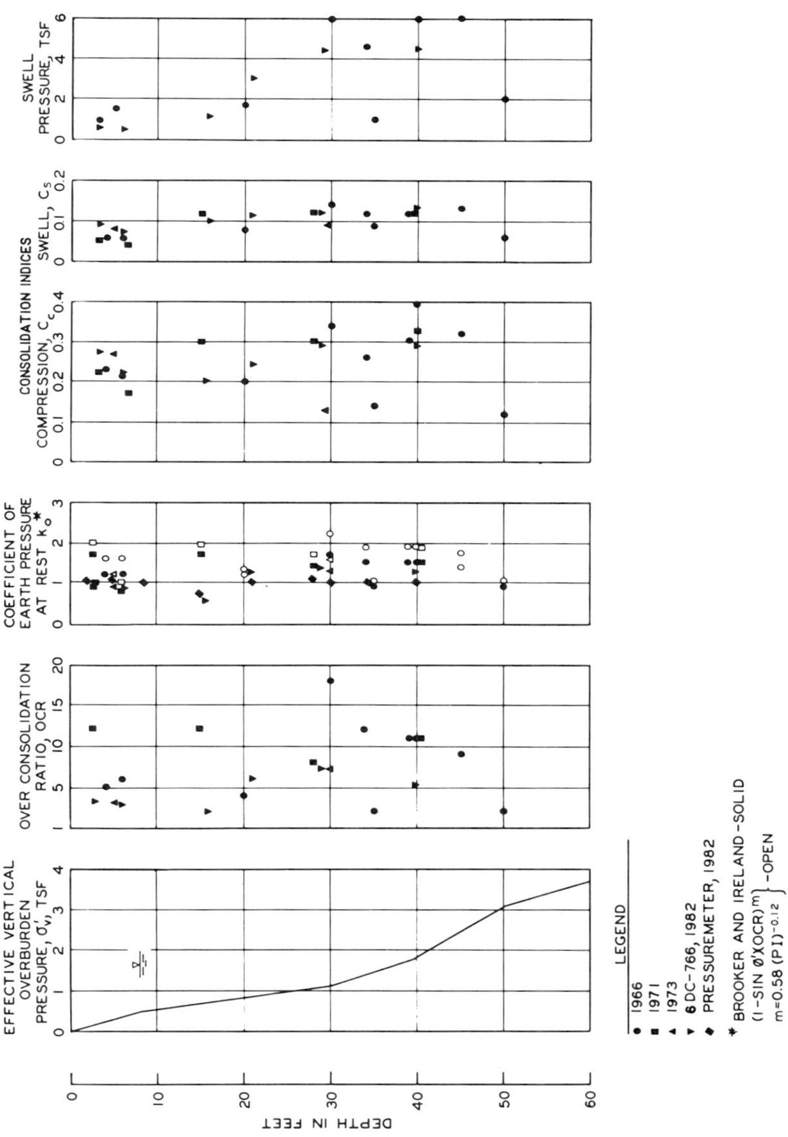

Figure 4. Consolidation data of LAFB test site soil

the initial modulus evaluated by the hyperbolic model. The undrained strength C_u evaluated from the limit pressure P^*_L by (Appendix F, 10).

$$C_u = \frac{P^*_L}{10} + 0.26 \text{ tsf } (25 \text{ kPa}) \quad (1)$$

provided strengths generally consistent with the other test results, Figure 3. Other equations for estimating undrained strength from pressuremeter data provided larger estimates than Equation 1. The coefficients of lateral earth pressure K_o, simply estimated from the total horizontal stress at rest P_{OH} determined from the corrected pressure-volume curve of the test results, are about 1.

<u>Heave Profile</u>. Figure 5 shows the accumulated heave profiles observed in July 1974 and September 1981 due to moisture accumulation. The soil heave profile from the freestanding benchmarks PSTBM shown by the heavy solid lines in Figure 5 indicate the least measured vertical swell in the test site.

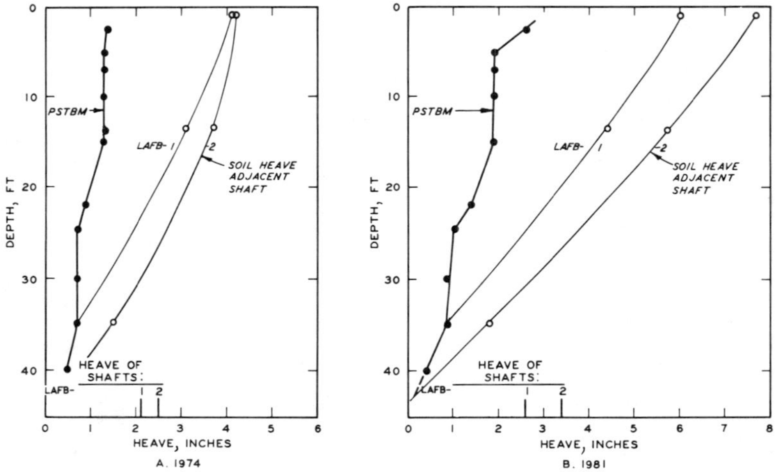

Figure 5. Heave profiles at LAFB test site in July 1974 and September 1981

The maximum heave at the ground surface is about 3 in. (76 mm) in 1981. The freestanding benchmarks were a minimum of 28 ft (8.5 m) from any area which was deliberately flooded. The thin solid lines in Figure 5, which indicate the soil heave profile adjacent shafts LAFB-1 and LAFB-2, show that LAFB-1 and LAFB-2 had heaved considerably more than the PSTBM by 1981. This heave is about 6 and 7.7 in. (152 and 196 mm) for soil at the ground surface near LAFB-1 and LAFB-2, respectively. Soil near the base of shaft LAFB-2 at 35 ft (10.7 m) had heaved about twice that of the 35-ft (10.7-m) benchmark PSTBM-10.

Shaft heaves measured at the shaft top in July 1974 and September 1981 are shown at the bottom of Figure 5. Shafts LAFB-1 and LAFB-2 had heaved significantly after the end of flooding from 1967 through 1981. These heaves of 2.6 and 3.4 in. (66 and 86 mm) are considerably more than 0.8 and 1.8 in. (20 and 46 mm) of soil heave observed near the shaft bases, respectively, so that the shaft stretched 1.6 in. (76 mm). Shaft LAFB-2 may have heaved more than LAFB-1 because LAFB-2 was located within the covered area. Because shafts LAFB-1 and LAFB-2 are underreamed, they must have fractured at one or more locations along the shaft lengths. The uplift thrust tensile resistance of the shafts should have been exceeded since fracture appears to have occurred. According to Equation 6-7 in TM 5-818-7 (7), the (negative) uplift thrust Q_{us} for 60,000 psi (414 MPa) steel is

$$Q_{us} = - \frac{A_s D_s^2}{C} \qquad (2)$$

where

A_s = percent steel

D_s = shaft diameter, ft (m)

$C = 0.03\ ft^2/ton\ (0.28\ m^2/MN)$

The maximum capacity to resist uplift thrust for LAFB-1 with $A_s = 2\%$ and $D_s = 1.5$ ft should have exceeded 150 tons (1.5 MN). Since the 34-ft (10.4 m) length of steel can only stretch 0.7 in. (18 mm) before rupture, the shaft had ruptured and the steel band broken, probably near the base at the neck of the underream.

Theoretical Models

Vertical axial loads are resisted by skin friction along the shaft - soil interface and by the bearing capacity of the soil beneath the shaft base. The ultimate capacity Q_u of the shaft is commonly given by

$$Q_u = Q_{su} + Q_{bu} \qquad (3)$$

$$Q_{su} = \pi D_s \int_o^L f_s\, dL \qquad (4)$$

$$Q_{bu} = q_{bu} A_b \qquad (5)$$

where

Q_{su} = ultimate mobilized skin resistance, tons (MN)

Q_{bu} = ultimate end bearing resistance, tons (MN)

D_s = shaft diameter, ft (m)

$\bar{f}_s$ = full mobilized skin friction, tsf (MPa)

dL = increment of shaft length L, ft (m)

q_{bu} = ultimate base resistance pressure, tsf (MPa)

A_b = base area, ft^2 (m^2)

The ultimate capacity Q_u is normally reduced to an allowable bearing capacity Q_a for design, usually by selection of suitable factors of safety. Procedures for evaluation of Q_a are discussed in the literature (10, 19).

End Bearing Capacity. Shaft foundations normally fail by punching shear in which the shear surface is not well defined and failure is progressive with continuing downward movement or punching of the soil. The end bearing capacity q_{bu} for deep foundations may be given by

$$q_{bu} = cN_c + \sigma_v' N_q \tag{6}$$

where

c = cohesion, tsf (MPa)

σ_v' = effective vertical overburden pressure, tsf (MPa)

N_c, N_q = bearing capacity factors

Factor N_c equals 9 and N_q equals 1 for total stress or undrained analysis (9). σ_v' is usually ignored to compensate for the weight of the shaft. The cohesion c equals the undrained shear strength C_u.

The bearing capacity factor N_q for effective stress or drained analysis has been evaluated by a variety of useful procedures. Cohesion is normally ignored. Two procedures after the work of Terzaghi for local and general shear failure are:

$$\text{Local (18):} \quad N_q = (1 + \tan\phi')\, e^{\tan\phi'} \tan^2\left(45 + \frac{\phi}{2}\right) \tag{7a}$$

$$\text{General (2):} \quad N_q = \frac{e^{\frac{270-\phi'}{180}\pi \tan\phi'}}{2\cos^2\left(45 + \frac{\phi'}{2}\right)} \tag{7b}$$

The bearing capacity factor N_q by Prandtl (13) provides similar, perhaps slightly larger, values than Equation 7b for general shear failure. Settlements required to achieve ultimate end bearing vary widely, but practical estimates for drilled shafts are up to 25% of the shaft diameter (18).

Skin Friction. The development of skin friction f_s in deep foundations depends on the relative displacement between soil and shaft. The full mobilized skin friction $f_{\bar{s}}$ occurs at relative displacements much less than those usually required to achieve full end bearing and often less than 0.5 in. (13 mm). Cylindrical shear models are considered most appropriate for analysis of skin friction resistance as indicated by Equation 4. The magnitude of ultimate skin friction for practical design applications may be computed by the same cylindrical shear model for applied structural loads, pullout loads, downdrag forces from consolidating soil, and uplift thrust from swelling soil. The maximum uplift thrust developed in swelling soil is considered in this paper.

Methodology for evaluating $f_{\bar{s}}$ by undrained analysis is

$$f_{\bar{s}} = \alpha C_u \tag{8}$$

where α is an empirical factor relating soil-shaft adhesion factor to the undrained shear strength C_u . α should be approximately 0.5 for the stiff clays of this site (15, 16); however, α for modeling uplift thrust from expansive soil may be larger than 0.5 and could approach 1.0 because the soil expands tightly against the shaft perimeter over the full length of the swelling soil.

Methodology for evaluating the full mobilized skin friction by drained analysis is given by

$$f_{\bar{s}} = \beta \sigma'_v \tag{9}$$

where β is the lateral earth pressure and friction angle factor. Among the variety of models available for evaluating β , the model (4)

$$\beta = K_o \tan \delta \tag{10}$$

is commonly used where K_o is the coefficient of lateral earth pressure. In practice the most appropriate soil-shaft friction angles δ appear to be several deg ($\approx$5 deg) less than ϕ' , or about 0.8 times the angle ϕ' measured from results of R tests (10).

The uplift thrust from swelling of expansive soil is expected to increase the coefficient of lateral earth pressure toward the passive coefficient K_p which could be modeled by the Rankine passive state ignoring cohesion and taking $\delta = \phi'$. The uplift thrust Q_{us} may also be modeled by (8, 11)

$$Q_{us} = 1.3\pi D_s \int_0^L \sigma_s \tan\phi'_{residual} dL \tag{11}$$

where

σ_s = swell pressure, tsf (MPa)

$\phi'_{residual}$ = residual effective friction angle, deg

A possible explanation for the factor 1.3 in Equation 11 is that measured swell pressures may be less than actual swell pressures.

Load-Displacement Behavior. The load-displacement behavior may be estimated by a variety of techniques including the Randolph and Wroth (14) elastic method, the Poulos (12) trilinear method, finite element method, and load transfer functions. A detailed comparison of load test results with finite element method and load transfer functions calculation is discussed elsewhere (10).

A simple parametric analysis was performed with a computer program AXILTR (10) using load transfer functions to approximate a trilinear load-displacement curve for shafts in general similar to the Poulos (12) method. The soil surrounding the shaft was assumed homogeneous. The soil beneath the base may be of different strength than that along the shaft. The ultimate load capacity was calculated by Equation 3. The ultimate displacement ρ_u in inches at Q_u may be approximated by (10)

$$\rho_u = \frac{2.75D_b}{2.5c_b^{0.25} + 1.1\phi_b^{0.3}} - \frac{4.4D_s}{10c_b + 0.44\phi_b^{1.07}} \tag{12}$$

where

ρ_u = ultimate displacement, in.

D_b = base diameter, ft

D_s = shaft diameter, ft

c_b = cohesion of soil beneath base, tsf

ϕ_b = angle of internal friction of soil beneath the base, deg

The concrete modulus was assumed 216000 tsf (21000 MPa) and the shaft skin friction was reduced 40% to increase conservatism. The cohesion

$c_b = 0$ for drained analysis and $\phi_b = 0$ for undrained analysis. The undrained strength of soil beneath the base $C_{ub} = c_b$ should be greater than 0.5 tsf. The effective friction angle $\phi'_b = \phi_b$ should be greater than 10 deg. The displacement in inches at 1/2 of Q_u is approximately

$$\rho_{u/2} = \frac{0.13 D_b^{1.7}}{D_s^{0.55}} \tag{13}$$

Results and Analysis of Vertical Load Tests

Shaft LAFB-1. The vertical load test was performed in May 1982 using the original anchor shafts constructed in 1966. A calibrated, electronically operated loading jack with 200 ton (2 MN) capacity was used to apply the loads. Displacements were measured by two extensometer gages sensitive to 0.001 in. (0.025 mm) positioned on each side of the top of the shaft and mounted on a wood frame independent of the shaft. Load tests were conducted in accordance with standard test procedure (1).

Results of the vertical load test shown in Figure 6 indicated

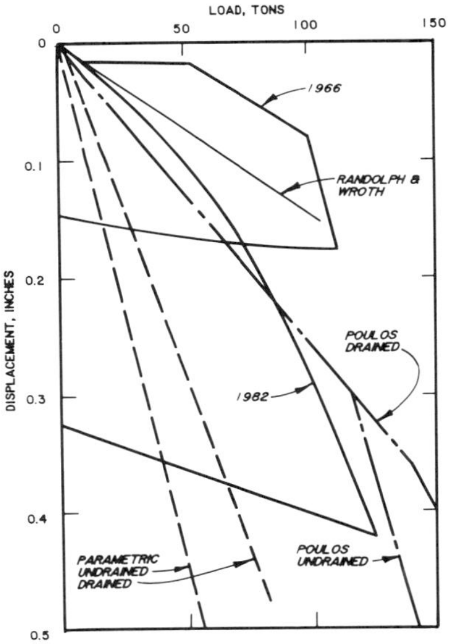

Figure 6. Comparison of observed with calculated vertical load-displacement behavior by Randolph and Wroth, Poulos, and parametric methods, LAFB-1

about twice as much displacement in 1982 at similar loads than observed in 1966. Little end bearing was mobilized in 1966 load test. At 130 tons (1.3 MN) the anchor shafts failed and terminated the test. Some additional skin resistance was possible because the maximum uplift thrust could have exceeded 150 tons (1.5 MN) calculated by Equation 2. The increased displacements observed in 1982 are attributed to (a) closure of cracks in the shaft, particularly at the neck of the bell and (b) loss of support beneath the shaft base because of shaft heave from uplift thrust of swelling soil. The high tensile strains observed below 15 ft (4.6 m) (10) may have caused the shaft to fracture in one or more locations below 15 ft (4.6 m) of depth during the years following 1966. One of these locations is indicated to be the neck of the underream from study of the loss in shaft stiffness observed from the strain gage data at this location. A 2-in. (50-mm) diameter core of LAFB-1 indicated no fracture in upper 12-ft (3.7 m) of shaft. The core boring exited shaft at 12 ft (3.7 m).

The load distribution with depth shown in Figure 7 indicates less

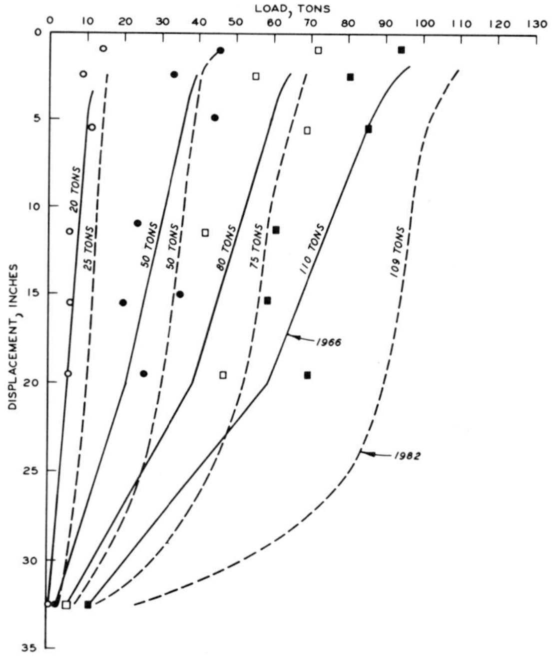

Figure 7. Load distribution behavior for vertical load test of shaft LAFB-1

skin friction above 20 ft (6 m) and more below this depth in 1982 than observed in 1966. Only 14 of the original 22 gages were operational in 1982. Appropriate concrete moduli were difficult to determine because of the large tensile strains observed in the shaft and possible fracture. The load-distribution curves for both 1966 and 1982 load tests show that most of the skin resistance is carried by soil below 20 ft (6.1 m) where soil shear strength is greatest.

Vertical load-displacement behavior analyzed by the Randolph and Wroth (14), Poulos (12), and parametric models is shown in Figure 6. Parameters used to evaluate this behavior include

	Drained	Undrained
E_c = 216,000 tsf (21000 Mpa)	N_q = 13	α = 0.45
E_s = 18.2z tsf (5.72z MPa)	ϕ'_{shaft} = 20 deg	C_{ushaft} = 0.8 tsf (76.6 kPa)
ν_s = 0.3	ϕ'_{base} = 27 deg	C_{ubase} = 1.5 tsf (143.7 kPa)

where ~z = depth in feet (m). The effective friction angles were reduced about 5 to 7 deg.

The elastic Randolph and Wroth model slightly underestimated the 1982 displacements, but remained within 0.1 in. (2.5 mm) up to 100 tons (1.0 MN) of load. The Poulos model, in which the elastic portion depends on the soil modulus E_s , indicated substantially greater displacements than observed although still less than 1 inch (25 mm).

Shaft LAFB-2. The vertical load test was performed in November 1982 using a 1,250-ton (12.5 MN) capacity loading frame supported by two 3-ft-(0.9-m)-diam shafts 36 ft (11.0 m) long with 7.5-ft (2.3 m) diam bells. Each anchor shaft contained ten 1.375-in. (35 mm) diam, 150,000 psi (1035 MPa) high-strength bars equally spaced around a 28-in. (710 mm) diameter ring. Vertical loads were applied with a calibrated and electronically operated hydraulic loading jack of 1,200 ton (12 MN) capacity. Displacements were measured by two dial gages sensitive to 0.001 in. (0.025 mm) positioned on each side of the test shaft and mounted on a wood frame independent of the shaft. Backup ruler and wire gages were also in position. The load test was conducted in accordance with a standard test method (1). Details of this test are provided elsewhere (10).

The shaft experienced an intermediate plunging failure at 250 tons (2.5 MN) causing rapid displacement to 1.5 in. (38 mm), Figure 8. This intermediate failure was attributed to applied loads exceeding the maximum skin resistance and the presence of a void and/or soft soil beneath the base. Since LAFB-2 had heaved 1.6 in. (41 mm) more than the soil had heaved near the vicinity of the base, a gap could have occurred in the shaft near the base and/or a void beneath the base. The strain gage data (10) had shown that the shaft had compressed about 1 in. (25 mm) during the load test, mostly below 20 ft (6.1 m), so that 0.6 in. (15 mm) of possible space may have existed beneath the base. The shaft held an additional 130 tons (1.3 MN) after the intermediate plunging failure, which is attributed to the ultimate base capacity. The failure load of 390 tons (3.9 MN) to 400 tons (40 MN) was maintained for 8 hr with a creep rate of 0.001 in./min (0.025 mm/min). Loads exceeding 400 tons (4.0 MN) significantly increased the creep rate and could not be maintained.

The load distribution curve, Figure 9, shows a maximum bearing capacity of 130 tons (1.3 MN) and maximum skin friction resistance of 250 tons (2.5 MN). About 10% of the load was taken in end bearing for

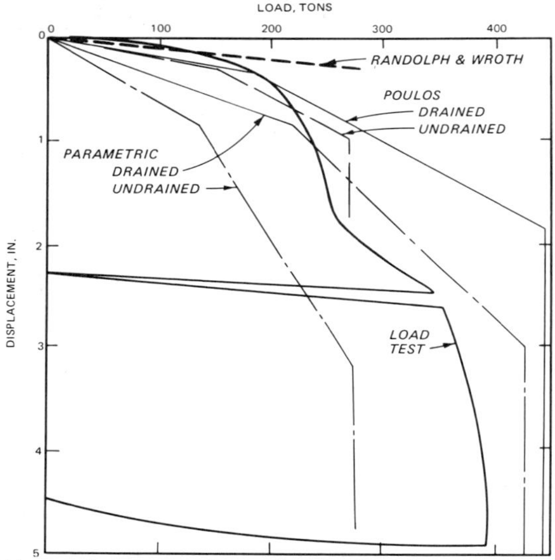

Figure 8. Comparison observed with calculated load-displacement behavior by Poulos (12), and Randolph and Wroth (14), and parametric models, LAFB-2

applied loads less than 250 tons (2.5 MN), indicating that any void that occurred beneath the base due to shaft uplift was largely eliminated by heaving of underlying soil. Most skin friction resistance was developed below 16 ft (4.9 m). Some load shedding of the skin friction resistance from 0 to 16 ft (4.9 m) occurred during the 8-hr period at maximum load of 390 tons (3.9 MN). Some of the shedded load was absorbed between 21 and 27 ft (6.4 and 8.2 m) and the remainder taken by end bearing. Strain gage readings above 11 ft (3.4 m) were not reasonable for applied loads less than 300 tons (3.0 MN).

An analysis of the skin friction distribution with depth is shown in Figure 10 for applied loads from 0 to 250 tons (2.5 MN). The maximum measured skin resistance is 0.6 tsf (59.5 kPa) from 0 to 16 ft (4.9 m) and 1.6 tsf (153.2 kPa) from 16 to 35 ft (4.9 to 10.7 m). The α factor at maximum skin resistance relative to the undrained shear strength is about 0.8 and 1.0 from 0 to 16 ft (4.9 m) and 16 to 35 ft (4.9 to 10.7 m), respectively. These α factors are larger than those expected in nonswelling soil, indicating intimate contact at

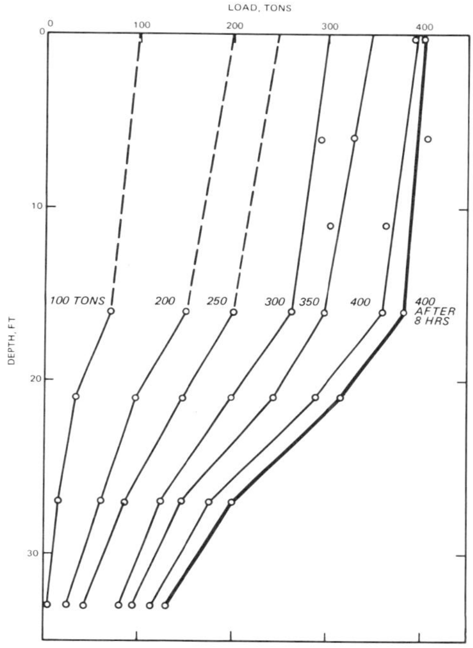

Figure 9. Load distribution behavior for vertical load test of shaft LAFB-2

the soil-shaft interface and substantial pressure acting on the interface.

Estimates of vertical behavior were made by the Randolph and Wroth (14), Poulos (12), and parametric models as shown in Figure 8 using the same parameters as for shaft LAFB-1. The Randolph and Wroth model agrees well with the early portion of the load-displacement curve for loads less than 150 tons (1.5 MN). The Poulos model provides reasonable correlation with load test results up to 200 tons (2 MN), but the drained analysis underestimates displacements above 200 tons (2 MN) which is attributed to a too large base friction angle. The Poulos undrained analysis underestimates the skin resistance. Results of the parametric analysis for undrained soil tend to overestimate displacements. Skin friction resistance by the parametric model tended to be underestimated while end bearing resistance was overestimated.

End bearing capacity Q_{bu} for undrained analysis using $N_c = 9$ is 170 tons (1.2 MN), which exceeds the actual end bearing capacity by 40 tons (0.4 MN). An effective bearing capacity factor $N_c = 7$

approximately simulates end bearing. Q_{bu} for drained analysis using ϕ' = 27 deg and the general shear model for N_q grossly overestimates end bearing at 278 tons (2.8 MN). N_q approximated by local shear failure from Equation 7a more closely simulates actual end bearing resistance. The strength parameters C_u and ϕ' are average measured values.

Skin resistance was underestimated for α = 0.45 and should have been 1.0 for undrained analysis consistent with results in Figure 10. Skin resistance for drained analysis is properly calculated using the measured effective friction angle from results of R tests and coefficient of lateral earth pressure K approaching 3.0. The earth pressure distribution measured by earth pressure cells and shown in Figure 11 indicates lateral pressures about 3 times the vertical pressure. The swell pressure model given by Equation 11 using $\phi'_{residual}$ = 9 deg and 1966 original swell pressures also lead to skin frictions comparable with those measured during the 1982 load tests.

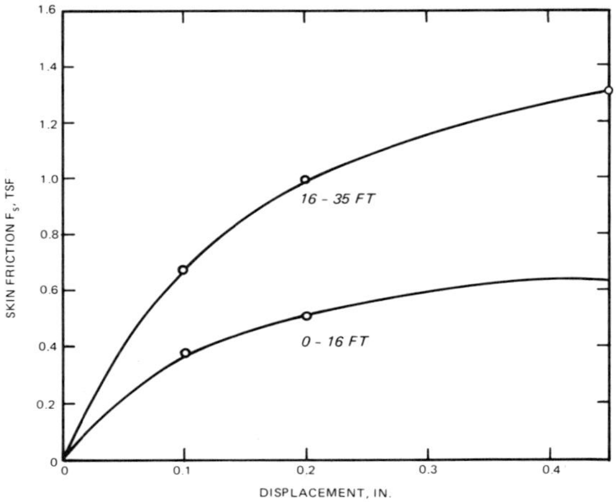

Figure 10. Skin friction-displacement curve for vertical load test of shaft LAFB-2 up to 250 tons (2.5 MN) load

Conclusions

The test shafts were displaced up more than soil heave at the base of the shafts due to uplift thrust and fractured, probably near the neck of the bell. The α factor for skin friction resistance by total stress or undrained analysis in expansive soil is larger than may occur in nonswelling soils and can approach 1.0 over the entire length of the shaft subject to lateral thrust from swelling soil. Maximum skin friction by effective stress or drained analysis was reasonably estimated by the swell pressure and residual friction angle model using original swell pressures or the cylindrical shear model

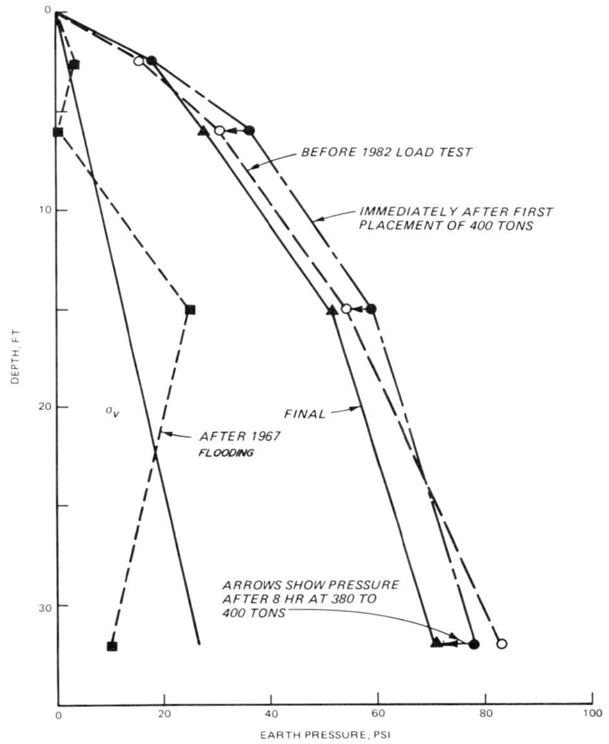

Figure 11. Measured earth pressure distribution for shaft LAFB-2

using measured effective friction angles and passive coefficients of lateral earth pressure calculated by the Rankine model.

End bearing capacity for LAFB-2 was less than expected and approximately given by $N_c = 7$ for undrained analysis and N_q given by local shear failure for drained analysis. The reduced end bearing capacity is attributed to shaft heave which lifted the base off the soil contributing to a possible void and soil heave/softening beneath the shaft base.

Drilled shaft foundations for supporting permanent structures in expansive soil should be loaded near the allowable bearing capacity assuming a minimum factor of safety consistent with good engineering practice. The amount of reinforcement steel required to resist uplift thrust caused by swelling soil should be based on the maximum shear strength of the adjacent soil. Reinforcement should be full length extending into any existing underream and well secured in the underream. Elements of deep foundations should not penetrate through perched water tables into deeper desiccated soil if practical.

Acknowledgement

This study was supported by a grant from the Office, Chief of Engineers, Washington, D. C.

Appendix 1. - References

1. American Society for Testing and Materials (ASTM) Specifications. 1981. "Piles Under Axial Compressive Load," ASTM D 1143, Part 19, Philadelphia, Pa.

2. Bowles, J. E. 1968. Foundation Analysis and Design, McGraw-Hill, New York, pp 607-609.

3. Brooker, E. W. and Ireland, H. O. 1965. "Earth Pressures at Rest Related to Stress History," Canadian Geotechnical Journal, Vol 11, No. 1, pp 1-15.

4. Chandler, R. J. 1968. "The Shaft Friction of Piles in Cohesive Soils in Terms of Effective Stress," Civil Engineering and Public Works Review, Vol 63, pp 48-51.

5. Duncan, J. M. and Chang, C. Y. 1970. "Nonlinear Analysis of Stress and Strain in Soils," Journal of the Soil Mechanics and Foundations Division, Proceedings of the American Society of Civil Engineers, Vol 96, No. SM5, pp 1629-1653.

6. Gardner, W. S. 1977. "Soil Property Characterization in Geotechnical Engineering Practice," Geotechnical/Environmental Bulletin, Vol 10, No. 2, Woodward-Clyde Consultants, San Francisco, CA.

7. Headquarters, Department of the Army. 1983. "Foundations in Expansive Soils," TM 5-818-7, Washington, D. C.

8. Jobes, W. P. and Stroman, W. R. 1974. "Structures on Expansive Soil," Technical Report M-81, Construction Engineering Research Laboratory, Champaign, IL.

9. Johnson, L. D. 1981. "Field Test Sections on Expansive Soil," 4th International Conference on Expansive Soils, Denver, Colorado, American Society of Civil Engineers, New York, NY.

10. Johnson, L. D. 1984. "Methodology for Design and Construction of Drilled Shafts in Cohesive Soils," Technical Report GL-84-4, U. S. Army Engineer Waterways Experiment Station, Vicksburg, MS.

11. O'Neill, M. W. and Poormoayed, N. 1980. "Methodology for Foundation on Expansive Clays," Journal of the Geotechnical Engineering Division, Proceedings of the American Society of Civil Engineers, Vol 106, No. GT12, pp 1345-1357.

12. Poulos, H. G. 1972. "Load-Settlement Prediction for Piles and Piers," Journal of the Soil Mechanics and Foundations Division, Proceedings of the American Society of Civil Engineers, Vol 98, No. SM9, pp 879-897.

13. Prandtl, L. 1920. "Uber die Harteplastischer Korper," Nachr. der Ges. der Wiss., Gottingen, Germany.

14. Randolph, M. F. and Wroth, C. P. 1978. "Analysis of Deformation of Vertically Loaded Piles," Journal of the Geotechnical Engineering Division, Proceedings of the American Society of Civil Engineers, Vol 104, No. GT12, pp 1465-1488.

15. Reese, L. C., Touma, F. T., and O'Neill, M. W. 1976. "Behavior of Drilled Piers Under Axial Loading," Journal of the Geotechnical Engineering Division, Proceedings of the American Society of Civil Engineers, Vol 102, No. GT5, pp 493-510.

16. Stewart, J. P. and Kulhawy, F. H. 1980. "Behavior of Drilled Shafts in Axial Uplift Loading," Geotechnical Engineering Report 80-2, School of Civil and Environmental Engineering, Cornell University, Ithaca, NY.

17. USAEDFW. 1968. "Investigations for Building Foundations in Expansive Clays," Vols I and II, U. S. Army Engineer District, Fort Worth, TX.

18. Vesic, A. S. 1977. "Design of Pile Foundations," National Cooperative Highway Research Program, Synthesis of Highway Practice No. 42, Transportation Research Board, National Research Council, Washington, D. C.

19. Wright, S. J. and Reese, L. C. 1979. "Design of Large Diameter Bored Piles," Ground Engineering, Vol 12, No. 8, pp 17-24, 51.

EVALUATION OF P-Y RELATIONSHIPS IN COHESIONLESS SOILS

By Jack M. Murchison,[1] A.M., ASCE, and Michael W. O'Neill,[2] M., ASCE

INTRODUCTION

In present-day practice laterally loaded piles are often analyzed using a finite difference model. The model uses beam theory to represent the pile and uncoupled, nonlinear load transfer functions, called p-y curves, to represent the soil (12, 16). This model makes possible relatively simple and straightforward computations of pile-head flexibility and of stresses along the pile. However, because of the complex stress conditions imposed on the soil by placement and subsequent loading of the pile, continuum effects not modeled by the method, and nonlinear effects, complete and comprehensive theoretical derivations of p-y curves have not yet been developed.

This paper considers four semi-empirical procedures for construction of p-y curves in cohesionless soils. Each was developed to fit data from a particular lateral load test or a specific set of tests on similar soils. Few, if any, systematic studies have been conducted to assess their universal validity. In this paper, the relative accuracy of these procedures is studied by applying each to a variety of conditions defined in a data base consisting of full-scale lateral load tests in very loose to very dense cohesionless soils and of piles with several cross-sectional shapes and a large range of diameters.

METHODS FOR CONSTRUCTION OF P-Y CURVES IN COHESIONLESS SOILS

The four procedures considered in this paper, denoted Method A - Method D, are described briefly below.

Method A

Method A is the p-y method for sand presently incorporated into API RP2A (1). It was introduced by Reese, Cox, and Koop in 1974 (17),

---

[1]Design Engineer, Bridge Division, Texas Department of Highways and Public Transportation, Austin, Texas 78701

[2]Professor, Department of Civil Engineering, University of Houston, Houston, Texas 77004

who describe it in detail. A brief synopsis is given here. P-y curves are constructed for desired depths. Each curve consists of three segments: two straight lines with a parabola between as shown in Fig. 1. The diameter of the pile is denoted D. The value of K_i (lb/ft^2 or N/m^2), the initial slope, is determined by multiplying k (lb/ft^3 or N/m^3) times depth, where k is a modulus of lateral soil reaction. The ultimate soil resistance p_u is determined from the lesser value given by Eqs. 1 and 2, modified by an empirical adjustment parameter, which differs for static and cyclic loading and varies with pile diameter and depth.

$$p_u = \gamma z[D(K_p - K_a) + zK_p \tan\phi \tan\beta] \qquad (1)$$

$$p_u = \gamma Dz(K_p^3 + 2K_oK_p^2 \tan\phi + \tan\phi - K_a) \qquad (2)$$

where

p_u = ultimate soil resistance per unit of depth,
z = depth,
γ = unit weight of soil (buoyant or non-buoyant, as appropriate),
K_a = Rankine active coefficient,
K_p = Rankine passive coefficient,
K_o = at-rest earth pressure coefficient,
ϕ = angle of internal friction,
β = $45^o + \phi/2$.

The value of p_m (beginning of second linear segment of curve) is a certain percentage (determined from empirical charts) of p_u, while the values for y_m and y_u are ratios of the pile diameter. The point (y_k, p_k) is determined from an empirical relationship involving y_m, y_u, p_m, and p_u. The procedure is somewhat complicated to apply manually but can be programmed for efficient development of p-y curves.

In this study, k was determined consistently from standard penetration test blow counts. First, the blow count was used to estimate the angle of internal friction, ϕ, using the correlation in Fig. 2 (7). Next, k was estimated from ϕ using Fig. 3 (13).

Method B

Method B, a modification of the API method, was introduced by Bogard and Matlock in 1980 (3), among others. By realizing that some terms in the formulation of p_u can be taken as constants with little error, they were able to simplify Method A. Method B is used by many geotechnical design offices.

The ultimate soil resistance is calculated in the same way as in Method A, except that the terms have been grouped to form constants which vary with ϕ, as shown in Eqs. 3 and 4. The factor p_u is taken as the minimum of the values given by Eqs. 3 and 4.

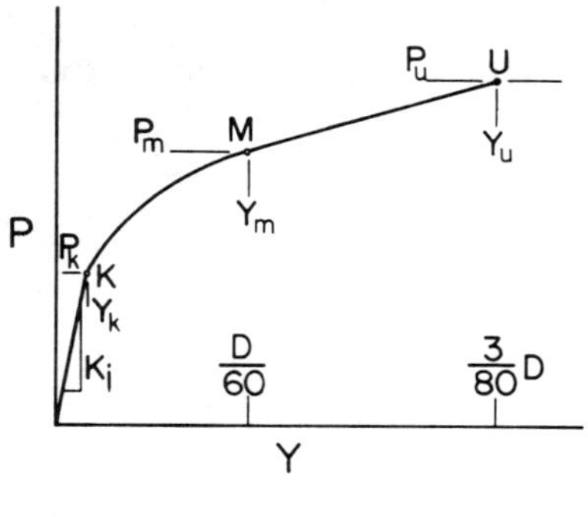

Fig. 1. Method A p-y Curve

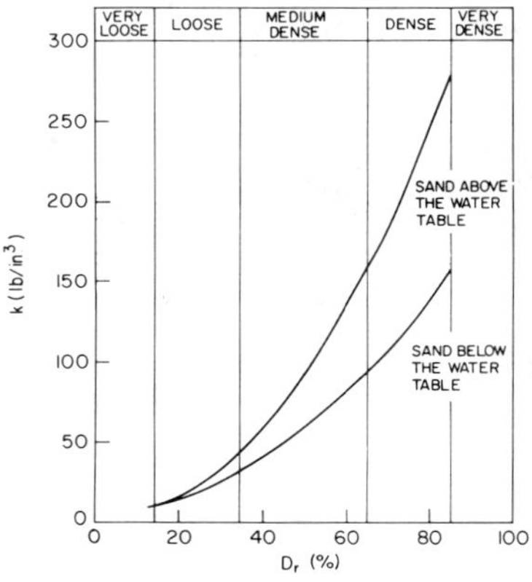

Fig. 3. k vs. Relative Density (Modified After Ref. 13)

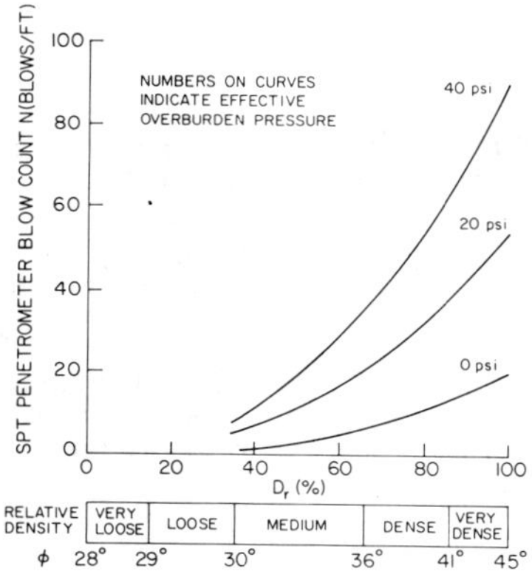

Fig. 2. SPT Blow Count vs. Friction Angle and Relative Density (Modified After Ref. 7)

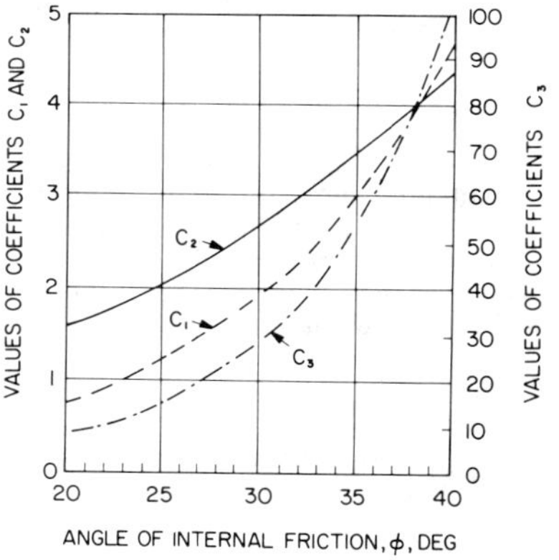

Fig. 4. Factors for Method B (After Ref. 3)

$$p_u = (c_1 z + c_2 D)\ \gamma z \tag{3}$$

$$p_u = c_3\ D\gamma z \tag{4}$$

The parameters c_1, c_2, and c_3 are evaluated in Fig. 4.

The p-y curves otherwise are the same as for Method A. Charts with non-dimensionalized values of p for corresonding values of y (Tables 1 and 2) have been developed to make it unnecessary to calculate the p values from the different segments of the Fig. 1 curve. The deflections are chosen to give the critical points on the Method A curve. Note that it is not necessary to compute k. Tables 1 and 2 make Method B probably the quickest of the four methods to use manually. As with Method A, Method B sets p_u (Eqs. 1 and 2, modified per Method A) as the limit on the resistance of the soil to lateral deflection.

Table 1. Values of p/p_u for Static Loading (3)

		Normalized Deflection y/D						
		0.001	0.002	0.005	0.010	0.0167	0.0375	1.00
z/D	0	0.89	1.10	1.46	1.80	2.10	2.90	2.90
	1	0.76	0.91	1.14	1.36	1.54	2.02	2.02
	2	0.45	0.56	0.74	0.91	1.06	1.46	1.46
	3	0.24	0.32	0.45	0.59	0.72	1.07	1.07
	4	0.10	0.15	0.26	0.39	0.52	0.90	0.90
	$\geq$ 5	0.09	0.14	0.24	0.37	0.50	0.88	0.88

Method C

Method C is interpreted from research performed by Scott (19). It differs from the previous criteria in at least two important aspects. First, the p-y curve is idealized by two straight line segments, which simplifies the calculations involved. The initial segment of the curve is similar to the other methods, because a subgrade modulus, k, times the depth, z, defines the slope. The other segment is empirically assigned a slope of kz/4, which highlights the second important difference from the other methods. Because the upper segment has a constant non-zero slope, the method assumes that as deflection increases, the soil resistance increases linearly with no limit. The ultimate soil resistance concept is therefore not applied.

Table 2. Values of p/p_u for Cyclic Loading (3)

		Normalized Deflection y/D						
		0.001	0.002	0.005	0.010	0.0167	0.0375	1.00
z/D	0	0.16	0.21	0.31	0.41	0.50	0.71	0.71
	1	0.39	0.47	0.59	0.71	0.81	1.07	1.07
	2	0.45	0.52	0.63	0.73	0.82	1.04	1.04
	3	0.29	0.36	0.47	0.59	0.69	0.96	0.96
	4	0.13	0.19	0.30	0.32	0.55	0.90	0.90
	> 5	0.14	0.20	0.31	0.43	0.55	0.88	0.88

The force per unit length p_k that exists at the beginning of the quasi-plastic line segment is given by Eq. 5, where σ_o' is the effective lateral stress in the soil ($K_o\sigma_v'$) and D is the pile diameter.

$$\Omega = \frac{\sigma_o' D}{p_k} = \frac{1}{\pi}\left(\frac{1}{\sin^2\phi} + \frac{1}{3 - 4\nu}\right)^{0.5} \tag{5}$$

Values of $1/2\ \sigma_o'/p_k$ are graphed in Fig. 5. The corresponding displacement at the beginning of the quasi-plastic range is given by

$$y_k = \frac{p_k}{E_s} \tag{6}$$

where $E_s = kz$, and k is determined from Figs. 2 and 3. The complete p-y curve is shown in Fig. 6. At the present time, there is no specific recommendation for the extension of Method C to cyclic loading conditions.

Method D

Method D was originally formulated by Parker from his study of small-diameter pipe piles (15) and reformulated by O'Neill and Murchison (14). The continuous hyperbolic tangent function is used to describe the p-y curves.

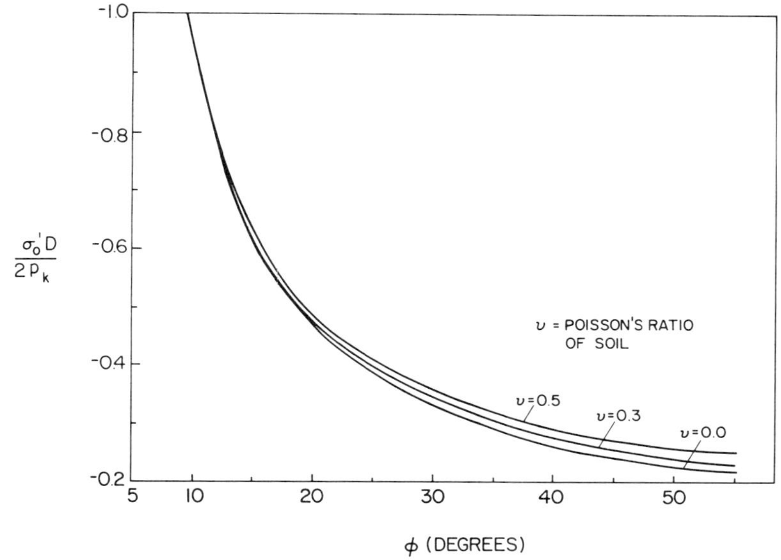

Fig. 5. Design Curve for Method C (Ref. 19)

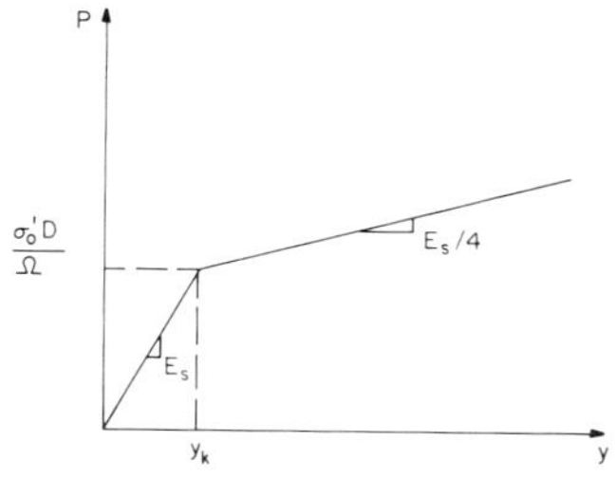

Fig. 6. Method C p-y Curve (Ref. 19)

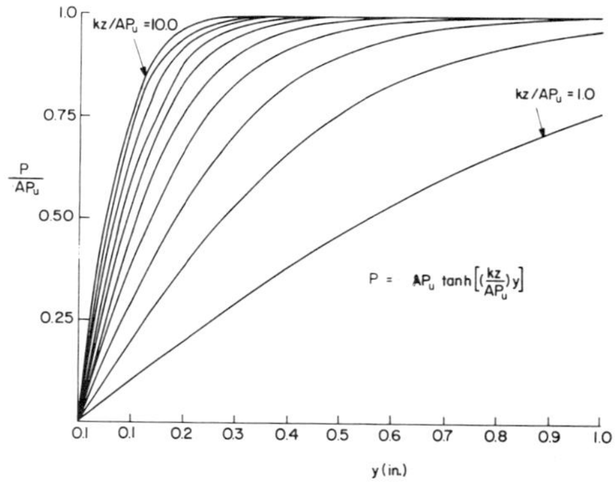

Fig. 7. Method D p-y Curve

The equation for the p-y curve is

$$p = \eta A\, p_u \tanh \left[\left(\frac{kz}{A\eta p_u}\right) y\right] \tag{7}$$

where p_u is the unmodified ultimate soil resistance (Eqs. 1 and 2). The empirical adjustment factors (A) are:

$$\begin{aligned} A &= 0.9 && \text{for cyclic loading} \\ A &= 3 - 0.8\, z/D \geq 0.9 && \text{for static loading} \end{aligned} \tag{8}$$

and η is a factor used to describe pile shape, taken to be equal to 1.5 for uniformly tapered piles, such as timber piles, and for H piles and 1.0 for circular, prismatic piles (14). The term kz is the product of lateral subgrade modulus and depth as used in Method A. Characteristic dimensionless p-y curves for this method are graphed in Fig. 7. Note that this method, unlike Method C, provides for a limiting value of p.

Uniformly tapered piles can be shown to develop p_u values that are higher than those for circular, prismatic piles by considering the obliquity of the resultant soil force on the pile, which is affected both by the angle of taper and the angle of internal friction. Figure 8 shows this effect for a 12-in. (305-mm) diameter pile at a depth of two diameters. An increase in p_u of 15-20 per cent is seen in the usual range of ϕ (30-35°). An additional 30-35 per cent increase in p_u is assumed to occur because of added sand densification produced by pile taper, leading to the η-value of 1.5 referenced previously. The assignment of $\eta = 1.5$ for H-piles is empirical, as it was necessary to use that factor to fit the available data.

The characteristic shape of the Method D curve is similar to that of the Method A curve, as is illustrated in Fig. 9. This is of interest to the designer who develops the curve manually, because the p-y curve is a single analytical function instead of three distinctly different functions.

DATA BASE

A data base was developed against which the various procedures could be tested. It consisted of only full-scale tests at sites where, in the judgment of the authors, soil conditions, loading conditions, and structural properties of the pile were adequately defined. After a comprehensive literature review, 14 static and cyclic tests at 10 sites characterized by cohesionless soils were selected. The site conditions varied from very loose clayey sand to very dense clean sand. Pile types included both driven piles and drilled piers. Driven piles were of four types: prismatic steel pipe, prismatic precast concrete,

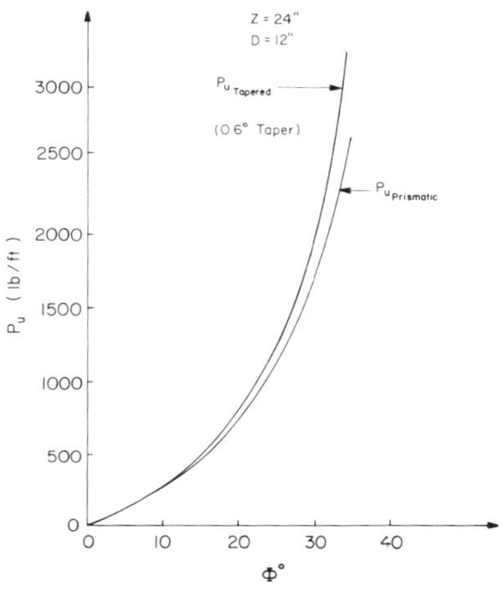

Fig. 8. Effect of Taper and Soil Friction Angle on p_u

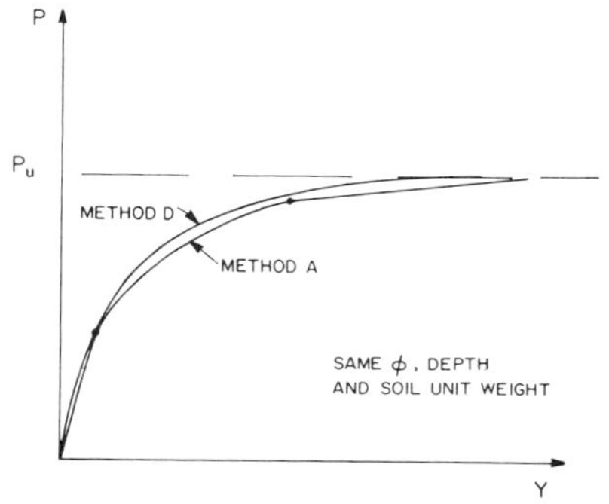

Fig. 9. Comparison of Shapes of Method A and D p-y Curves

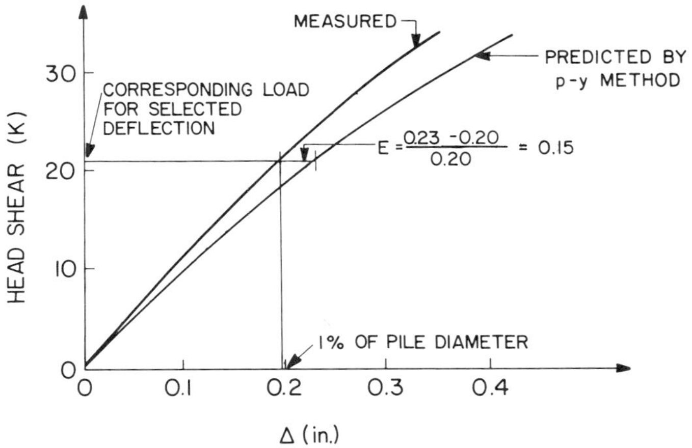

Fig. 10. Example of Error Calculation

prismatic H, and tapered timber. Diameters of the piles and piers varied from 2 in. (51 mm) to 48 in. (1.22 m). The condition of buoyancy varied, although a majority of the tests were conducted in a submerged sand. All tests were conducted on free-headed piles.

Table 3 provides a summary of the tests that were considered. Further details are available in Ref. 14.

Table 3. Data Base Synopsis

Test (Ref.)	Type	Pile	Average ϕ (degrees)	Soil
Mustang Island (4)	Cyclic Static	Steel Pipes 24" diam.	38	SM
Arkansas River (10)	Cyclic Static	Precast Conc. 20" width Steel Pipe 16" diam.	42	SP
Mason-Bishop (11)	Static	H-Pile 16" width	35	Med. Sand
Parker-Reese (15)	Static	Steel Pipe 2" diam.	40	Med. Sand
Padre Island (Unpublished)	Static	Steel Pipe 20" diam.	37	Med. Sand
Bhushan (2)	Cyclic Static	Drilled Pier 36" and 48" diam. Drilled Pier 42" diam.	36	SM
Robinson (18)	Static	Tapered Timber 12" butt	34	Sandy Silt
Leicht and Reese (9)	Static	Precast 14" width	23	SC
Monolith M6 (20)	Cyclic Static	Tapered Timber 12" butt	35	Med. Sand
Pile T3 (Unpublished)	Cyclic Static	H-Pile 14" width	35	Med. Sand

COMPARISON OF METHODS

Procedure for Comparison

The comparison of the methods was accomplished by computing pile-head deflections and bending moment diagrams for a range of pile-head loads for each test in the data base using each p-y curve construction method. A widely available finite difference algorithm, BMCOL 28 (12), was employed in this process. In order to make unbiased calculations, p-y curves were input for each procedure consistently at each of the depths and deflections indicated in Tables 1 and 2, and discrete increment lengths on the pile (the major numerical approximation in the finite difference algorithm) were taken to be 0.50 D. Continuous relationships between applied load and head deflection, applied load and maximum moment, and applied load and depth to maximum moment were then fitted to the discrete output. It was decided that standard statistical procedures would not be appropriate to compare the relative accuracy of the methods because (a) the nature of the problem is such that the distribution of errors (e.g., normal, log-normal) could not be established, and (b) the size of the data base was small. Instead, errors in the computation procedures (and thus in the p-y curve methods) were evaluated numerically in each of six categories in the manner explained below. The categories were:

a. predicted versus measured head deflection (static),
b. predicted versus measured maximum moment (static),
c. predicted versus measured depth to maximum moment (static),
d. predicted versus measured head deflection (cyclic),
e. predicted versus measured maximum moment (cyclic),
f. predicted versus measured depth to maximum moment (cyclic).

Not all tests could be evaluated in all categories, as some of the test piles were not instrumented to measure moment or subjected to an appropriate number of load cycles to achieve stable cyclic degradation. Method C has no cyclic component, so Method C was not evaluated for its ability to predict cyclic behavior.

Errors (absolute value of (predicted value minus measured value)/measured value) were calculated at head deflection values of 1.0, 2.5, 5.0, 7.5, and 15 per cent of the pile diameter for the static tests and at 1.0, 2.5, and 5.0 per cent for the cyclic tests. Figure 10 demonstrates the error (E) calculation. Weighting factors were applied to the computed errors as follows: (a) E was multiplied by 2.0 if the predicted deflections or moments were smaller than the measured values (to penalize an unconservative prediction), and (b) E was (again) multiplied by 2.0 for pipe piles, 1.5 for non-pipe driven piles, and 1.0 for drilled piers. The latter weighting factors were used in order to assess which method was best suited to driven pipe piles while not excluding evidence from tests on piles of other types.

Results of Comparison

The resulting weighted errors were summed over all values of deflection for all tests, and the resulting cumulative errors are tabulated in Tables 4 and 5.

Table 4. Cumulative Weighted Errors for Static Tests

Method	Pile-Head Deflection (Δ)	M_{max}	Depth to M_{max}
A	74.8 (1.00)	16.5 (1.00)	12.7 (1.00)
B	70.1 (0.94)	16.0 (0.97)	11.4 (0.90)
C	83.1 (1.11)	12.8 (0.78)	*8.2 (0.65)
D	*51.8 (0.69)	*9.6 (0.58)	8.7 (0.69)
Number of data points per method	(36)	(22)	(22)

*Indicates lowest error

The numbers in parentheses are the cumulative errors normalized with respect to Method A (API Method).

Table 5. Cumulative Weighted Errors for Cyclic Tests

Method	Pile-Head Deflection (Δ)	M_{max}	Depth to M_{max}
A	17.3 (1.00)	4.9 (1.00)	0.8 (1.00)
B	18.3 (1.06)	4.4 (0.90)	*0.6 (0.75)
D	* 8.3 (0.48)	*4.0 (0.82)	0.8 (1.00)
Number of data points per method	(15)	(8)	(5)

*Indicates lowest error

The numbers in parentheses are the cumulative errors normalized with respect to Method A (API Method).

Tables 4 and 5 indicate that Method D is the most accurate of the four methods over the data base considered. Method C is seen also to provide improved accuracy in the computation of moments compared to the current API method (Method A) in the static case.

Further analyses were conducted with Method D. Plots of unweighted error versus head deflection were made for the individual tests for (a) head deflection (static), (b) maximum moment (static), and (c) head deflection (cyclic). These plots are shown as Figs. 11-13. The errors appear to be relatively independent of the head deflection. A Student-t distribution was applied to each of the figures to determine the average error (6, 8). For all of the categories, the average error is near zero. A 95 per cent confidence band was determined and plotted for the average error, implying that there is 95 per cent confidence that the average error will fall between the two bounds. In addition, a confidence band of 90 per cent on the actual errors for the individual tests has been plotted, which implies that there is 90 per cent confidence that the method will not be in error outside of the indicated curves for a specific pile.

These 90 per cent confidence bands are unfortunately rather wide. One could reasonably expect not to underpredict static pile-head displacement by more than about 70 per cent or static maximum moment by more than about 20 per cent with Method D, although overpredictions in both cases can be higher. Somewhat narrower 90 per cent confidence bands exist for cyclic head deflection (Fig. 13) and maximum moment under cyclic loading (not shown due to lack of sufficient data) than were observed for static loading. This may be due to the fact that there were less available test data for cyclic loading or that errors in characterizing soil are less significant under cyclic loading. Confidence bands for depth to maximum moment were similar to those for value of maximum moment in both static and cyclic data sets.

It is noted that the tests marked 1 and 2 in Figs. 11 and 13 were not included in the calculation of confidence limits. Test h (Fig. 11) was on a pile in a mixed silt and sand profile. The other two tests were excluded due to either questionable measurements or non-virgin soil conditions. Test g, a test on a large-diameter drilled pier, was included; however, it should be noted that no attempt was made in the prediction process to include effects of either vertical shear or base resistance produced by shaft rotation (5, 21).

An additional exposition of individual errors, comparing Methods A and D, is presented in scatter-diagram form in Fig. 14.

Sensitivity of Method D to Parameter Variation

In order to evaluate the effects of errors in the soil parameter inputs for Method D, a hypothetical cyclically loaded, free-headed steel pipe pile, 24 in. (610 mm) O.D., with an EI value of 6×10^{10} lb-in^2 (1.7×10^5 kN-m^2), penetrating 50 ft (15.3 m) into a submerged sand deposit was analyzed. The sand had a nominal angle of internal friction ϕ of 35 degrees and a total weight γ of 100 pcf (15.7 kN/m^3). Baseline calculations were made with the above parameters, and ϕ, γ,

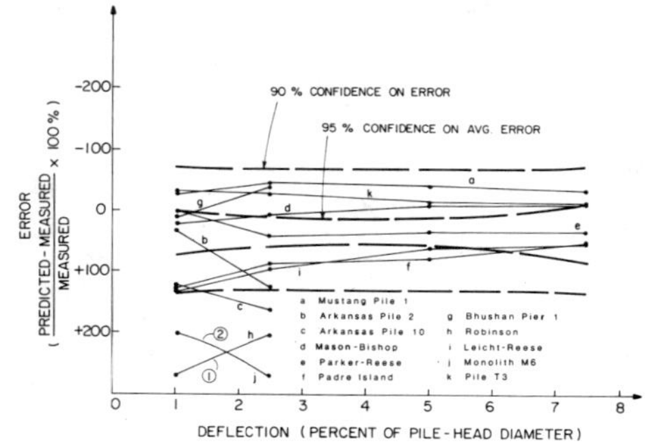

Fig. 11. Error in Head Deflection vs. Deflection, Method D, Static

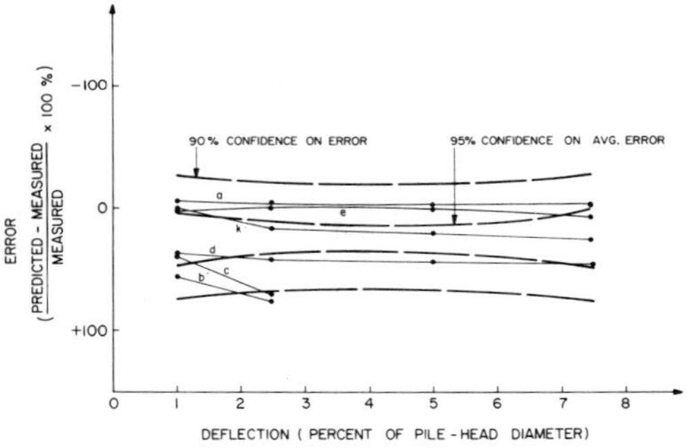

Fig. 12. Error in Maximum Moment vs. Deflection, Method D, Static

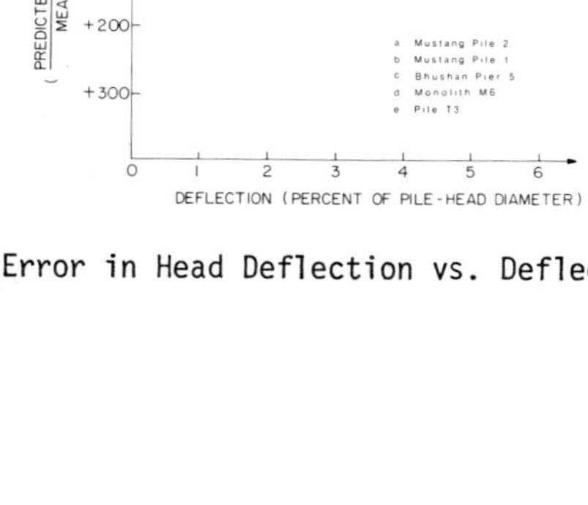

Fig. 13. Error in Head Deflection vs. Deflection, Method D, Cyclic

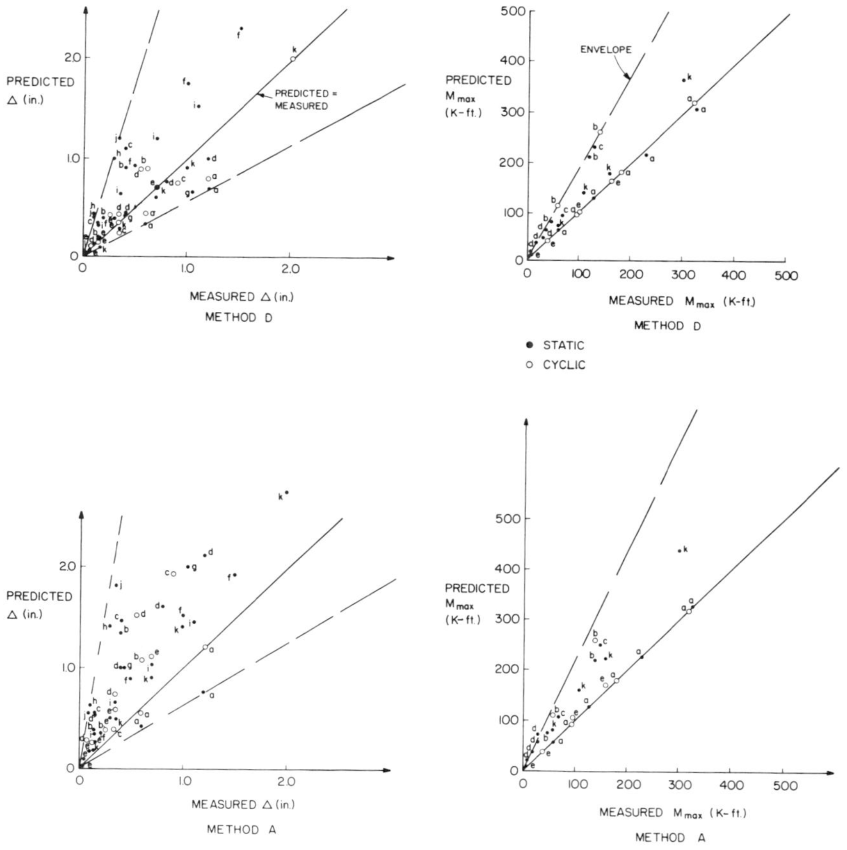

Fig. 14. Comparison of Errors in Methods A and D

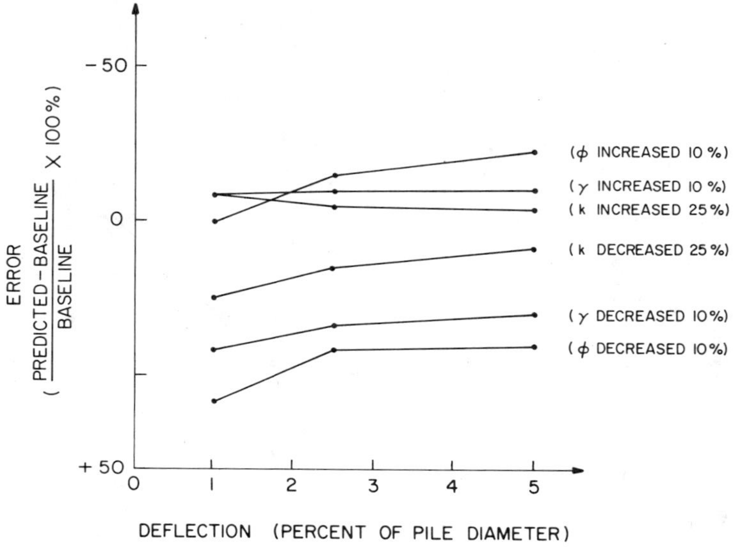

Fig. 15. Errors in Computed Head Deflection by Method D, Hypothetical Pile

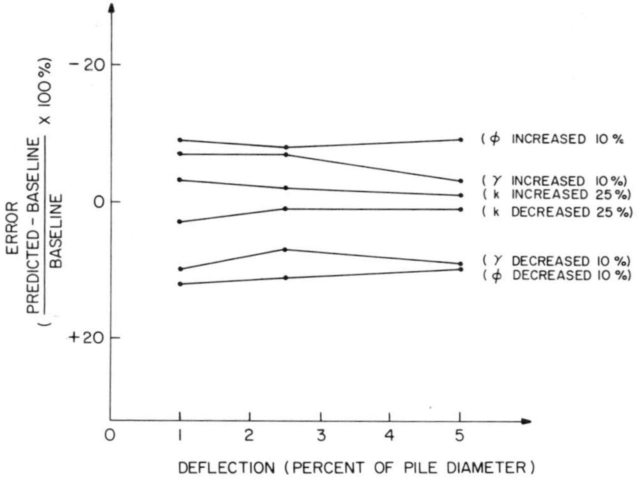

Fig. 16. Errors in Computed Maximum Moment by Method D, Hypothetical Pile

and k were then varied independently, as shown on Figs. 15 and 16. Head deflections (Fig. 15) are seen to be more sensitive to parameter variation than are maximum moments. Outputs are seen to be most sensitive to ϕ and least sensitive to k. The magnitudes of these errors suggest that errors in soil property evaluation at the various test sites could have partially produced the errors described in the previous section, since the errors in ϕ and γ implied on Figs. 15 and 16 are reasonable extreme bounds. However, comparing Fig. 15 with Fig. 13, in which the 90 per cent confidence limits lie outside the band of calculations in Fig. 15, it is likely that other factors, not included in the p-y curve formulation in Method D, are operative. This observation suggests that further study into the fundamental mechanisms of lateral pile-soil interaction in cohesionless soil is warranted.

CONCLUSIONS

1. Of the four methods studied, the bilinear method (Method C) is the simplest to use and is at least as accurate as the present API method (Method A). Method B is also simpler than Method A and is at least as accurate. The hyperbolic method (Method D) was the most accurate of the methods, although it is less simple to apply than Method C.

2. In none of the methods did accuracy appear to be influenced by the magnitude of pile diameter.

3. Errors evidenced with all methods, including Method D, are high, suggesting that further research into fundamental mechanisms of lateral pile-soil interaction is warranted.

4. The data base was small due to the unavailability of appropriately documented full-scale test data. The small data base undoubtedly had some influence on the numerical error computations. Further high-quality field tests, especially on instrumented, large-diameter piles, are needed to enlarge the data base and to permit future reassessment of procedures for analyzing laterally loaded piles in cohesionless soil.

ACKNOWLEDGEMENT

The study described herein was sponsored by the American Petroleum Institute (Project PRAC 82-41). API is not responsible for any conclusions or other items represented as fact, as these are solely the responsibility of the authors. The authors are grateful for API's permission to publish these results.

APPENDIX - REFERENCES

1. API Recommended Practice for Planning, Designing, and Constructing Fixed Offshore Platforms, American Petroleum Institute, Washington, D.C., Thirteenth Edition, 1982.

2. Bhushan, K., Lee, L.J., and Grime, D.B., "Lateral Load Tests on Drilled Piers in Sand," Drilled Piers and Caissons, ASCE, 1981, pp. 114-131.

3. Bogard, D., and Matlock, H., "Simplified Calculation of P-y Curves for Laterally Loaded Piles in Sand," Unpublished Report, The Earth Technology Corporation, Inc., Houston, Texas, April 1980.

4. Cox, W.R., Reese, L.C., and Grubbs, B.R., "Field Testing of Laterally Loaded Piles in Sand," Preprints, Sixth Annual Offshore Technology Conference, Vol. 2, Paper No. 2970, Houston, Texas, May 1974.

5. DiGioia, A.M., Jr., Davidson, H.L., and Donovan, T.D., "Laterally Loaded Drilled Piers, A Design Model," Drilled Piers and Caissons, ASCE, 1981, pp. 132-149.

6. Freund, J.E., and Miller, I., Probability and Statistics for Engineers, Second Ed., Prentice-Hall, Englewood Cliffs, N.J., 1977.

7. Gibbs, H.J., and Holtz, W.G., "Research on Determining the Density of Sands by Spoon Penetration Testing," Proceedings, Fourth International Conference on Soil Mechanics and Foundation Engineering, London, Vol. 1, 1957, pp. 35-39.

8. Kennedy, J.B., and Neville, A.M., Basic Statistical Methods for Engineers and Scientists, Harper and Row, New York, 1974.

9. Leicht, T.J., and Reese, L.C., "Load Deformation Characteristics of Pile Foundations," Research Report GR 81-17, Geotechnical Engineering Center, The University of Texas, Austin, Texas, August 1981.

10. Mansur, C.I., Hunter, A.H., and Davisson, M.T., Pile Driving and Loading Tests--Lock and Dam No 4, Arkansas River and Tributaries, Fruco and Associates, St. Louis, Missouri, September 1964.

11. Mason, H.G., and Bishop, J.A., "Measurement of Earth Pressure and Deflection Along the Embedded Portion of a 40-Ft. Steel Pile," Special Technical Publication No. 154-A, American Society of Testing and Materials, 1953, pp. 1-21.

12. Matlock, H., and Haliburton, T.A, BMCOL28, Hudson Matlock Publisher, Austin, Texas, June 1964.

13. Meyer, B.J., and Reese, L.C., "Analysis of Single Piles Under Lateral Loading," Research Report No. 244-1, Center for Transportation Research, The University of Texas, Austin, Texas, December 1979.

14. O'Neill, M.W., and Murchison, J.M., "An Evaluation of P-y Relationships in Sands," Research Report No. GT-DF02-83, Department of Civil Engineering, University of Houston, Houston, Texas, May 1983.

15. Parker, F., Jr., and Reese, L.C., "Experimental and Analytical Study of Behavior of Single Piles in Sand Under Lateral and Axial Loading," Research Report No. 117-2, Center for Highway Research, The Univerity of Texas, Austin, Texas, November 1970.

16. Reese, L.C., "Laterally Loaded Piles: Program Documentation," Journal of the Geotechnical Engineering Division, ASCE, Vol. 103, No. GT4, April 1977, pp. 287-305.

17. Reese, L.C., Cox, W.R., and Koop, F.D., "Analysis of Laterally Loaded Piles in Sand," Proceedings, Sixth Annual Offshore Technology Conference, Vol. 2, Paper No. 2080, Houston, Texas, May 1974.

18. Robinson, K.E., "Horizontal Subgrade Reaction Estimated from Lateral Loading Tests on Timber Piles," Behavior of Deep Founations, ASTM, STP 670, Raymond Lundgren, Ed., American Society for Testing and Materials, 1979, pp. 520-536.

19. Scott, R.F., "Analysis of Centrifuge Pile Tests: Simulation of Pile Driving," Research Report, American Petroleum Institute OSAPR Project 13, California Institute of Technology, Pasadena, California, June 1980.

20. Stevens, J.B., Holloway, D.M., Moriwaki, Y., and Demsky, E.C., "Pile Group Response to Axial and Lateral Loading,' Symposium on Deep Foundations, ASCE, 1979, pp. 396-419.

21. Vallabhan, C.V.G., and Alikhanlou, F., "Short Rigid Piers in Clays," Journal of the Geotechnical Engineering Division, ASCE, Vol. 108, No. GT10, October 1982, pp. 1255-1272.

EVALUATION OF P-Y RELATIONSHIPS IN COHESIVE SOILS

By Sal M. Gazioglu,[1] A.M. ASCE, and Michael W. O'Neill,[2] M. ASCE

INTRODUCTION

The present-day practice of analyzing laterally loaded driven and bored piles often requires modeling of soil response in terms of a family of uncoupled, non-linear unit load transfer functions. These functions, commonly called p-y curves, are input into computer codes that use finite difference or finite element representations of beam behavior (9, 16). With this method of analysis, the forces, bending moments, and displacements in the pile can be predicted if the relationship describing the p-y curves is accurately defined along the pile. Because of the complexity of stress conditions developed in the soil during pile installation and subsequent loading, complete and comprehensive theoretical derivations of p-y curves have not been yet developed.

This paper considers three semi-empirical procedures for constructing p-y curves in cohesive soils that utilize data from laboratory tests. One method, called the Soft Clay Method, was developed based on analysis of data from lateral load tests from two soft clay sites and from models (10). The second method, called the Stiff Clay Method, was derived from full-scale test data from a stiff clay site (17). These two methods differ appreciably in complexity from each other. A third (new) method is described in this paper that integrates relative pile-soil stiffness and soil degradability effects to represent the response of all cohesive soils in a consistent manner. A relatively broad data base is used to compare the three methods.

METHODS FOR CONSTRUCTION OF P-Y CURVES IN COHESIVE SOILS

Three procedures for developing p-y curves for driven piles under a free-water surface, denoted the Soft Clay Method, the Stiff Clay Method, and the Integrated Clay Method, are described in this section in summary form. The first two methods are often used in the design of piles for offshore structures. The third method is introduced in this

[1]Project Supervisor, D'Appolonia Consulting Engineers, Inc., Houston, Texas 77042.

[2]Professor and Acting Chairman, Department of Civil Engineering, University of Houston, Houston, Texas 77004.

paper as a possible means of improving the accuracy of analysis for driven pipe piles.

Soft Clay Method

The Soft Clay Method for p-y curve generation is the method presently suggested in American Petroleum Institute guidelines RP2A (1). It was based on a careful analysis of the behavior of instrumented full-scale pipe piles tested at two submerged soft clay sites (one lacustrine site, which was slightly overconsolidated, and one marine site, which was approximately normally consolidated), and on a limited series of model tests (10).

The Soft Clay Method employs an expression to describe the variation of the ultimate soil resistance, p_u, with depth:

$$p_u = N_p \; c \; D \tag{1}$$

where:

p_u = ultimate soil resistance,
N_p = ultimate lateral soil resistance coefficient,
c = undrained soil shear strength, and
D = pile diameter

The value of the ultimate lateral soil resistance coefficient, N_p, is a function of depth below the ground surface and reaches a limiting value which corresponds to plane strain flow failure of the soil around the pile. The variation prescribed for N_p is

$$N_p = 3 + \frac{\bar{\sigma}}{c} + J \frac{x}{D} \leq 9 \tag{2}$$

where:

$\bar{\sigma}$ = effective overburden stress at depth x,
J = an empirical constant with an approximate value of 0.5 for soft offshore clays of the Gulf of Mexico and a value of 0.25 for stiffer clays, and
x = depth below ground surface.

The depth x at which surface failure transforms to plane strain failure ($N_p = 9$) is termed x_{cr}. Above x_{cr}, c is taken as an average value between the depth of interest and the surface. Below x_{cr}, c is taken as the value at the depth of interest.

The characteristic shape of the p-y curves is defined by a power function that was fitted to the shapes of the experimental p-y curves:

$$p = 0.5\ p_u \left(\frac{y}{y_c}\right)^{0.33} \tag{3}$$

where:

p = lateral soil resistance, in units of F/L,
y = pile deflection, and
y_c = reference deflection of the pile.

The reference pile deflection, y_c, is defined as:

$$y_c = 2.5\ \varepsilon_c\ D \tag{4}$$

where:

ε_c = major principal strain at one half the maximum deviator stress in a UU triaxial compression test.

The limiting effects of cyclic loading are accounted for by restricting the ultimate soil resistance to $0.72p_u$ at a deflection of $3y_c$ or greater for depths equal to or greater than x_{cr}. For depths less than x_{cr}, the ultimate soil resistance is further reduced by the ratio of actual depth to critical depth at large deflections. Figure 1 presents construction details for static and cyclic p-y curves using the Soft Clay Method. The cyclic curve is intended to represent an envelope to actual p-y curves after cyclic degradation of the soil has reached a stable condition, which is assumed to exist.

Stiff Clay Method

The Stiff Clay Method as used here refers to a preliminary criterion proposed by Reese et al (17) based on a series of lateral load tests on instrumented pipe piles at a site consisting of heavily overconsolidated, jointed clay that had been ponded several months prior to installing and testing the piles. The method, outlined in Fig. 2, follows the same general logic as the Soft Clay Method, with several notable differences. First, the maximum resistance is defined as p_c and is evaluated by:

$$p_c = A\ [2\ c\ D + \bar{\sigma}\ D + 2.83\ c\ x] \leq 11\ A\ c\ D, \text{ or} \tag{5}$$

$$p_c = B\ [2\ c\ D + \bar{\sigma}\ D + 2.83\ c\ x] \leq 11\ B\ c\ D \tag{6}$$

Equation 5 pertains to static loading, and Equation 6 pertains to the cyclic envelope. The factors A and B represent empirical corrections

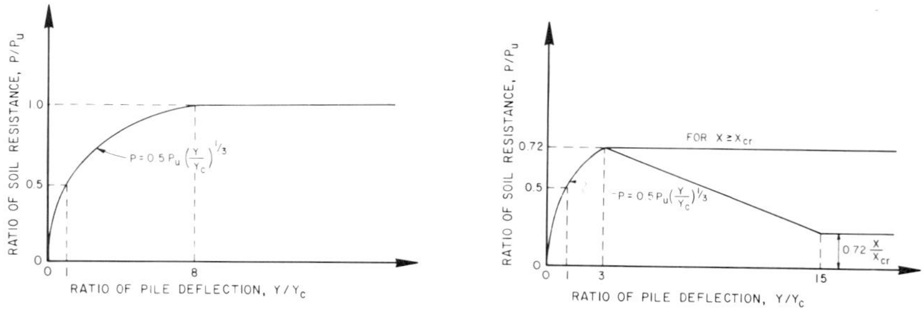

Static

Cyclic

Fig. 1. Soft Clay Method (10)

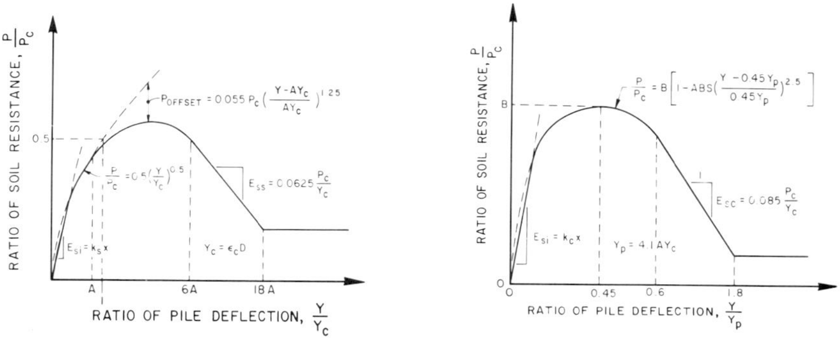

Static

Cyclic

Fig. 2. Stiff Clay Method (17)

to expressions for p_u that were developed analytically. They are evaluated in Fig. 3.

The p-y curves are constructed using a series of line segments. The initial portions of both the static and cyclic curves are straight lines that are defined by:

$$E_{si} = k\ x \tag{7}$$

where:

E_{si} = initial subgrade modulus,
k = a parameter depending upon whether loading is static or cyclic and evaluated in Table 1, and
x = depth.

Table 1. Recommended Values of k (17)

	Average Undrained Shear Strength (tons/ft^2)		
	0.5-1.0	1.0-2.0	2.0-4.0
k_s (lbs/ in^3)	500	1000	2000
k_c (lbs/ in^3)	200	400	800

k_s is for static loading; k_c is for cyclic loading.

Note: 1 tsf = 95.8 kPa; 1 pci = 271 kN/m^3.

In the static curves the initial straight line portion is followed by a segment of a parabola, followed by an offset from that parabola, as shown in Fig. 2. The offset begins at y/y_c = A (from Fig. 3) and terminates at y/y_c = 6A. Note that $y_c = \varepsilon_c D$. In the cyclic case the parabola is as shown in Fig. 2, terminating at y/y_c = 0.6 with no offset. Both the static and cyclic curves then have negative-sloping straight lines, as shown in Fig. 2, to the values of y indicated. The remaining branches depict constant p for increasing y.

Extreme degradation occurs in both the static and cyclic curves in the Stiff Clay Method. This may be due to the presence of a transient moisture condition in the heavily overconsolidated soil from which the method was derived. Once a gap developed adjacent to a test pile, free water from the surface pond may have been imbibed by the moisture-deficient clay through its joint structure, producing a softening effect that would otherwise not be present in moisture-stable clays.

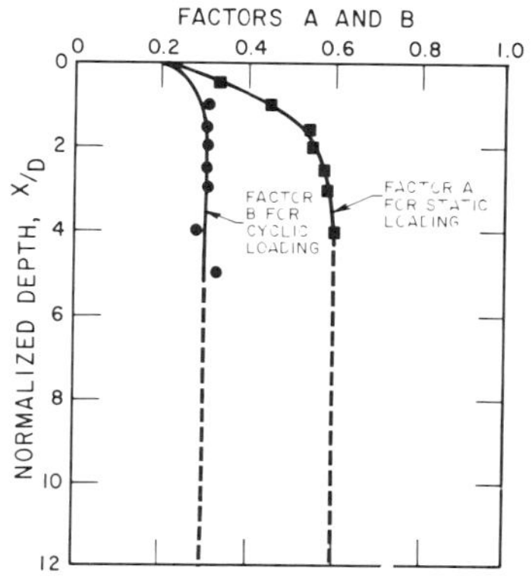

Fig. 3. Ultimate Resistance Reduction Factors for Stiff Clay Method (17)

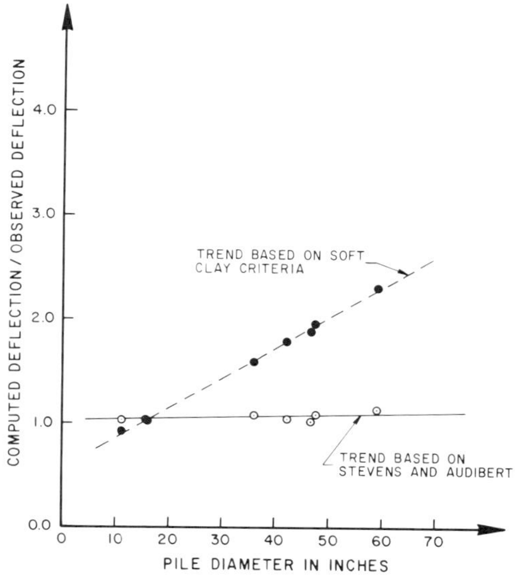

Fig. 4. Effect of Making Reference Deflection a Function of $D^{0.5}$ (18)

Integrated Clay Method

A new p-y curve construction method, termed the Integrated Clay Method (13), was developed in order to address several concerns, outlined below:

1. Division of methods into "soft" and "stiff" clay categories for design purpose is artificial. Factors such as soil ductility and dispersion characteristics may be more significant factors than consistency. It is desirable to have only one method that incorporates appropriate factors that apply to all clays. At least one unified method, applicable to all clays, already exists (19). However, that method was not studied because choices of parameters must be made a priori that essentially convert it to either a stiff-clay-like method, or a soft-clay-like method, based on the judgment of the user.

2. Studies by Stevens and Audibert (18) among others, have suggested that the reference deflection, y_c, used to construct p-y curves according to the Soft Clay Method, is not linearly dependent on pile diameter, as suggested in Eq. 4, but instead is approximately defined by Eq. 8, which involves the square root of pile diameter:

$$y_c = 8.9 \ \varepsilon_c \ D^{0.5} \tag{8}$$

where D and y_c are both expressed in units of inches. Stevens and Audibert analyzed several load tests using the Soft Clay Method but where Eq. 8 was substituted for Eq. 4, with the excellent results indicated in Fig. 4.

3. Most present methods, including the Soft and Stiff Clay Methods, contain no term that represents continuum effects. While p-y curve construction methods derived experimentally from load tests contain such effects implicitly for the deflected pile shape imposed in the test, there is no assurance that a stiffer or more flexible pile would produce the same unit soil response.

The Integrated Clay Method is a new semi-empirical method applicable to all clays which includes empirical terms that reflect soil ductility (Concern 1), non-linear dependence on pile diameter (Concern 2), and relative pile-soil stiffness (Concern 3). Wang (20) conducted pinhole dispersion tests and pile-soil-scour-potential (PSSP) tests on samples of clays from four sites, three of which had been at or near the locations of well-documented load tests on piles under the water surface. As can be seen in Table 2, no correlation appears to exist between these measures of soil dispersion characteristics and observed degradability of the soil under cyclic lateral loading. Therefore, a soil dispersion expression was not included in the Integrated Clay Method.

The following computations are made in the Integrated Clay Method. Details are given in Ref. 13.

1. Assess Critical Pile Length, L_c. If a pile is "long," there

Table 2. Soil Degradability Effects (20)

Undisturbed Sample	Soil Classification: Pinhole Dispersion Test	Soil Classification: PSSP Test	Degradability Under Cyclic Lateral Pile Loading (Submerged)
Manor Clay	Non-dispersive	Intermediate scour susceptibility	Major
North Sea Clay	--	Significant scour susceptibility	Unknown
Sabine Clay	--	Significant scour susceptibility	Minor
Houston Clay	Non-dispersive	Significant scour susceptibility	Intermediate

exists a penetration L_c beyond which pile-head response is essentially unaffected by further penetration. Based upon finite element studies by Randolph (15), L_c can be evaluated for a quasi-elastic soil from Eq. 9:

$$L_c = 3[EI/E_s D^{0.5}]^{0.286} \tag{9}$$

where EI is the flexural stiffness of the pile and E_s is an operational soil modulus, perhaps representing a secant Young's modulus at a deviator stress level of one-half of the deviator stress at failure in undrained triaxial compression. Values used in the study reported in this paper are based on correlations with the average UU triaxial compression shear strength between the surface and depth L_c and are tabulated in Table 3. Note that assessment of L_c is an iterative process if the shear strength varies with depth.

2. <u>Assess Reference Deflection, y_c.</u> The reference deflection y_c is evaluated from Eq. 10:

$$y_c = A' \varepsilon_c D^{0.5} \left(\frac{EI}{E_s}\right)^{0.125} \tag{10}$$

where E_s is the value of soil modulus from Table 3 at the depth of interest, and A' is a constant to be determined. Equation 10 was derived by assuming that y_c could be expressed in the form

Table 3. Soil Modulus vs. Undrained Shear Strength for Integrated Clay Method

Undrained Shear Strength, c (lb/ft^2)	Soil Modulus, E_s (lb/in^2)
< 500	50
500 - 1000	50 - 150
1000 - 2000	150 - 450
2000 - 4000	450 - 1500
4000 - 8000	1500 - 5000
> 8000	5000

Note: 1 psf = 47.9 Pa; 1 psi = 6.89 kPa

$$y_c = A' \varepsilon_c D K_r^n \quad (11)$$

where K_r is a relative stifness (continuum term modified from a definition by Poulos (14) for long piles defined by

$$K_r = EI/E_s L_c^4 \quad (12)$$

Equation 10 results from combining Equations 9, 11, and 12.

The factor A' was determined through an optimization technique based on modeling seven well-documented, full-scale static and cyclic load tests (that were conducted under water) with the complete criterion. Table 4 summarizes the results. The A'-factors appear generally to increase with increasing degree of overconsolidation and to increase somewht with increasing load. The two extremely large average values of A' (Houston (small pile) and Manor) are believed to be anomolous. At Manor, as has already been discussed, unstable moisture conditions probably existed, and at Houston, it is believed that the shear strength profile selected for analysis modeled the soil near the surface as being too stiff. Based on the data available and upon the judgment of the authors, A' was selected to be a constant for all soils equal to 0.8. It is possible that future additions to the data base will show A' to be a function of overconsolidation ratio, sensitivity, and possibly other factors.

3. Formulate Ultimate Soil Resistance, p_u. The ultimate soil resistance is determined using Eq. 13:

$$p_u = F N_p c D \quad (13)$$

where F is a reduction factor from Table 5, based on soil ductility and

Table 4. Results of A'-Factor Optimization Study

Site Location	Pile Head Condition	Lateral Load (kips)	Optimized A'-Factor	Average A'-Factor	Consolidation of Site Soil
Sabine, TX	Free	2.9 8.5 12.6	0.40 0.50 0.50	0.467	Approx. NC
Lake Austin, TX	Free	3.0 11.0 17.2	0.70 0.86 0.90	0.820	Slightly OC
Houston, TX (48.00-in.)	Free	71.0 129.0 170.0	0.50 0.60 1.08	0.727	OC
Houston, TX (10.75-in.)	Free	4.0 9.6 14.5	0.96 2.90 4.40	2.433	OC
Manor, TX (25.25-in.)	Free	11.0 40.0 66.0	1.50 3.00 4.00	2.833	Heavily OC
Sabine, TX	Restrained	4.8 12.5 17.1	0.40 0.60 1.00	0.667	Approx. NC
Harvey, LA	Restrained	1.2 3.8 5.5	0.30 0.31 0.51	0.373	Approx. NC

Notes: 1 k = 4.45 kN; 1 in. = 25.4 mm.
References to tests and pile properies are given in Table 7.

form of loading (static or cyclic), and the other factors are as defined previously. N_p has the following variation with depth:

$$N_p = 3 + 6(x/x_{cr}) \leq 9 \tag{14}$$

and

$$x_{cr} = L_c/4 \tag{15}$$

Table 5. Soil Degradability Factors, F

Factor	Loading Condition	UU Triaxial Compression Failure Strain		
		<0.02	0.02-0.06	>0.06
F_s	Static	0.50	0.75	1.00
F_c	Cyclic	0.33	0.67	1.00

This definition of x_{cr} follows the logic advanced by Briaud et al (2) based on pressuremeter data (that x_{cr} depends on relative pile-soil stiffness) and not the principles used in the Soft and Stiff Clay Methods, which do not consider relative stiffness. Table 6 shows that the methods based on relative stiffness (Pressuremeter and Integrated Clay) produce smaller values of x_{cr} for a given pile as the soil becomes stronger, while the soft and stiff clay methods produce larger values of x_{cr} with increasing soil strength.

Table 6. Comparison of Critical Depth, x_{cr}

Method	D (Inches)	EI (lbs-in^2)	x_{cr} in inches, c in lbs/in^2: 3	15	30
Soft Clay	12	$8.5\text{x}10^9$	116.1	137.4	140.6
	24	$7.5\text{x}10^{10}$	194.6	262.8	274.8
	48	$7.5\text{x}10^{11}$	293.9	483.2	525.5
Stiff Clay	12	$8.5\text{x}10^9$	36.6	37.8	38.0
	24	$7.5\text{x}10^{10}$	73.2	75.7	76.0
	48	$7.5\text{x}10^{11}$	146.4	151.4	152.0
Pressuremeter	12	$8.5\text{x}10^9$	67.2	30.0	15.6
	24	$7.5\text{x}10^{10}$	96.0	40.8	19.2
	48	$7.5\text{x}10^{11}$	144.0	57.6	19.4
Integrated Clay	12	$8.5\text{x}10^9$	118.8	63.4	44.9
	24	$7.5\text{x}10^{10}$	200.5	106.9	75.8
	48	$7.5\text{x}10^{11}$	350.8	187.1	132.6

Note: 1 in. = 25.4 mm; 1 psi = 6.89 kPa

4. Construct the p-y Curves. Figure 5 describes graphically the construction of static and cyclic p-y curves using the Integrated Clay Method. Note that the curves are similar to those of the Soft Clay Method, except that cyclic degradation begins when p/p_u reaches 0.5 (not 0.72). The decision to include non-zero residual static surface resistances is based on the fact that most loading conditions are load-controlled, in which it is virtually impossible to degrade the soil resistance totally from any non-zero peak value.

It is observed that the initial slopes of the curves for both the Soft Clay and Integrated Clay Methods are infinite. Since p-y curves are normally input into computational algorithms in discrete form, this causes no particular problem as long as the first non-zero point is input at a sufficiently large deflection. It is recommended that the first non-zero value of y (and corresponding p) be inut at $y \geq 0.33$ per cent of the pile diameter for analyses not involving extremely low-displacement pile response (such as piles supporting machine foundations), where it should not be assumed that any of these p-y methods apply.

DATA BASE

Full-scale field tests from various clay-profile sites were selected for comparison of observed pile behavior with pile behavior predicted by the three p-y methods considered for this study. The tests included in the data base were selected such that the soil properties, pile properties, and details of loading were defined reasonably accurately. Some soil properties, such as failure strain levels, were assumed based on the available information in the absence of high-quality stress-strain relationshps. Tests considered were limited to round steel piles, usually in a submerged condition. Table 7 provides a summary of all static and cyclic tests included in the data base. Further details may be found in Ref. 13. In general, soil characterization was based on UU triaxial test data supplemented by other undrained test data, when available. A wide range in pile sizes and soil consistency is observed in the data base. All piles can be classified as "long" piles. Cyclic tests were normally conducted under controlled displacement conditions to a number of cycles such that pile-head displacement was stable at the steady state. Accurate moment measurements were made in the first five tests. In the remainder of the tests, the piles were either uninstrumented or instrumented too sparsely to provide well-defined moment diagrams over a wide range of loads.

COMPARISON OF METHODS

Procedure for Comparison

The load tests in the data base were modeled for all three methods using a widely available algorithm, BMCOL28 (9). In order to generate

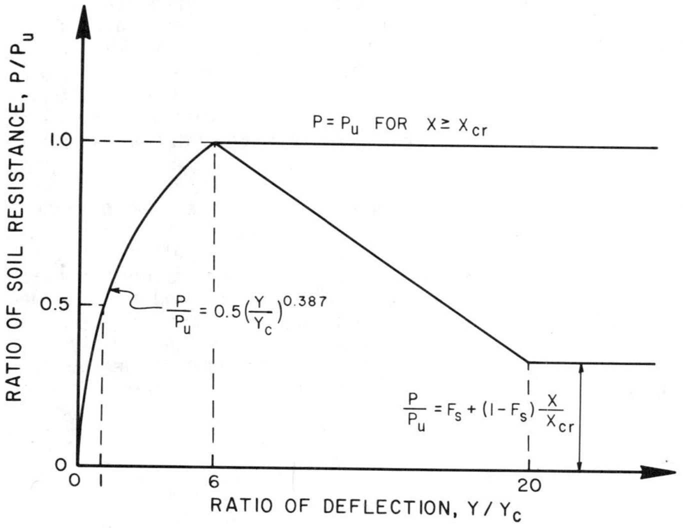

Static

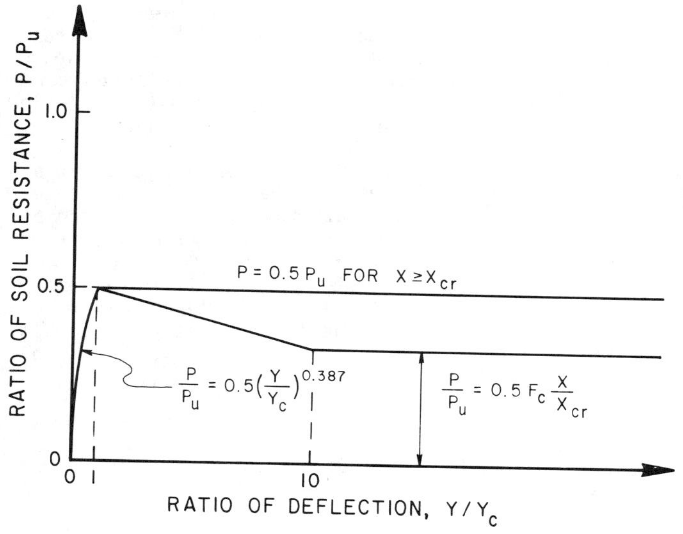

Cyclic

Fig. 5. Integrated Clay Method

Table 7. Data Base Synopsis

Site (Ref.)	No. of Tests	Pile Head Conditions	Soil	Pile Properties
Sabine (10)	2 Static 2 Cyclic	Free Restrained	Soft Clay	12.75-inch O.D. Steel Pipe Piles
Austin (10)	1 Static 1 Cyclic	Free	Medium Stiff Clay	12.75-inch O.D. Steel Pipe Pile
Houston*	2 Static 2 Cyclic	Free	Stiff Clay	10.75-inch O.D. Steel Pipe Pile, 48-inch O.D. Steel Pipe Pile
Manor (17)	3 Static 1 Cyclic	Free Restrained	Very Stiff Clay	25.25-inch O.D. Steel Pipe Pile 6.625-inch O.D. Steel Pipe Piles
Harvey (11)	1 Static 1 Cyclic	Restrained	Soft Clay	6.625-inch O.D. Steel Pipe Piles
Hamilton (4,5) (Flooded Site) Hamilton (4,5) (Dry Site)	4 Static 4 Static	Free Free	Medium Stiff Clay Stiff Clay	4.50-inch, 8.62-inch, 12.75-inch, 16.00-inch O.D. Steel Pipe Piles
Manfredonia (7)	1 Static 1 Cyclic	Free	Very Soft Clay	48-inch O.D. Steel Pipe Piles
Chalmette (8)	1 Static 1 Cyclic	Free	Very Soft Clay	Octagonal Pile with Diameter Varying from 33 inches to 22 inches
Biwako-Ohashi Bridge (6)	1 Static	Free	Soft Clay	59-inch O.D. Steel Pipe Pile
St. Gabriel (3)	1 Static	Free	Soft Clay	10-inch O.D. Steel Pipe Pile Filled with Concrete

*Unpublished data

consistent and unbiased results, p-y curves were computed and input for each p-y method consistently at depths of integer multiles of the pile diameter, D, up to a depth of 6D and then at depths of even integer multiples of D up to a depth of 12D. Additional p-y curves were input at x_{cr}, a depth of 20D and the bottom of the pile. Eight to ten values of p and y were input for each curve to define accurately the characteristic shape of the p-y curve for each method.

Four different magnitudes of lateral load were prescribed to generate plots of lateral load versus deflection and maximum moment. Continuous relationships between lateral load and pile-head deflection and lateral load and maximum moment were then fitted to the discrete output. The measured pile deflections and maximum moments were also plotted along with the plots generated using the p-y methods to assess the ability of each method to predict the measured values accurately.

Errors in predicted behavior were assessed using the procedure described by Murchison and O'Neill (12). Definitions of errors for two cases of behavior are described graphically in Figs. 6 and 7. In "Case 1" behavior the predicted and measured relationships generally conformed to one another. Most tests fell into this category. In "Case 2" behavior the p-y method (usually the Stiff Clay Method) would predict premature failure. Errors in pile-head deflection, E_y, and errors in maximum moment, E_m, were calculated at shear loads corresponding to pile-head deflections of 1, 5, 10, and 15 per cent of D, where possible. For tests on piles with diameters exceeding 24 in. (0.61 m), large displacements were not achieved, so in those tests E_y and E_m were computed at loads corresponding to deflections of 0.1, 0.5, 1.0, and 1.5 per cent of D. Error evaluation protocol established the errors using Case 1 definitions unless errors were undefined by Case 1. In that event, Case 2 definitions were used.

Cumulative errors for all tests and all discrete loads were evaluated using Eqs. 16 and 17.

$$(\overline{E}_y)_j = \sum_{\substack{i=1\\j=1}}^{m,n} (w|E_y|)_{i,j} \tag{16}$$

and

$$(\overline{E}_m)_j = \sum_{\substack{i=1\\j=1}}^{m,n} (w|E_m|)_{i,j} \tag{17}$$

where :

$\overline{E}_y$ = Cumulative errors in deflection for all static or cyclic tests for one method,
$\overline{E}_m$ = Cumulative errors in maximum moment for all static or cyclic tests for one method,
j = Subscript denoting a particular method,

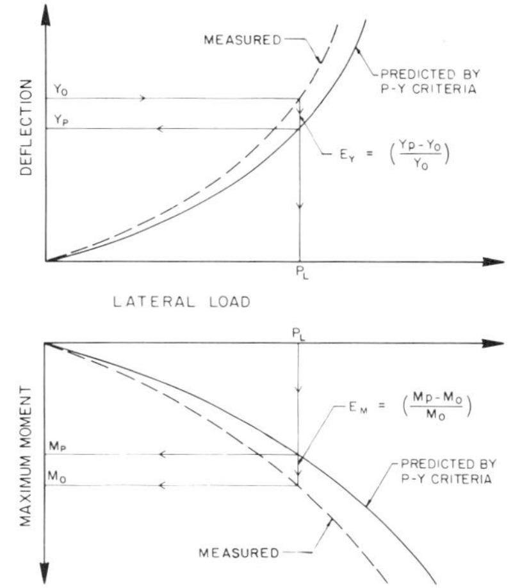

Fig. 6. Case 1 Error

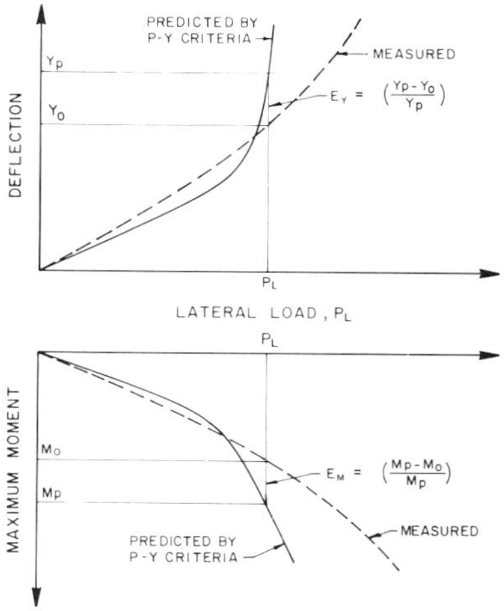

Fig. 7. Case 2 Error

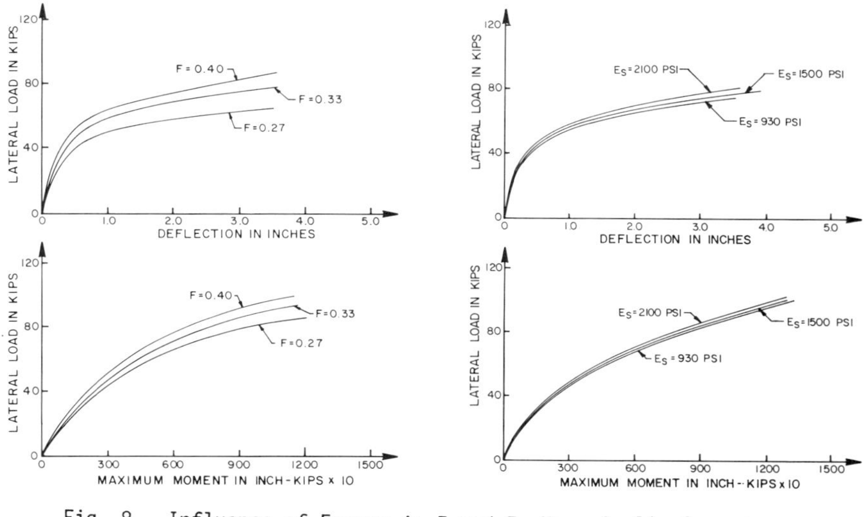

Fig. 8. Influence of Errors in F and E_S Upon Cyclic Curves
(1 in. = 25.4 mm; 1 k = 4.45 kN; 1 psi = 8.89 kPa)

i = Subscript denoting lateral load point,
m,n = Total number of lateral load points considered in evaluating the cumulative error for the indicated method (3 or 4 per test times the number of tests) for static and cyclic tests, respectively, and
w = A weighting factor.

Weighting factors were applied to the calculated errors as follows:

1. w = 2 if the error was unconservative (predicted deflection or moment that was too small),

2. w = 4 if Case 2 behavior occurred, and

3. w = 0.5 if the pile was not a prismatic pipe pile (Chalmette test only).

In the event that more than one weighting factor applied to a given point, the appropriate factors were multiplied to give w for Eqs. 16 and 17.

Results of Comparison

The cumulative weighted errors for the three methods considered are summarized in Tables 8 and 9. Errors for individual tests can be found in Ref. 13. In most cases, the Integrated Clay Method produced the lowest average (cumulative) error across the data base. However, the Soft Clay Method yielded the lowest average error for pile-head deflection under cyclic loading. Table 10 compares normalized errors for the Integrated Clay Method with errors computed from the Soft Clay Method on soft clay sites and from the Stiff Clay Method on stiff clay sites. The Integrated Clay Method yielded lower average errors than either of the two consistency-specific criteria, except for pile-head deflection under cyclic loading at soft clay sites, where the Soft Clay Method is clearly superior.

Average errors (unweighted), E_a, and 95 per cent confidence limits were evaluated for the Integrated Clay Method, assuming the errors to be normally distributed. The results of these calculations are summarized in Table 11. The average errors for head deflection (for all tests), defined as predicted value minus measured value divided by measured value, tend to be slightly negative (underprediction) and to become more negative with increasing displacement under static loading, whereas the errors became more positive (overprediction) with increasing deflection under cyclic loading, suggesting that the method is not completely optimized for the data base that was considered. Average errors in maximum moment were almost random with respect to displacement and were relatively small. Although locations of measured and predicted maximum moment were not treated statistically, errors in predicting the location of maximum moment below grade were observed to be small.

Table 8. Cumulative Weighted Errors for Static Tests (All Sites)

Method	Pile-Head Deflection	Maximum Moment
Soft Clay	54.8 (1.00)	10.0 (1.00)
Stiff Clay	77.8 (1.42)	18.4 (1.84)
Integrated Clay	*44.2 (0.81)	*8.4 (0.84)
Number of data points per method	(83)	(32)

*Indicates lowest error

The numbers in parentheses are the cumulative errors normalized with respect to the Soft Clay Method (API Method).

Table 9. Cumulative Weighted Errors for Cyclic Tests (All Sites)

Method	Pile-Head Deflection	Maximum Moment
Soft Clay	*15.4 (1.00)	11.9 (1.00)
Stiff Clay	44.8 (2.91)	29.3 (2.46)
Integrated Clay	21.3 (1.38)	*10.6 (0.89)
Number of data points per method	(36)	(28)

*Indicates lowest error

The numbers in parentheses are the cumulative errors normalized with respect to the Soft Clay Method (API Method).

Table 10. Cumulative Errors in Integrated Clay Method Normalized by Cumulative Errors in Soft and Stiff Clay Methods for Soft and Stiff Clay Sites

Site Consistency	Static Deflection	Static M_{max}	Cyclic Deflection	Cyclic M_{max}
Soft Clay	0.81	0.74	1.49	0.86
Stiff Clay	0.73	0.73	0.79	0.91

Numbers of data points:

Soft, Static: 48 (deflection), 16 (moment)
Soft, Cyclic: 24 (deflection), 16 (moment)
Stiff Static: 36 (deflection), 20 (moment)
Stiff Cyclic: 12 (deflection), 12 (moment)

The confidence limits on both head deflection and maximum moment are seen to be relatively uniform with displacement of the pile head under static loading conditions. Confidence limits are tabulated as percentages of average computed values. Interpretation of Table 11 suggests that, under static conditions, one could reasonably not expect to underpredict or overpredict true displacement by more than about 70 per cent or underpredict or overpredict true maximum moment by more than about 30 per cent. Under cyclic conditions, the respective errors are larger, especially at high displacements. The existence of confidence limits exceeding -100 per cent (cyclic loading, high deflection) is probably due to an insufficiency of data at certain deflections. The confidence limits in static deflection and moment are of the same general magnitude as were found in a parallel study in cohesionless soils (12). The trend in cohesionless soils, however, is for predictions to be more accurate under cyclic loading than under static loading, not less accurate, as occurs in cohesive soils.

SENSITIVITY OF INTEGRATED CLAY METHOD TO PARAMETER VARIATION

The effect of varying input parameters to the Integrated Clay Method was studied for several hypothetical soil profiles (13). Figure 8 illustrates the effect of varying F and E_s for a situation in which D = 24 in. (610 mm), length = 600 in. (15.2 m), EI = 7.5 x 10^{10} lb-in^2 (2.15 x 10^5 kN-m^2), c = 4000 psf (192 kPa), ε_c = 0.005, F (correct) = 0.33, and E_s (correct) = 1500 psi (10,300 kPa). The p-y curves are rather insensitive to E_s but considerably more sensitive to F. This observation suggests that it is more important to define post-peak degradation soil characteristics than to arrive at the correct relative pile-soil stiffness (which E_s is used to determine).

Additional critical evaluation of the Integrated Clay Method suggests that the value of p/p_u at which cyclic degradation begins may

Table 11. Confidence Limits for Integrated Clay Method

Loading Condition	Deflection (Per Cent of D)	Head Deflection E_a (Per Cent)	Head Deflection 95 Per Cent Confidence Limits (Per Cent +/-)	Maximum Moment E_a (Per Cent)	Maximum Moment 95 Per Cent Confidence Limits (+/-)
Static	1	- 1.6	71.9	+ 6.9	41.6
	5	- 11.6	66.3	- 1.9	30.0
	10	- 15.6	62.4	- 2.1	27.2
	15	- 15.6	55.7	- 1.7	28.8
Cyclic	1	- 16.0	82.9	- 4.3	48.9
	5	- 6.2	90.3	- 5.6	24.4
	10	+ 1.7	112.6	- 0.2	48.5
	15	+ 16.1	133.6	+ 3.0	62.4

need to be slightly higher than indicated in Fig. 5 (0.60 ±) in order to reduce the displacement errors at high displacements. As has already been stated, it is also logical that A' (Eq. 10) may be more appropriately expressed as a function of soil sensitivity, overconsolidation ratio, and/or construction method, rather than as a universal constant.

CONCLUSIONS

1. Of the three methods studied, the Integrated Clay Method yielded the most accurate predictions of pile deflection under static loading and of maximum moment under static or cyclic loading when applied to a relatively broad data base. However, the Soft Clay Method provided better predictions of pile displacements under cyclic loading, especially at soft clay sites.

2. The significant improvement realized by the Integrated Clay Method for static loading seems to suggest that a reference displacement characterized by both the square root of pile diameter and a relative stiffness term is desirable compared to reference displacements that are simply linearly dependent on pile diameter.

3. The Integrated Clay Method can probably be improved by redefining the soil resistance value at which cyclic degradation begins to be slightly greater than 0.5 p_u, by developing A' (Eq. 10) as a

function of appropriate soil and construction variables, and perhaps by reconsidering the degradation factors, F. It is unlikely that these changes can be accomplished without performance of more high-quality full-scale or centrifuge tests on laterally loaded piles.

4. The confidence in predicting deflections and moments with the method that has the highest average accuracy (Table 11) is unfortunately rather poor, suggesting that further research into fundamental mechanisms of lateral pile-soil interaction is warranted.

ACKNOWLEDGEMENT

The study described here was sponsored by the American Petroleum Institute (Project PRAC 82-41). API is not responsible for any conclusions or other items represented as fact, as these are solely the responsibility of the authors. The authors are grateful for API's permission to publish these results.

APPENDIX - REFERENCES

1. American Petroleum Institute, API RP 2A, "Recommended Practice for Planning, Designing, and Constructing Fixed Offshore Platforms," Washiongton, D.C., Thirteenth Edition, 1982.

2. Briaud, J-L, Smith, T.D., and Meyer, B.J., "Using the Pressuremeter Curve to Design Laterally Loaded Piles," Proceedings, 15th Offshore Technology Conference, Paper No. 4501, May, 1983, Vol. 1, pp. 495-502.

3. Capazzoli, L., "Test Pile Program at St. Gabriel, Louisiana," Louis J. Capazzoli and Associates, Baton Rouge, Louisiana, July, 1968.

4. Gill, H.L., "Soil Behavior Around Laterally Loaded Piles," Report R-571 Naval Civil Engineering Laboratory, Port Hueneme, California, April, 1968.

5. Gill, H.L., and Demars, K.R., "Displacement of Laterally Loaded Structures in Nonlinearly Responsive Soil," Report R-670, Naval Civil Engineering Laboratory, Port Hueneme, California, April, 1970.

6. Ishii, Y., and Fujita, K., "Field Test on the Lateral Resistance of Large Diameter Steel Pipe Piles and Its Application to the Aseismic Design of Pile Bent-Type Pier," Proceedings, Third World Conference on Earthquake Engineering, New Zealand, 1965.

7. Jamiolkowski, M., and Marchetti, S., "Determinazione del Modulo di Reazione Orizzontale del Terreno Attaverso Misure Inclinometriche Effettuate Durante Prove di Carico Orizzontale Sul Pali," Associazone Geotechnica Italiana, XI Convegno di Geotechnica, Milano, March, 1973.

8. Lee, P.Y., and Gilbert, L.W., "Behavior of Laterally Loaded Pile in Very Soft Clay," Proceedings, 11th Offshore Technology Conference, Paper No. 3401, May, 1979, Vol. 1, pp. 387-396.

9. Matlock, H., and Haliburton, T.A., "A Program for Finite-Element Solution of Beam-Columns on Nonlinear Supports," Austin, Texas, June, 1964.

10. Matlock, H., "Correlations for Design of Laterally Loaded Piles in Soft Clays," Preprints, 2nd Annual Offshore Technology Conference, Paper No. 1204, May, 1970, Vol. 1, pp. 577-588.

11. Matlock, H., Ingram, W.B., Kelley, A.E., and Bogard, D., "Field Tests of the Lateral-Load Behavior of Pile Groups in Soft Clay," Proceedings, 12th Offshore Technology Conference, Paper No. 3871, May, 1980, Vol. IV, pp. 163-174.

12. O'Neill, M.W., and Murchison, J.M., "An Evaluation of P-y Relationships in Sands," A Report to the American Petroleum Institute, Report PRAC 82-41-1, University of Houston, May, 1983.

13. O'Neill, M.W., and Gazioglu, S.M., "Integrated Formulation of P-y Relationships in Clays," A Report to the American Petroleum Institute, Report PRAC-82-41-2, University of Houston, April, 1984.

14. Poulos, H.G., "Behavior of Laterally Loaded Piles: I-Single Piles," Journal of the Soil Mechanics and Foundations Division, A.S.C.E., SM4, May, 1971, pp. 711-751.

15. Randolph, M.F., "The Response of Flexible Piles to Lateral Loading," Geotechnique, 1981, Vol. 31, No. 2, pp. 247-259.

16. Reese, L.C., and Sullivan, W.R., "Documentation of Computer Program COM624," Geotechnical Engineering Center, Department of Civil Engineering, University of Texas, August, 1980, 297 pp.

17. Reese, L.C., Cox, W.R., and Koop, F.D., "Field Testing and Analysis of Laterally Loaded Piles in Stiff Clay," Proceedings, 7th Offshore Technology Conference, May, 1975, Vol. II, pp. 671-690.

18. Stevens, J.B., and Audibert, J.M.E., "Re-Examination of P-y Curve Formulations," Proceedings, 11th Offshore Technology Conference, Paper 3402, May, 1979, Vol. I, pp. 397-403.

19. Sullivan, W.R., Reese, L.C., and Fenske, C.W., "Unified Method for Analysis of Laterally Loaded Piles in Clay," Numerical Methods in Offshore Piling, Institution of Civil Engineers, London, 1980, pp. 135-146.

20. Wang, S.T., "Development of a Laboratory Test to Identify the Scour Potential of Soils at Piles Supporting Offshore Structures," Unpublished Master's Thesis, The University of Texas at Austin, August, 1982.

LATERAL LOAD TESTS OF PILES IN SLOPING ROCK FILL

Gerald M. Diaz[1], Bartlett W. Patton[2], Members ASCE
George L. Armstrong[3], Mohammad Joolazadeh[4],
Associate Members ASCE

ABSTRACT

A new container wharf constructed at the Port of Los Angeles features vertical prestressed concrete piles to provide both vertical and lateral support. A rockfill embankment, used to allow a steeper under-wharf slope and a reduced wharf width, resulted in the piles being embedded in the rockfill slope. Lateral load tests were performed on two production piles to evaluate the lateral load performance of the piles embedded in the sloping rockfill. This paper presents the results of the load tests for designers and researchers who are evaluating similar installations, to compare their predictions of lateral performance to that obtained in the tests. Back-calculated p-y curves based on the pile lateral load tests are compared to p-y curves calculated for cohesionless soils with a level surface. Conclusions are presented regarding the reduction of lateral resistance, p, necessary to account for the effects of the sloping rockfill, in lateral resistance verses deflection (p-y) analyses.

INTRODUCTION

The Port of Los Angeles, embarking on a major expansion program, commissioned a study (1) of structural systems for containership berthing facilities. The study objective was to identify a versatile and economical system for new containership facilities. Berths 216-218 were selected as the prototype for the study.

Preliminary evaluations indicated that retained earth and gravity structures currently in use in many areas of the world were not economically suited to the design seismic loads, soil conditions, and proposed draft at the Port of Los Angeles. Conventional open field pile-supported wharves were indicated as the most suitable alternative. Batter piles have traditionally been used to resist lateral loads in the Los Angeles Harbor, but in general, each new design has included more batter piles, which contributed to more difficult and

---

1 & 2. Principal Engineers, Harding Lawson Associates, Tustin, CA 92680

3 & 4. Project Engineers, Harding Lawson Associates, Tustin, CA 92680

costly construction. The use of all vertical piles was proposed to facilitate construction and to create a structure that would be compliant while resisting lateral loads.

The wharf section is shown in Figure 1. The rock embankment used is part of standard Port of Los Angeles practice to reduce the width and cost of the wharf structure by allowing a steeper slope than could be achieved in the natural soils. The structure is supported on vertical prestressed concrete piles; batter piles are not used. The piles in the back (landward) row (Row F) have the least eccentricity of applied horizontal load and therefore are the most effective in resisting lateral loads. To develop sufficient lateral capacity, the piles in this row are closely spaced. The piles are embedded in the sloping rockfill.

DESIGN

Previous research for lateral load (p) versus pile deflection (y) relationships has been based on piles founded in sands and clays with a level surface. Therefore, to evaluate the lateral load capacity of vertical piles embedded in a sloping rock fill, it was necessary to modify the p-y relationships to reflect the properties of the rockfill and to reduce the lateral resistance (p) to account for the sloping surface.

Initally, p-y curves were generated using the procedures proposed by Reese, Cox, and Koop (4) for cohesionless soils on level ground. The rock fill at Berth 216-218 consisted of quarry run material specified as follows:

Maximum rock size	500 pounds
Maximum fines (passing No. 4 sieve)	20 percent

The generalized soil profile used for analysis of the back row of piles was:

DEPTH	SUBSURFACE CONDITIONS
0 to 23 feet	Quarry Run Rockfill (GP) $\phi = 45^{o}$ (2, 3) $c = 0$
23 to 57 feet	Silty Sand (SM-SP) $\phi = 35^{o}$ $c = 0$
57 to 76 feet	Silt (ML) $\phi = 25^{o}$ $c = 0$ psf
Below 76 feet	Sand and Silty Sand (SM-SP) $\phi = 37^{o}$ $c = 0$

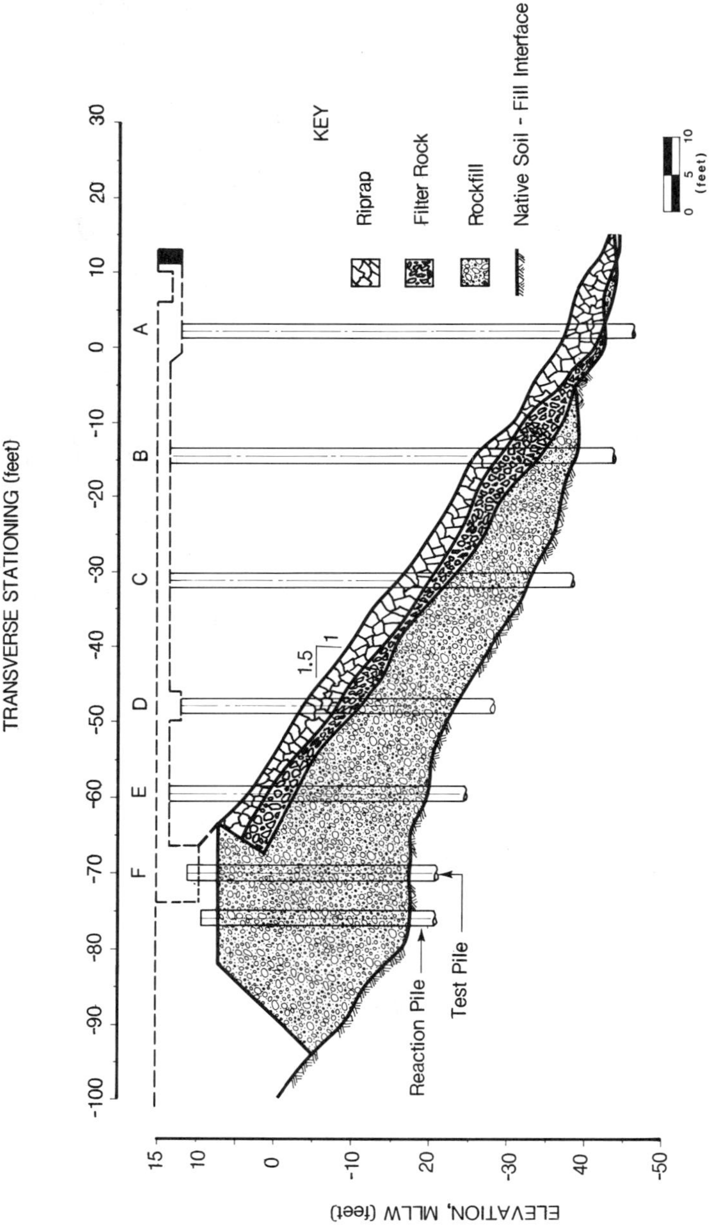

Figure 1 - Wharf and Actual Slope Cross-Section

The generated p-y curves were adjusted above a limiting depth of 25 feet (the depth below which the passive mode of soil failure no longer controls [4]), to account for the effects of the slope using the following reduction factor:

$$\beta = \frac{K_{ps} - K_{as}}{K_p - K_a} \qquad (1)$$

where: β = Lateral resistance slope correction factor, applied to soil lateral resistance (p) values

K_{ps} = Coulomb passive earth pressure coefficient (vertical wall, sloping backfill)

K_{as} = Coulomb active earth pressure coefficient (vertical wall, sloping backfill)

K_p = Coulomb passive earth pressure coefficient (vertical wall, horizontal backfill)

K_a = Coulomb active earth pressure coefficient (vertical wall, horizontal backfill)

The resulting p-y curves were used in the computer program COM622 (5) to estimate the pile lateral behavior when subjected to the design vertical and lateral loads. The analysis indicated that 24-inch (61cm) octagonal prestressed concrete piles spaced on five foot (1.5m) centers in the back (landward) row (Row F in Figure 1) would meet the structural design criteria for lateral capacity and deflection.

Because no test data was found in the literature for similar design conditions, (i.e. laterally loaded vertical prestressed piles embedded in a sloping rockfill) pile lateral load tests were recommended and authorized. Because of the proposed construction schedule and to save the cost of constructing a test rock embankment and test piles, the pile lateral load tests were performed using two production piles.

PRODUCTION PILE CONSTRUCTION

The wharf foundation construction involved four phases:

- Dredging the natural soils to a slope ratio of 2-1/4:1 (horizontal to vertical)

- Pre-jetting a pilot hole at each pile location to within a few feet of the design tip elevation, then driving the piles to refusal in the dense sand below an elevation of approximately -75 feet

- Installing stay lath to maintain the position of the piles

- Placing the rock embankment around the piles by dumping from skips.

Figures 2 and 3 illustrate the pile installation and rock placement.

Figure 2 - Pile Installation

Figure 3 - Rockfill Placement

LATERAL LOAD TESTS

Load Test Setup

Two instrumented production/test piles were driven in the back row. An instrumented reaction pile was driven approximately six feet (1.8 m) (center to center) behind each test pile. All test and reaction piles were 80 feet (24.4 m) long, and were installed using the same procedures as for the production piles.

The construction schedule and plan precluded placing the level backfill behind the rockfill (and reaction piles) before the pile lateral load tests were performed. Accordingly, the original plan to obtain data from the reaction pile for the level-surface condition for comparison to the production pile for sloping surface condition was not possible.

The actual cross-section at the test pile location is shown in Figure 1. The test pile setup is shown on Figure 4.

Instrumentation for each test and reaction pile included:

1. Two dial gauges with strokes of one and two inches (25 and 50 mm), and a precision of 0.001 inch (0.025 mm) to measure pile head deflections parallel to the axis of loading.

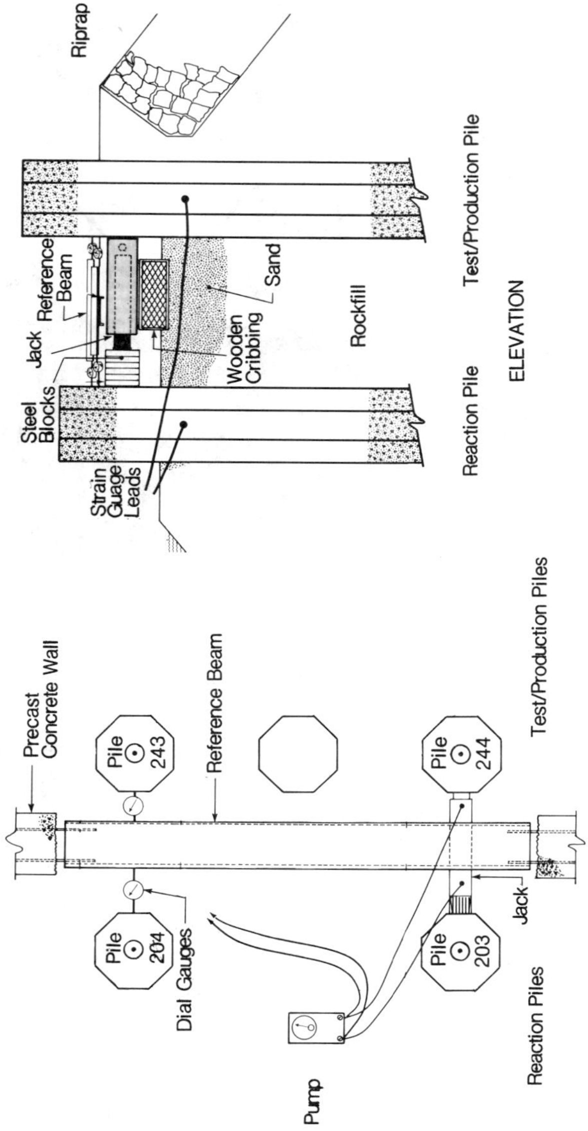

Figure 4 - Test Pile Setup

2. 2.75-inch (69.9 mm) diameter, self-aligning inclinometer casing located along the central axis of each pile. The casing extended from the pile butt to within ten feet (3.05 m) above the pile tip and was oriented to provide inclinometer readings parallel and perpendicular to the axis of loading.

3. Twenty strain gauges, ten attached to the reinforcing steel and ten embedded in concrete on both sides of the pile's neutral axis at 6, 12, 18, 24, and 48 feet (1.8, 3.7, 5.5, 7.3, and 14.6 m) from the pile butt.

After the piles were cast the strain gauge wire leads were cut by the contractor to ease placing the piles in the driving leads. This necessitated splicing all of the strain guage wire leads.

The piles were loaded by a hydraulic jack with a 20-inch (508 mm) stroke. Hydraulic pressure was supplied by an electric pump with a manual control valve. Pressure was measured by an in-line pressure gauge calibrated the week of the test.

Test Procedure

Initial dial gauge, strain gauge and inclinometer readings were made prior to the loading of the piles. Each pair of test piles was loaded by jacking between the test pile and the reaction pile in steps of 8, 16, 24, and 32 tons (71, 142, 214, 285 kN). After each load step, the piles were cyclically loaded in the same direction as the inital static load.

Each load step was applied at a rate of approximately two tons (18 kN) per minute. Deflections of the pile heads were recorded at one-half minute intervals during loading. Digital recordings of the strain gauge output were made at each load step. Each load step was maintained until the pile head movement stabilized, then dial gauge, strain gauge and inclinometer readings were obtained. The load then was relaxed at a rate of approximately two tons (18 kN) per minute. The pile head was allowed to rebound until the pile head movement stabilized, at which time final instrument readings were recorded.

The cyclic loads were then applied and varied from zero to the desired load step. Loads were applied and relieved at a rate of two tons (18 kN) per minute. Following load relaxation, the piles were allowed to stabilize for two minutes, at which time the dial gauge readings were recorded. These procedures were repeated 20 times. During the cyclic loading phase, the strain gauge and inclinometer readings were recorded at the start of each load step, at the 20th load cycle, and after relaxation of the 20th load cycle.

After cyclic loading at 32 tons (285 kN) , the piles were loaded at a rate of two tons (18 kN) per minute until the deflection limits of the jack were reached or failure of the test system occurred. The maximum lateral loads reached were 64 and 68 tons (570 and 606 kN) for the two test piles. Dial gauge readings were recorded at each two-ton load step; inclinometer and strain gauge data were read at each eight-ton load step.

Figure 5 illustrates head deflection versus lateral load for the range of loads applied.

ANALYSIS AND RESULTS

Pile Deflection

The stabilized pile head deflection for a specific test load, defined by the dial gauge data, was used to back-calculate a coefficient of linear variation of soil modulus with depth (k) using the procedures formulated by Reese and Matlock (6). Twelve k values, developed from analysis of the head deflections from 12 test loads, were used to calculate the piles' deflected shapes. These deflected shape plots were then compared with the inclinometer data to check the accuracy of the analytical procedure. The agreement was judged to be good.

It was noted the piles had deflected during installation and subsequent rock placement and settlement. Accordingly, the position and shape of the piles, as indicated by the dial guage and inclinomenter readings before the initial test loading, was taken as the "zero" position.

Pile Moments

The strain gauge data was used to calculate the pile moments induced by the lateral loads. The data analysis assumed an uncracked section using the formula:

$$M = \frac{E \epsilon I}{y} \qquad (2)$$

where: M = Pile moment
E = Pile modulus of elasticity
ϵ = Strain (measured by strain gauges)
I = Pile moment of inertia
y = Distance from neutral axis to strain gauge

While some of the strain gauge data produced reasonable results, most of the strain gauge data resulted in wide scatter and unrealistic values and therefore, was deemed inconclusive. Moments calculated from the most reasonable data are compared to moments calculated using the computer program COM624 (7) in Figure 6. The movement data shown is for a free-head condition and lateral loads of 8 and 24 tons. The unreliable strain gauge data was probably due to moisture infiltrating the field splices.

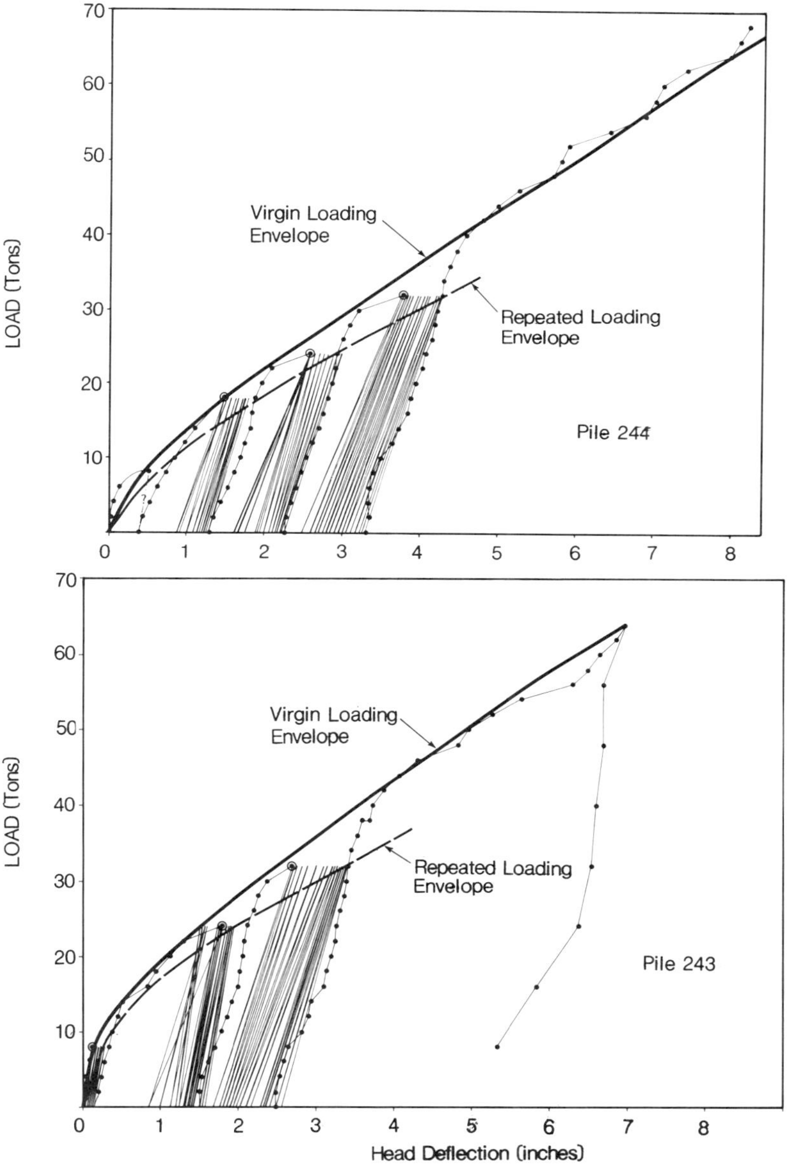

Figure 5 - Head Deflection vs. Lateral Load

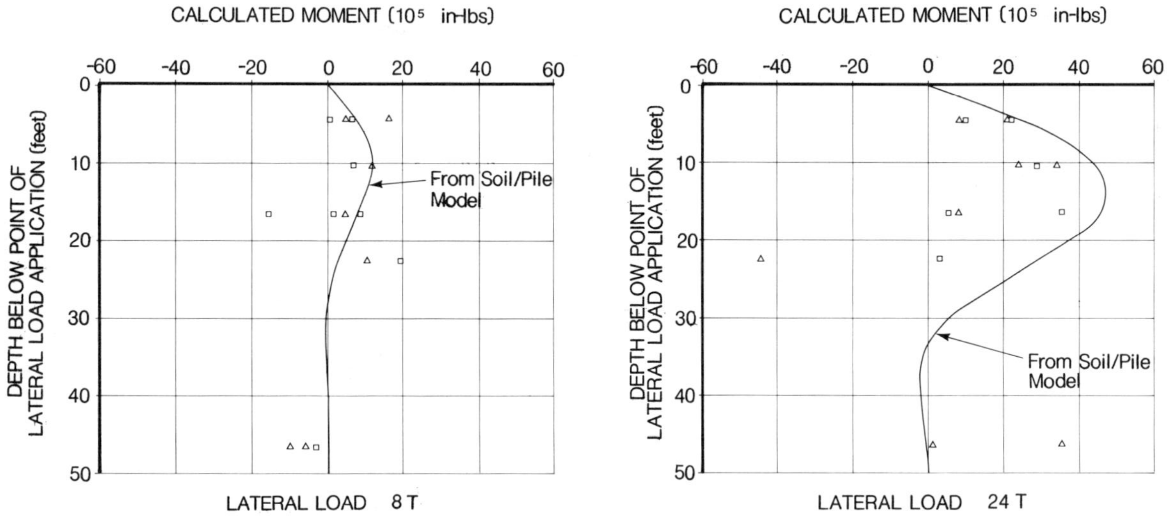

Notes:
1. △ Measured Moment, Pile 244 strain guage data
2. □ Measured Moment, Pile 243 strain guage data
3. The above moments assume an uncracked section, Free head, F row pile

Figure 6 - Comparison of Moments Calculated from Strain Guage Data to Moments Calculated Using COM624

Soil-Pile Model Development

The relationship between the stiffness coefficient k and lateral load was developed by evaluating the k values from the stabilized head deflections from each test load. The k coefficient was found to vary with the pile lateral load.

Soil resistance (p) and pile deflection (y) values due to the various lateral loads were calculated with depth using the k values corresponding to the lateral loads. This data was then plotted to formulate initial p-y curves.

Initial pile deflection, shear and moment data was calculated using the computer program COM624. Based on a comparison of this calculated data with the test data, the input p-y curves were modified and new deflection, shear and moment data were calculated. Subsequent iterations were performed until the calculated deflection data reasonably fit the test data. The final calculated head deflections are shown in comparison to the measured head deflections on Figure 7. The corresponding p-y curves are shown in Figure 8. In comparing the new p-y curves shown on Figure 8, to those originally used in the design, the new p-y curves were found to be "stiffer" near the surface and "softer" at greater depths. Figure 9 shows the comparison of pile deflections and moments calculated for design loading conditions on the basis of:

a) P-Y curves developed for cohesionless soil with a level surface.

b) P-Y curves developed from the pile lateral load test.

Lateral Resistance Correction Factor

The relationship between lateral resistance (p), calculated for level surface conditons, and p determined from the pile lateral load test, versus depth for different pile deflection (y) values is shown as a lateral resistance correction factor on Figure 10.

As would be expected, the lateral resistance correction factor approaches unity at depths greater than 20 feet (6m). That is, the effect of the slope on the lateral resistance of the pile diminishes as the horizontal distance to the slope increases. By comparing the deflections and moments, calculated with and without the correction factor, it was found that at depths of 30 to 40 feet (9.2 to 12.2 m) the slope has no effect. Thus the correction factor is shown to approach unity at 40 feet (12.2 m). Similarly it would be expected that the correction factor would be very small near the surface; however, it was found to approach or even exceed unity at the surface. It is believed this can be explained as follows. First, the test piles were set back three to five feet (0.9 to 1.5 m) from the top of the slope, thus the lateral resistance at shallow depth was not affected by the slope. Secondly, the unconfined rockfill at the

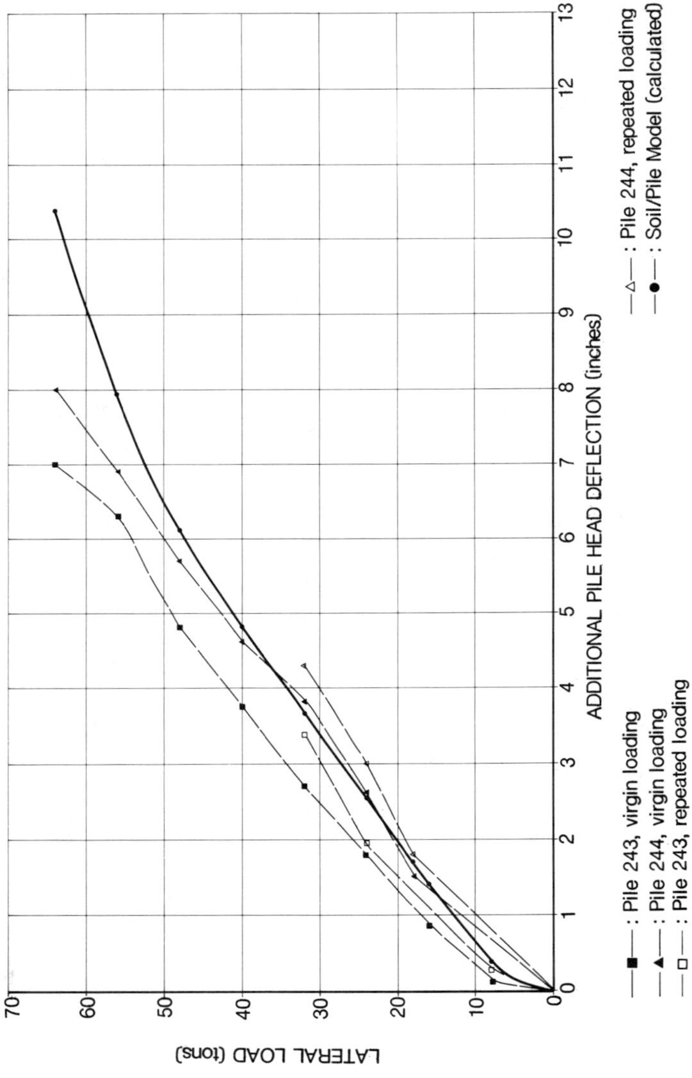

Figure 7 - Measured and Calculated Head Deflection

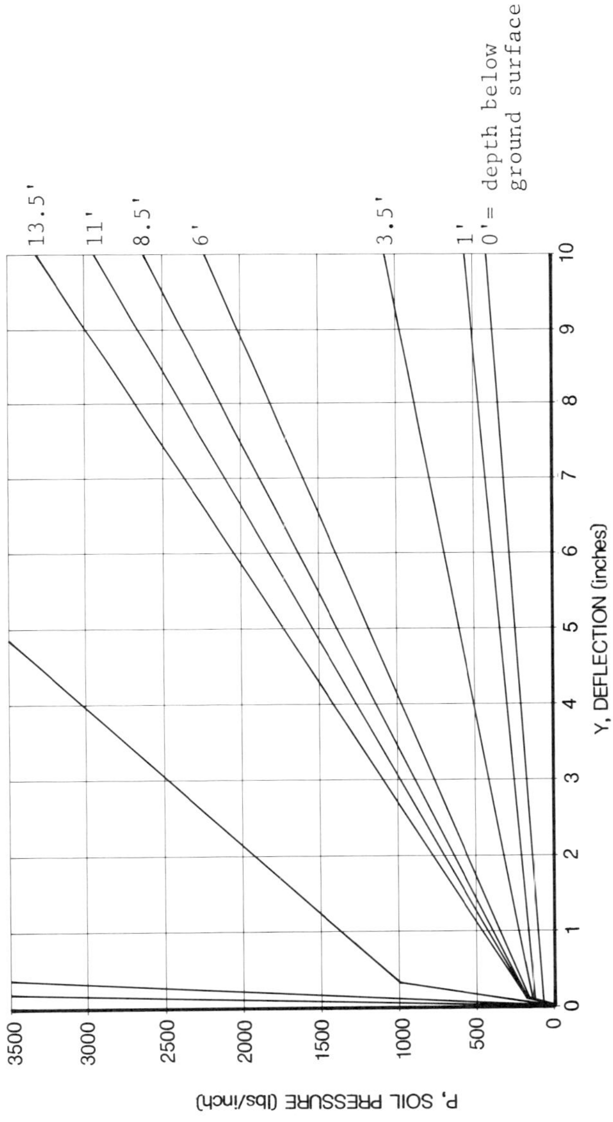

Note: Based on 24" prestressed reinforced concrete piles, tip elevation - -70 feet, MMLW

Figure 8 - Soil/Pile Model p-y Curves, Sloping Rockfill

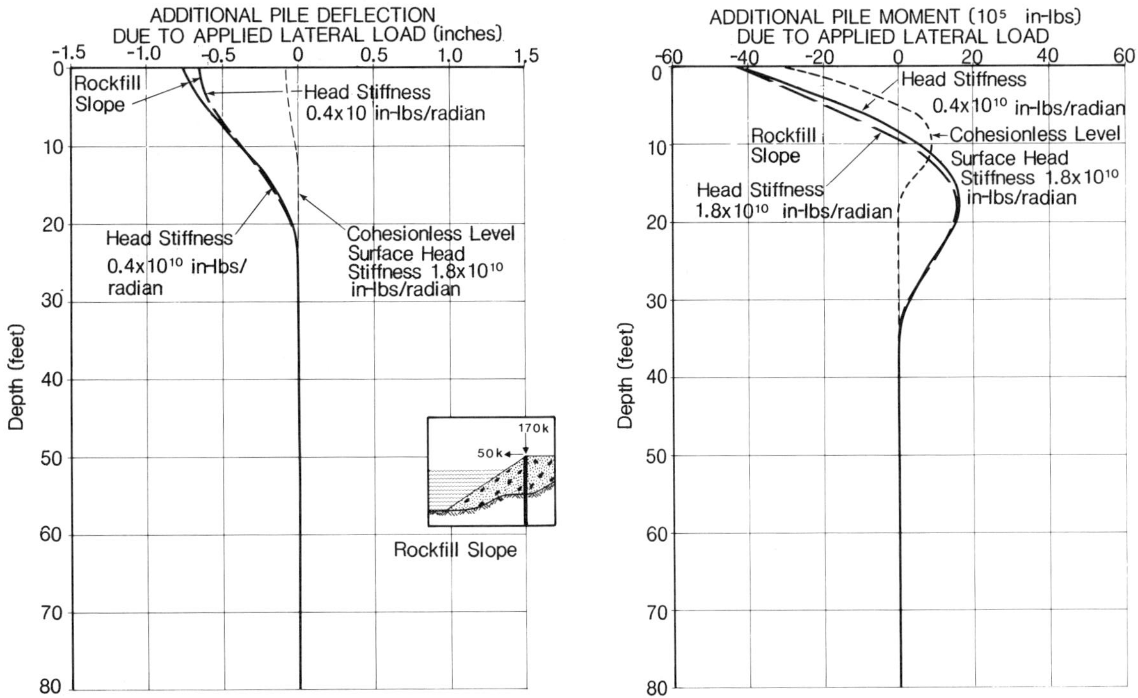

Notes: 1. The above values are applicable for semi-fixed head, F row piles subjected to loads indicated above

2. The deflection and moments above are due to the applied 50^k lateral load and are in addition to any moment and of the lateral load

3. 24 Octagonal Prestressed Concrete Pile I=18.180 in^4, E=4x106 PSI, A=477 in^2.

FIGURE 9 - Calculated Pile Deflections and Moments under Design Conditions surface provides more resistance to lateral load than sand as it would be necessary to "lift" the surface rocks over one another for lateral movement to occur; however, the unconfined sand, on which the comparison model was based, would offer very little lateral resistance at the surface.

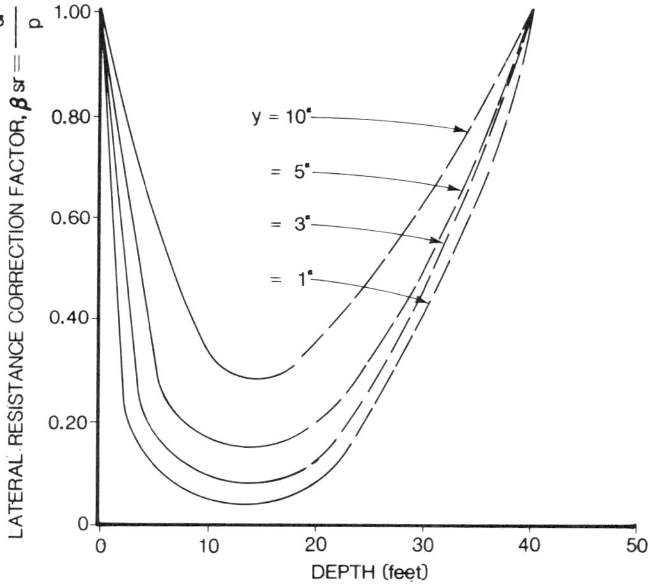

$$\beta_{sr} = \frac{P_{sr}}{p}$$

Psr = Lateral resistance for sloping rockfill for test conditions

p Soil lateral resistance for cohesionless soils on level surface after Reese, Cox, Koop (1974)

Figure 10 - Lateral Resistannce Correction Factors

CONCLUSIONS

The lateral load tests provided field data for the behavior of laterally loaded piles embedded in sloping rockfill. The lateral load tests confirm that vertical prestressed piles embedded in a sloping rock fill can be used to resist lateral loads for the conditions described. Figure 10 indicates correction factors for lateral resistance that can be considered for future projects with similar design and construction conditions including pile type and size, rock quality, placement, and slope.

As noted, the piles were deflected due to installation, rock placement and rock movement prior to the lateral load testing. The deflections and moments presented in this paper are based on the pre-test position of the piles being the "zero" position. Accordingly, an allowance for the effects of construction deflection and the resulting induced moments should be added to the calculated deflections and moments.

ACKNOWLEDGEMENTS

The pile lateral load test was funded as part of the project costs by the owner, the Port of Los Angeles (POLA), who granted permission to publish these results. The project was designed by Daniel, Mann, Johnson and Mendenhall (DMJM) who provided design details and project photographs. Guy F. Atkinson Construction Company was the construction contractor and installed the test piles and provided the test setup.

The writers wish to thank Mr. Edward Gorman, POLA Chief Harbor Engineer and his staff for their help in making the test possible and their assistance during the course of the work; Mr. Dick Bayles, Guy F. Atkinson Project Manager and his crews for their cooperation during the test; and Mr. John Warwar, DMJM Project Manager, and Dr. Donald W. Quigley, Harding Lawson Associates Principal Engineer, for their helpful comments.

REFERENCES

1. Daniel, Mann, Johnson and Mendenhall,: "State of the Art Report, Containership Berthing Facilities" for Los Angeles Harbor Department, April 27, 1981

2. Leps, Thomas M.: "Review of Shearing Strength of Rockfill," Journal of The Soil Mechanics and Foundation Division, ASCE, Vol. 96, No. SM4, July 197, pp 1159

3. Marachi, N. Dean, Chan, Clarence K., Seed, H. Bolton: "Evaluation of Properties of Rockfill Materials," Journal of the Soil Mechanics and Foundation Division, ASCE, Vol. 98, No. SM1, January 1972, pp 95

4. Reese, L.C.; Cox, W.R.; Koop, F.D.: "Analysis of Laterally Loaded Piles in Sand", Offshore Technology Conference, OTC 2080, 1974

5. Reese, L.C.: "Laterally Loaded Piles: Program Documentation" Journal of Soil Mechanics and Foundations Division, ASCE Vol. 103, No. GT4, April 1977, pp L87

6. Reese, L.C.; Matlock, H.: "Non-dimensional Solutions for Laterally Loaded Piles with Soil Modulus Assumed Proportional to Depth", Proceedings: Eighth Texas Conference on Soil Mechanics and Foundation Engineering, Special Publication No. 29, Bureau of Engineering Research, The University of Texas, Austin, 1956

7. Reese, L.C.; Sullivan, W.R.: "Documentation of Computer Program COM624, Analysis of Stress and Deflections for Laterally Loaded Piles Including Generation of p-y Curves", Geotechnical Engineering Center, Department of Civil Engineering, The University of Texas, Austin, 1980

A FLAT-DILATOMETER STUDY OF LATERAL SOIL RESPONSE

E.Sabri Motan[1],A.M.ASCE, and Muhammed A. Gabr[2]

Abstract

The effect of flat dilatometer penetration on the lateral stress conditions in soil around the dilatometer blade, and the soil response against dilatometer diaphragm deflection were investigated in a laboratory study. A specially constructed aluminum block, containing a dilatometer diaphragm and a strain-gage-equipped leaf cantilever to continuously record lateral movement of the diaphragm, was used during the tests. The vertically positioned block was laterally forced inside sand specimens prepared at different relative densities to simulate the lateral separation undergone by the soil during dilatometer penetration. Data obtained reflected the change in lateral in-situ stresses in response to penetration. The cantilever beam deflection sensor made it possible to obtain the non-linear lateral pressure-deflection behavior of soils for realistic analyses of pile deflections under lateral loading. Also, the effect of repetitively expanding the diaphragm on the pressure-deflection curve was considered. Experience gained in this study is expected to lead to improvements in dilatometer design and confidence in extrapolating the dilatometer data to the design and analysis of laterally loaded piles.

Introduction

A variety of penetration devices developed primarily as a result of a need to profile the subsoil conditions more accurately than conventional methods of drilling and intermittent sampling have, in general, met certain amount of success in achieving the purpose. One such piece of equipment, the flat dilatometer, was first introduced in the mid 1970's (10). It represents an attempt to obtain, in addition to the soil resistance during penetration, data related to the soil deformation characteristics with the

---

[1]Asst. Prof. of Civil and Env. Eng., Clarkson University, Potsdam, New York.

[2]Graduate Assistant, Clarkson University, Potsdam, New York.

help of an expandable, circular steel diaphragm attached on a 14 mm (0.550 in)-thick stainless steel blade. The blade is jacked into the soil using a penetrometer rig. A nylon tube that runs through the rods connects the dilatometer control unit with a chamber behind the diaphragm. The nylon tube contains an electrical wire. During the penetration, the diaphragm is kept flush with the surface of the blade by the lateral soil pressure acting on it. As soon as the desired depth is reached, the diaphragm is inflated by means of pressurized gas. The first reading taken is the pressure corresponding to the moment at which the outward movement of the diaphragm is initiated. This moment is indicated by the silencing of the buzzer on the control unit. The second pressure reading is taken when the diaphragm center has deflected by one millimeter (0.039 in) at which moment the buzzer is activated again. The difference between the two pressure readings -which must be corrected for diaphragm stiffness- and the corrected first reading are used to compute soil index parameters which correlate with in-situ soil type and characteristics. Previous applications of the flat dilatometer include profiling of the subsoil conditions and estimation of a number of soil parameters such as at--rest lateral earth pressure, overconsolidation ratio, coefficient of volume compressibility, along with liquefaction potential assessments in saturated sands (9,11,12). By virtue of its construction, the flat dilatometer is capable of obtaining lateral soil stiffness data in a nearly continuous manner in both cohesive and cohesionless soils, and thereby can be employed as an alternative to the methods currently used in assessing the soil response during lateral loading of pile foundations. Differences in the soil strains resulting from dilatometer penetration and pile driving, however, present difficulties in extrapolating the dilatometer data for use in the analysis of laterally loaded pile behavior. This study was aimed at i. investigating the lateral pressure buildup in the immediate vicinity of the dilatometer during penetration, and ii. obtaining the nonlinear soil response in the form of a continuous pressure-deflection curve during the inflation of the dilatometer diaphragm. The results presented are of preliminary nature. However, the data obtained - which should be substantiated by calibration studies on laterally-loaded full-scale piles - definitely indicate a potential for effective use of the flat dilatometer in securing field data for lateral pile analyses.

Lateral Soil Stiffness

Among a number of techniques and equipment available for the evaluation of the lateral soil stiffness are, in addition to the flat dilatometer, full-scale lateral pile loading tests (8,16,17,18,19), plate-loading tests (2,21), triaxial testing (7,22,23), consolidation test (5), pressuremeter (1,14), the Iowa stepped blade (6), standard penetration test (3,15,24), and empirical correlations with other soil properties (2,4,20). Because of the significant differences between the deformation modes imposed on soil during the tests and those which occur as a result of pile installation and subsequent lateral loading, none of the techniques listed above - with the exception of full-scale pile load testing - worked satisfactorily under all circumstances. Also, as is often the case, discontinuous profiling of the soil, as related to the lateral response parameters, increases the margin of uncertainty in the analysis. Therefore, short of conducting in-place full-scale lateral loading tests on prototype piles, much remains to the judgment of the engineer to extrapolate the field data - which is very often obtained in the form of standard or cone penetration test results - far enough to make reasonable estimates regarding the lateral soil stiffness parameters.

Recently, pressuremeter testing has gained popularity as a relatively accurate piece of equipment in evaluating the lateral soil stiffness. Pressuremeter test results can be readily applied to the analysis of piles cast in-place with no lateral displacement of the surrounding soil. However, in the case of driven piles the margin of error in the analysis may be significant.

The primary drawback of the flat dilatometer is that the results obtained yield soil stiffness values at an intermediate value of lateral soil disturbance, leaving one with the task of extrapolating the data backward to zero lateral strain condition to obtain the undisturbed soil stiffness and, forward - within reason - to estimate the pressure-displacement relations for driven piles of substantially larger cross sections than the dilatometer blade.

Based on the assumption of linear elasticity, Marchetti (11) suggested the use of

$$\frac{E}{1-\mu^2} = \frac{2D\Delta p}{\pi S_o} \qquad (1)$$

as an attempt to obtain the dilatometer modulus, $E/(1-\mu^2)$. In Equation 1, S_o is the deflection of the diaphragm center and Δp is the difference between two pressure readings corrected for the diaphragm rigidity: first reading taken immediately after the outward movement of the diaphragm initiated (p_0), and the second at 1 mm lateral deflection of the diaphragm center (p_1). D and μ are the diaphragm diameter and Poisson's ratio of soil, respectively.

Considering the generally non-linear stress-strain

response of soils, the modulus value calculated with the formula above is actually a secant modulus representing the soil stiffness in the immediate vicinity of the blade as the diaphragm is inflated.

Suitably placed displacement sensors inside the blade can be made use of in recording the displacement of the diaphragm center as a function of the inflation pressure. However, there is no assurance that this non-linear pressure-deflection curve will be applicable if the soil is penetrated by a driven pile of cross- section dimensions substantially larger than those of the dilatometer blade.

The coefficient of subgrade reaction has been shown to be a function of the dimensions of the loaded area, and presently this is the weak point of the lateral subgrade reaction approach to the analysis of laterally loaded piles. A subgrade modulus value obtained by the dilatometer should be modified using an appropriate factor in order to account for the size effect. The extent of the size effect is illustrated with an example later in the paper.

Ideally, a field calibration test can be conducted on a driven pile, instrumented with a number of dilatometer plates or stress cells mounted flush with the surface (13). Another individual dilatometer can be used to profile the soil in the vicinity of the pile and to establish correspondence between the lateral pile response obtained. This technique, however, by most standards is time-consuming and is relatively expensive to implement as a routine field test.

Experimental Study

A laboratory study was carried out under controlled conditions concentrating primarily on:

a. The shape of the pressure-deflection curve of the dilatometer diaphragm when inflated against soil, and
b. The effect of the dilatometer penetration on the lateral earth pressure conditions in the soil in contact with the blade.

The present design employed in the construction of the flat dilatometer yields two readings as described earlier. As to the shape of the pressure-deflection curve between zero- and one-millimeter diaphragm center deflections, Marchetti (11) makes the linearity assumption and calculates the dilatometer modulus by Equation 1 given previously. However, with a standard dilatometer, repetitive pressuring of the diaphragm yields gradually a decreasing p_1, indicative of inelastic behavior. The nature of the pressure-deflection curve for repeated inflation of the diaphragm will be considered later.

In order to characterize the nonlinearity in pressure-deflection behavior, a leaf-type cantilever beam deflection sensor, instrumented with a half-bridge strain gage arrangement, was used behind the dilatometer diaphragm. The cantilever deflection sensor had the maximum tip travel of approximately 1.13 mm (0.044 in.) and no significant creep was observed in deflection readings during the course of the experimental program. Both the cantilever beam and the diaphragm were mounted on an aluminum block, 150 mm (6 in.) long, 100 mm (4 in.) wide, and 25 mm (1 in.) thick. Experimental work was carried out in a steel bin filled with a sand compacted to relative densities of 15, 30, and 45 percent (Figure 1). A 25 millimeter (1 in.) diameter pipe section was attached to the back-side of the block and was extended out through a hole on the short side of the bin. Through the pipe, the dilatometer pressure line and the deflection sensor wires were taken out.

During the tests, to simulate increasing lateral separation of soil as a result of dilatometer penetration, the block was laterally pressed against the soil by a trailer jack mounted rigidly on a steel frame which, in turn, was welded to the short side of the bin. Overburden stress was simulated by applying a vertical force through a hydraulic jack on a rigid steel plate placed on the sand surface. The vertical distance between the pressure plate and the dilatometer diaphragms was assumed to be small enough so that the effects of side friction on the vertical stress at diaphragm level can be neglected. The pressure plate was purposely not extended over the aluminum block in order to avoid damaging the block as the vertical pressure is applied. The vertical force was measured by a load cell mounted between the steel plate and the hydraulic jack. Before the start of each test, overburden load was applied, and through a hole cut in the steel plate a standard dilatometer was penetrated to the same depth as the aluminum block, diaphragms facing each other. The minimum and maximum pressure readings were taken by the standard dilatometer after seven millimeters (0.275 in.) of lateral movement of the aluminum block. The inflation pressure was measured by a pressure transducer mounted on the dilatometer control box at a distance of 750 millimeters (30 in.) from the dilatometer block along the pressure line.

Initially, the pressure versus diaphragm center deflection curves were taken at 0, 2, 4, and 7-mm lateral movement of the aluminum block. One such group of curves (a) is given in Figure 2 which was redrawn by tracing over experimental curves. However, it was later discovered that inflating the diaphragm at 2- and 4- millimeter lateral penetration substantially decreased p_o taken at 7-millimeter penetration. Therefore, a new series of experiments was

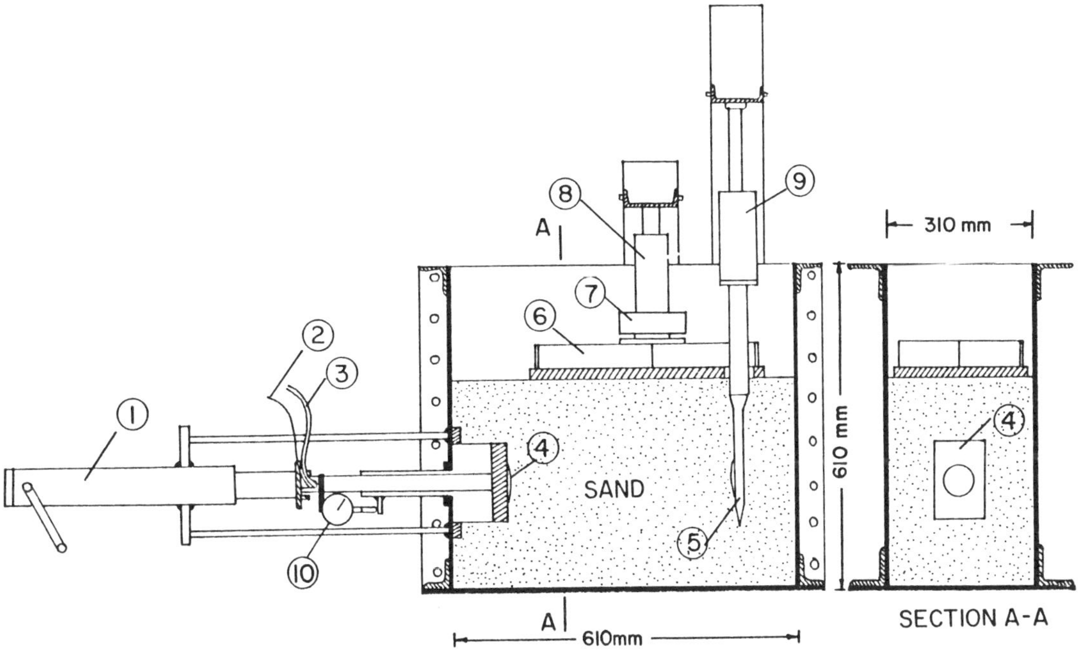

Figure 1. Experimental Setup (1. Trailer Jack, 2. Deflection Sensor Leads, 3. Pressure Line, 4. Aluminum Block, 5. Flat Dilatometer, 6. Pressure Plate, 7. Load Cell, 8. and 9. Hydraulic Jacks, 10. Dial Gage)

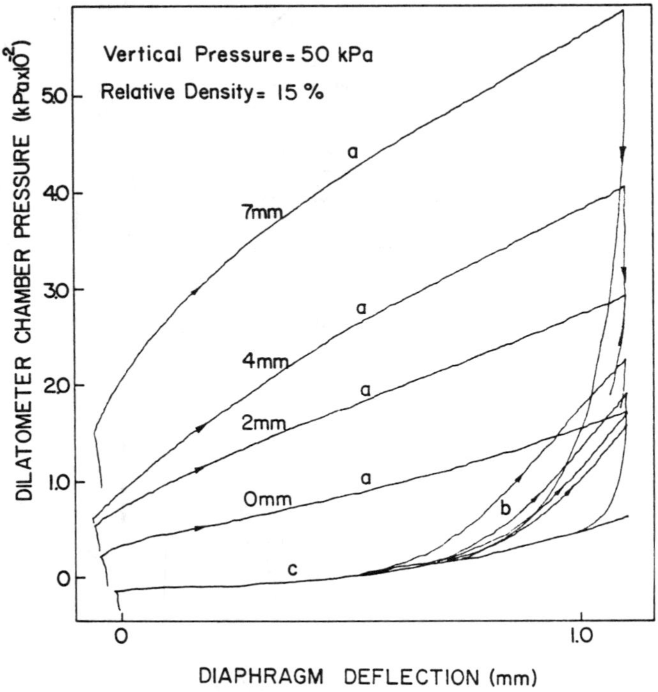

Figure 2. Dilatometer Chamber Pressure-Diaphragm Center Deflection Curves (a. Initial Inflation Curves b. Repeated Pressurizing Curves, c. Diaphragm Calibration Curve)

performed in which the inflation curves were obtained at zero- and 7-millimeter lateral penetration values only. Repetitive pressuring of the diaphragm - five times - was performed in the majority of tests at 7-millimeter lateral movement of the block (curve b in Figure 2). The calibration curve of the diaphragm was taken by inflating it against atmospheric pressure (curve c in Figure 2). The net pressure-diaphragm deflection curves can be obtained subsequently with reference to the calibration curve.

Experimental Results and Discussion

Pressure-Diaphragm Deflection Relations. An increasingly nonlinear behavior of soil was observed in curves relating the diaphragm inflation pressure to the diaphragm center displacement as the aluminum block was jacked laterally into the soil. Curve family (a) in Figure 2 illustrates this behavior. At all three initial relative densities, however, within zero to one-millimeter deflection range of the diaphragm center, no significant nonlinearity was observed in the pressure-deflection curves taken prior to the lateral movement of the aluminum block. This stands to indicate that, if the diaphragm inflation was started at at-rest earth pressure condition, the deflection range of the diaphragm would be inadequate to detect the nonlinearity in the soil stiffness which would definitely occur at larger deflections. If, however, the point of interest is specifically the soil stiffness for the small strain lateral response analysis of cast-in-place piles, a relationship between the soil modulus obtained by a standard dilatometer and the modulus corresponding to lateral loading beginning with at-rest earth pressure conditions will be convenient. One such relation as presented in Figure 3, indicates linear dependence between the two modulus values, which is relatively independent of the relative density and overburden stress conditions, although some scatter is present. The slope of the line could very well be a function of the soil type, more specifically the angle of internal friction. A relation of this type will be useful in estimating the Young's Modulus of the soil for possible use in formulas relating the modulus of elasticity of the soil to the lateral coefficient of subgrade reaction (22,23). Also, although not directly related to this study, for seismic analyses a plot of this nature is expected to yield a good indication of the in-situ shear modulus of the soil at small strains if a reasonable assumption with respect to the poisson's ratio is made. If the shear modulus at still smaller strains is needed, then one would have to make use of the continuous pressure-diaphragm deflection curve obtained by the cantilever deflection sensor.

For the majority of the experiments, the secant value of the dilatometer modulus calculated from the pressure-deflection curve using Equation 1 yielded reasonable average values for the modulus within zero to one-millimeter range of lateral soil displacements. However, the tests also indicated that the initial tangent dilatometer modulus can be as much as fifty percent higher than the secant modulus, pointing out the possibility of substantially lower actual modulus value if the inflation curve is extrapolated significantly beyond one-millimeter point, or substantially greater modulus at lateral displacements less than one millimeter.

During the experiments, p_o and p_1 values obtained with a standard dilatometer were found to be consistently less than the corresponding values obtained after laterally pressing the aluminum block toward the soil by seven millimeters. Figure 4 illustrates the difference for 30 kPa vertical pressure at initial relative densities of 15, 30, and 45 percent. This is interpreted as indicative of the effect of vertical shear deformations accompanying lateral separation of soil during the penetration of the standard dilatometer whereas no such action took place as the aluminum block was pressed laterally. Table 1 presents these readings for the tests conducted for comparison purposes.

Prior to the use of the dilatometer block in the experiments, based on the previous field experience with the standard dilatometer, the authors were under the impression that, since the pressure to reach one millimeter deflection decreases upon succesive cycling, degradation should occur in the soil. Continuous measurement of pressure and diaphragm deflection, however, indicated increased soil stiffness for repeated cycling of pressure, although beyond the second cycling of the pressure the increase observed in the stiffness was practically insignificant (curve group b in Fig. 2). Residual deformation continued to build up with continued cycling of pressure. The initial upward curvature of the cycling curves is thought to be due to receding of the diaphragm by a small amount from the soil following the depressuring of the chamber. Upon repressuring, once the contact is established, the curves became significantly steeper.

At this point, it is believed that the cycling curves can be used in estimating the buildup of irrecoverable deformations in soil surrounding piles laterally loaded in a repetitive manner. Further research is planned to investigate the potential.

<u>Pressure Buildup with Lateral Soil Separation</u>. The lateral pressure necessary to start the outward movement of the diaphragm on the block (p_o), plotted as a function of the lateral movement of the block yielded approximately a straight line for all three relative densities and vertical pressures. Figure 5 illustrates two such curves. The

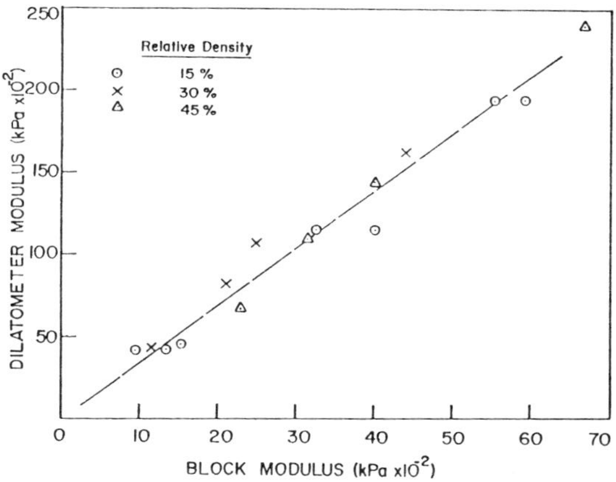

Figure 3. Standard Dilatometer Modulus versus At-Rest Condition Block Dilatometer Modulus

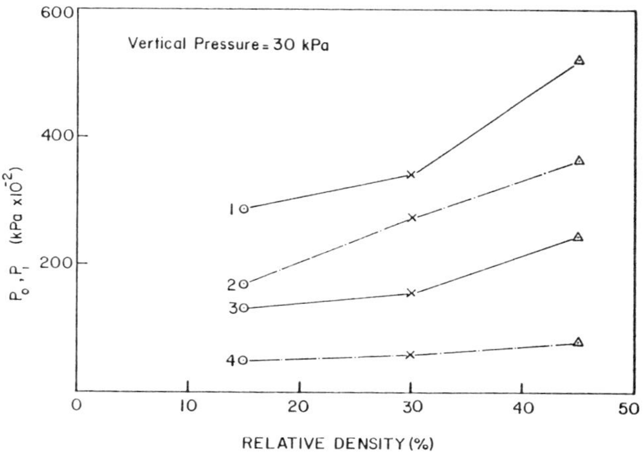

Figure 4. p_o and p_1 Values Obtained by Standard Dilatometer and the Aluminum Block as a Function of the Initial Relative Density (Lines 1 and 3 Aluminum Block; Lines 2 and 4 Standard Dilatometer)

Table 1. Dilatometer Block and Standard Dilatometer Diaphragm Inflation Data

Initial Relative Density (%)	Vertical Pressure (kPa)*	Dilatometer Block (At-Rest)		Dilatometer Block (7-mm)		Standard Dilatometer	
		p_0 (kPa)	p_1 (kPa)	p_0 (kPa)	p_1 (kPa)	p_0 (kPa)	p_1 (kPa)
15	12.5	10	35	80	190	38	147
	30	10	50	130	285	48	167
	50	75	180	350	610	93	393
	100	110	255	460	820	133	638
30	12.5	20	50	100	265	38	152
	27	30	85	155	340	58	273
	50	45	110	235	475	78	358
	100	80	195	395	695	148	572
45	12.5	5	65	145	375	68	243
	30	27.5	110	245	520	78	363
	50	45	150	300	625	123	498
	100	85	260	445	1050	218	843

*1 kPa = 0.145 lbs/in^2

slope of a continuous curve of p_o versus the lateral movement of the block should indicate the coefficient of subgrade reaction of the soil for an object separating the soil laterally; in this specific case, the flat dilatometer. Although the flat dilatometer is not capable of obtaining such relationships in field applications, it would be enlightening to see if the diaphragm pressurization data taken after seven millimeters (half-thickness of the dilatometer blade) of lateral penetration might be of any use as to the value of the slope of this curve. For this, consider the curves in Figure 5. With linearity assumption between zero- and seven-millimeter lateral separation values, the lateral coefficient of subgrade reaction would be

$$k_h = \frac{p_{o(7)} - p_{o(0)}}{d}$$

where d half-thickness of the dilatometer blade.

The numerical data for the upper curve yielded

$$k_h = \frac{245-27.5}{0.007} = 31{,}070 \text{ kN/m}^3 \ (114.3 \text{ lb/in}^3)$$

With the block width of 100 mm (4 in.), the subgrade reaction modulus becomes

$$K = 3{,}070 \times 0.1 = 3107 \text{ kN/m}^2 \ (450 \text{ lb/in}^2)$$

Similarly, a "Dilatometer Subgrade Reaction Coefficient" was defined by rearranging Equation 1 in the form:

$$\frac{\Delta p}{S_o} = \frac{\pi}{2D} \cdot \frac{E}{1-\mu^2} \qquad (2)$$

Both Δp and S_o are measured quantities during a dilatometer inflation test, and the ratio $\Delta p/S_o$ can be calculated without the knowledge of the terms on the right-hand side. The data obtained by a standard dilatometer for this particular test yielded approximately 285,000 kN/m^3 (1046 lb/in^3) for the dilatometer subgrade reaction coefficient. A comparison indicates that this value is an order of magnitude greater than the lateral coefficient of subgrade reaction obtained previously. One possible way of interpreting the difference is that, as the block penetrates laterally, yielding of the soil in the vicinity of the edges is taking place, whereas when the diaphragm is inflated, it is being forced toward the part of the soil which is gradually being densified by the lateral movement of the block without undergoing yielding action. Therefore, the subgrade reaction modulus obtained by inflating the diaphragm will be greater than the block subgrade reaction modulus. The difference is also believed to be partly a typical display of the dependence of the subgrade reaction on the contact area and the difference in the modes of deformation imposed by the blade in the form of lateral soil separation, and by the diaphragm as essentially axisymmetric loading of a thin plate. One can also compute, with linearity assumption, the lateral subgrade reaction modulus using p_o obtained by a standard dilatometer. For the same test considered above the standard dilatometer data yielded

$$k_h = \frac{78}{0.007} = 11,142 \text{ lb/in}^3 \text{ (41.0 lb/in}^3\text{)}$$

This value, also, is substantially lower than the dilatometer subgrade reaction coefficient calculated on the basis of the diaphragm inflation data.

Because of the disturbance in soil created upon penetrating the dilatometer, the average coefficient of subgrade reaction obtained by the dilatometer will be different from that of the block as illustrated above. Considering that the lateral subgrade reaction obtained by the block will be more representative as a design value in the case of cast-in-place piles, it is interesting to note that these two values, when plotted as a function of each other, yield approximately a straight line with evidently little dependence on the relative density (Figure 6). The slope of this line is defined herein as the "Disturbance Index" and represents the correction factor by which the standard dilatometer p_o value should be divided to obtain disturbance-free minimum pressure reading, and subsequently, the average lateral subgrade reaction coefficient. At this point, it is not known whether the relationship is significantly dependent upon the type of sand.

It is the opinion of the authors that the lateral subgrade reaction modulus calculated on the basis of p_o and 7-mm lateral separation, can be used in estimating the horizontal subgrade reaction coefficient for piles of different sizes. A case study on a 1370 mm (4.5 ft)-diameter and 4500 mm (15 ft)-long pier foundation yielded good agreement between the measured lateral pier-top displacement and the computed value obtained by using the subgrade reaction coefficient corrected according to Figure 6. In the analysis, the slope of the initial straight line portion of p-y curves (17) was computed on the basis of the corrected sugrade reaction coefficient. An extensive field calibration study, however, seems absolutely necessary in order to

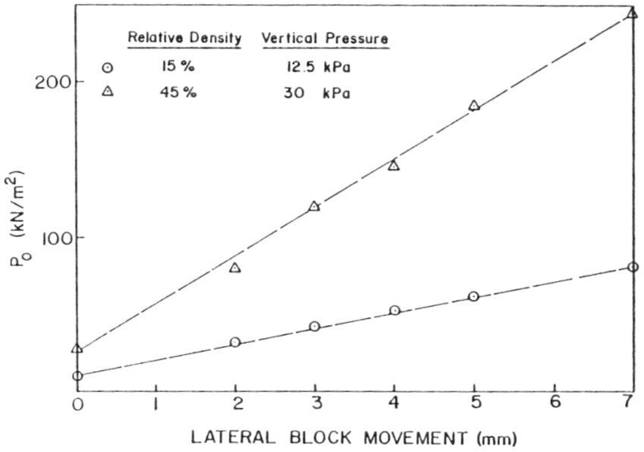

Figure 5. p_o versus Lateral Block Movement Curves Obtained by the Aluminum Block

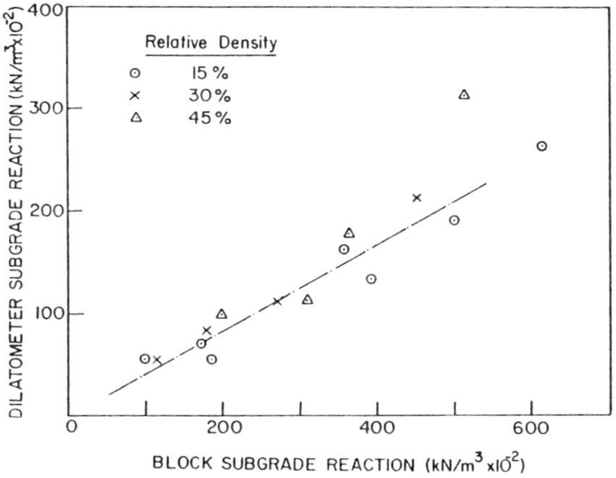

Figure 6. Relationship Between Average Values of Dilatometer and Block-Subgrade-Reaction Calculated over a Separation Distance of Seven Millimeters.

evaluate the potential of the flat dilatometer in securing data for use in pile analyses under lateral loading. In the case of piles constructed in sand, a sample of the soil can be tested in the lab using a setup similar to that in Figure 1 to obtain the value of the disturbance index for subsequent analyses.

The test results presented indicate that the direct use of the dilatometer subgrade reaction coefficient (as defined in Equation 2) in analyses will probably result in a substantial overestimation of the actual soil stiffness for the pile dimensions considered. It should, however, also be stressed that the laboratory study was conducted on sand. The clay behavior can be significantly different in this respect and may require a different scheme for the evaluation of the disturbance index as defined above.

Conclusion

As related to the usefulness of the dilatometer data in estimating the lateral subgrade coefficient of soils for lateral loading, the potential certainly exists. The results of this study point out a viable procedure for obtaining representative values of soil stiffness for piles constructed in-place with little disturbance to the soil in the immediate vicinity. However, the more general problem of relating the data to the actual soil stiffness as a function of the pile cross-section and the method of pile installation is presently the primary obstacle that has to be removed through further research if the flat dilatometer is to be a viable field instrument in this application category.

The limitations of the laboratory test setup in duplicating the field at-rest pressure conditions and the mode of deformations resulting from dilatometer penetration are recognized in this study. Refinements, however, are being made to improve the quality of the data substantially.

The results obtained, although preliminary in nature, point out some of the factors that have to be considered and the questions to be answered in order to realize the full potential of the flat dilatometer.

References

1. Baguelin, F. and Jezequel, J.F., (1975), "Further Insights on the Self-Boring Technique Developed in France," Proceedings ASCE Specialty Conference on In-situ Measurement of Soil Properties, Raleigh, N.C.

2. Broms, B.B., (1964), "Lateral Resistance of Piles in Cohesive Soils," Journal of the Soil Mechanics and Foundations Division, ASCE, Vol. 90, SM2.

3. D'Appolonia, D.J., D'Appolonia, E., and Brisette, R.F., (1970), "Discussion of Settlement of Spread Footings on Sand," Journal of the Soil Mechanics and Foundations Division, ASCE, V. 96, SM2.

4. Davisson, M.T., (1970), "Lateral Load Capacity of Piles," Highway Research Record, No. 333.

5. Francis, A.J., (1964), "Analysis of Pile Groups with Flexural Resistance," Journal of the Soil Mechanics and Foundations Division, ASCE, V. 90, SM3.

6. Handy, R.L., Remmes, B., Moldt, S., Lutenegger, A.J., and Trott, G., (1982), "In Situ Stress Determination by Iowa Stepped Blade," Journal of the Geotechnical Engineering Division, ASCE, V. 108, GT11.

7. Hughes, J.M.D. and Goldsmith, P.R., (1977), "Stress Path Interpretation of Triaxial Data," New Zealand Geomechanics News, No. 14.

8. Hughes, J.M.O., Goldsmith, P.R., and Fendall, H.D.W., (1979), "Full Scale Laterally Loaded Pile Test at the Westgate Freeway Site, Melbourne, Australia: Load Deflection Predictions and Field Results," Report No. 190, Department of Civil Engineering, University of Auckland, Auckland, New Zealand.

9. Lacasse, S. and Lunne, T., (1982), "Penetration Tests in Two Norwegian Clays," Proceedings of the Second European Symposium on Penetration Testing, V. 2.

10. Marchetti, S., (1975), "A New In-situ Test for the Measurement of Horizontal Soil Deformability," Proceedings, Conference on In Situ Measurement of Soil Properties, ASCE, V. 2,.

11. Marchetti, S., (1980), In Situ Tests by Flat Dilatometer, Journal of Geotechnical Engineering Division, ASCE, V. 106, GT3.

12. Marchetti, S., (1982), "Detection of Liquefiable Sand Layers by Means of Quasi-Static Penetration Tests," Proceedings of the Second European Symposium on Penetration Testing, Amsterdam, The Netherlands, Vol. 2.

13. Marchetti, S., (1982), Personal Communication.

14. Menard, L., (1965), "Rules for the Calculation of Bearing Capacity and Foundation Settlement Based on Pressuremeter Tests," Proc. 6th International Conference on Soil Mechanics and Foundation Engineering, Vol. 2.

15. Parry, R.H.G., (1971), "A Direct Method of Estimating Settlements in Sands from SPT values," Proc. Symposium on Interaction of Structures and Foundations, Midlands Soil Mechanics and Foundation Engineering Society, Birmingham, U.K.

16. Poulos, H.G., (1971), "Behavior of Laterally Loaded Piles: II-Pile Groups," Journal of the Soil Mechanics and Foundations Division, ASCE Vol. 97, SM5.

17. Reese, L.C., Cox, W.R., and Koop, F.D., (1974), "Analysis of Laterally Loaded Piles in Sand," Paper #UTC 2080, Presented at the 1974 Fifth Annual Offshore Technology Conference, Houston, Texas.

18. Reese, L.C. and Welch, R.C., (1975), "Lateral Loading of Deep Foundations in Stiff Clay," Journal of the Geotechnical Engineering Division, ASCE, V. 101, No. GT7.

19. Reese, L.C., Cox, W.R., and Koop, F.D., (1975), "Field Testing and Analysis of Laterally Loaded Piles in Stiff Clay," Paper #OTC 2312, Proceedings, 7th Offshore Technology Conference, Houston, Texas.

20. Skempton, A.W., (1951), "The Bearing Capacity of Clays," Building Research Congress, Institution of Civil Engineers, Div. 1.

21. Terzaghi, K., (1955), "Evaluation of Coefficient of Subgrade Reaction," Geotechnique, V. 5, No. 4.

22. Vesic, A.S., (1961), "Bending of Beams Resting on Isotropic Elastic Solid, Journal of the Engineering Mechanics Division, ASCE, V.87, EM2.

23. Vesic, A.S., (1961), "Beams on Elastic Subgrade and the Winkler's Hypothesis," Proc. 5th International Conference on Soil Mechanics and Foundation Engineering, Vol. 1.

24. Yoshida, I. and Yoshinaka, R., (1972), "A Method to Estimate Modulus of Horizontal Subgrade Reaction for a Pile,," Soils and Foundations, V. 12, No. 3.

PILE CAPS SUBJECTED TO LATERAL LOADS

Jai B. Kim[1], M.ASCE

Abstract

This paper reviews current pile cap design practices briefly and also presents the test results of full-scale lateral load tests on two pile caps with different reinforcement locations. The various design practices differ primarily in (a) the thickness of the pile cap, (b) the depth of the pile embedment into the cap, and (c) the requirement of the pile cap reinforcement and its location. Lateral load tests performed on two pile cap designs with 12 in. (305 mm.) pile embedments indicated that for a given lateral load, the lateral deflections and rotations of the cap with steel reinforcement 15 in. (381 mm.) above its base were approximately twice those of the other cap with steel reinforcement 3 in. (76 mm.) above its base. The reverse was true for bending stresses in the piles at the bases of the caps.

Introduction

Piles caps are a very important part of pile-supported foundations, yet little research has been conducted on this subject. Up to this time, most research efforts have been concentrated on the interaction of the pile and the surrounding soil. However, a few investigations have been directed towards an understanding of the interaction between the pile caps and the piles. With the advent of increased design stresses and also an increased use of high capacity piles, the connection that transfers column loads to pile/soil stratam has assumed considerable importance.

A review of available information (1-12) indicates that many different design procedures are in use for pile caps. These procedures differ in (1) the thickness of the pile caps, (2) the depth of the pile embedment into the caps, and (3) the requirement of the reinforcement and its location in the pile cap. One common practice is to place the reinforcement at the top of an embedded pile. This particular location is chosen because of (1) the ease in placing the reinforcement and casting the concrete, (2) the greater effectiveness in resisting punching shear, and (3) the better load distribution. Current design practices have evolved from experience and experimental studies dealing with only vertical concentric column loads.

However, pile foundations are usually subjected to various combinations of vertical and lateral loads. Therefore, the present investigation is conducted on two full-scale pile caps under lateral loads.

---

1 Professor and Chairman, Civil Engrg. Dept., Bucknell University, Lewisburg, PA. 17837

Pile Cap Design Practices for Concentric Vertical Loads

Pile cap design standards used by state transportation departments vary widely from state to state. The minimum design requirements differ in the thickness of the pile cap which varies from 2 ft. (0.61 m.) to 3 ft. 6in. (1.07 m.), pile embedments into the cap from 6 in. (152 mm.), to 24 in. (0.61 m.), and the reinforcement ratio and location. Some states require a moderate amount of reinforcement, while others require a very heavy amount of reinforcement. Reinforcement types also vary from straight to stirrups to looped bars and spiral reinforcements. Some states, including Pennsylvania, require reinforcement at the top of the embedded pile head. Others, such as Michigan, call call for reinforcement both at the bottom and top of the cap.

The AASHTO Standard Specifications for Highway Bridges treat pile caps in the same manner as a footing on soil. They recommend at least 12 in. (305 mm.) of pile embedment into the reinforced concrete cap for steel piles and 6 in. (152 mm.) for cast-in-place piles.

Since the required pile cap depth is usually governed by shear, the steel reinforcement is generally near the minimum amount of reinforcement by ACI Code 318-77 (2). The looped bars are intended to provide sufficient lateral confinement capacity of the concrete at the embedded tip of the pile.

Ohio Department of Highways conducted research on the strength of the connection between concrete caps and embedded end of steel H piles (5). The tests were conducted under vertical loads only. It was found from these tests that pile embedment into the cap for a given thickness of the cap does not appreciably affect the strength of the connection. In other words, the strength of the connection does not depend upon the pile embedment into the cap. Also, examination of the test specimens after the tests indicates that slab action plays some part in the failure of the cap. The strength of the cap increases steadily and consistently as the cap thickness increases, which shows that as the cap becomes thicker, its thickness is better able to cope with the slab action problems and work more effectively in resisting stresses due solely to the punching-through tendency of the pile. Some caps were equipped with two way reinforcement bars at the bottom of the cap intended to insure against failure due to slab action. It was thought that such bars would resist the punching-out force and it was expected that these specimens would resist a greater load than the comparable specimens of plain concrete. But the reinforced specimens failed earlier than the plain concrete specimens although the difference was not much.

Current practices for minimum clear edge distances from the side of the pile to the nearest edge of the pile cap vary from 4 in. (102 mm.) to 9 in. (229 mm.) Based on the results from two full-scale pile cap tests under concentric vertical loading, Mozer (7) states that steel pile embedment and edge distance should be such that the pile will bear against sound concrete with sufficient lateral confinement of the high bearing stresses at the embedded top of the pile. Mozer also points out the importance of maintaining adequate edge distance from the steel pile to the free edge of the concrete pile cap. Mozer states that the edge distance to 3/4 times the cap plate size or pile size appears to be reasonable using a concrete compressive strength of 3000 psi (76.2 Pa.). This is particularly true at the corners of the pile cap which provide least confinement of the pile bearing stresses.

An investigation by "The Research and Development Division of the Cement and Concrete Association" of England (9) included the layout of steel either in the form of a uniform grid or U-shaped bars bent over the piles, and the type of anchorage. The result showed that if the straight bars or bars with anchorage are used, bending the steel over the pile gives about a 15 percent increase in the ultimate load compared with the same steel arranged in a grid. It also showed that there is no significant increase in the strength where the bars with nominal hooks were used compared with straight bars. And if the steel was used in the form of a grid, the full anchorage did show some advantage. The anchorage may be achieved by welding the reinforcement bars to the embedded pile or by making holes through the piles and passing the reinforcing bars through these holes.

It is clear from the above discussion that there still exists a need for a proper understanding of pile cap behavior, especially under lateral loads.

Pile Caps Under Lateral Loads

Pile foundations are usually subjected to various combinations of vertical and lateral loads. Paduana and Yee (8) performed field tests on embankment vertical piles for bridges. The investigation included the determination of lateral load vs. deflection behavior of individual vertical piles. The results of the above investigation indicated that a fixed-head pile resisted approximately twice as much lateral load as a free-head pile at the same lateral deflection of 0.25 in. (6.4 mm.).

Based on the elastic theory of homogeneous material, two rigid concrete caps, each supported on two piles, one with a pinned head connection and the other with a fixed (or rigid) head connection, can be compared to two steel frames with similar connections. It is evident from basic theory that the frame with pinned head connections deflects four times as much as the frame with fixed head connections would under the same lateral loading. In actual conditions, it may not be possible to achieve 100% fixity at the connections. However, it would be possible to obtain a high percent of fixity by proper embedment of the pile, the cap, and favorable placement of the reinforcement in the cap. Even with less than 100% fixity, the nearly fixed (or rigid) connections can resist bending and lateral loads much more effectively than can pinned-head connections. Therefore, pile caps with a fixity as near to 100% as is feasible at the pile-to-pile cap connections would provide the greatest resistance to lateral loads and bending moments.

If the concrete pile cap is reinforced at its base, then the opening of any cracks developed at the base due to lateral loads and bending moments would be minimized, achieving more fixed (or rigid) connections between the pile and the concrete. However, the lack of steel reinforcement at the base of the concrete cap would allow the complete separation of the concrete and the pile as soon as cracks developed at the vicinity, resulting in less fixity of the connection. The present investigation is limited to full-scale lateral load tests on two concrete pile caps with two different steel reinforcement locations.

Program of Test

In this investigation two concrete caps were tested. The caps, designated as north cap and south cap, were identical except for the location of the reinforcement. Both caps were 62 in. (1.57 m.) in length, 26 in. (0.66 m.) in width and 18 in. (0.46 m.) in height. In the north cap, the reinforcement was located 3 in. (76 mm.) above the base of the cap and in the south cap, it was located 3 in. (76 mm.) above the top of the embedded pile top (15 in. (381 mm.) above the base). The same amount of reinforcement was provided in each cap consisting of 4-#9 deformed bars. The caps were cast side by side, 24 in. (609 mm.) apart, as shown in Fig. 1.

Each cap was supported by two HP8 x 36 steel piles 4 ft. (1.22 m.) long, spaced 3 ft. (0.91 m.) apart. The piles were embedded 12 in. (305 mm.) into the caps. The lower ends of the piles were welded to two W36 x 300 beams which provided a near rigid support to the piles.

Two strain gages were installed on each pile at the base of the cap. Dial gages were attached to all bottom corners of the caps. Lateral loads were applied against the loading platform (see Fig. 2) by means of two hydraulic rams.

Test Procedure

Initially the lateral load was raised to 10 kips(4.45 kN)/pile, and thereafter loads were monitored in increments of 5 kips(22.3 kN)/pile. The dial gage readings were taken twice for each increment of load at an interval of fifteen minutes to observe the deflections of the caps in horizontal and vertical directions. Strain readings in the piles at the base of the caps were taken twice for each increment of load.

During the course of the test, the caps were carefully observed for any sign of distress or cracks. The test was terminated when the south cap failed with large cracks at its base.

Test Results

The failure of the south cap, which has reinforcement 3 in. (76 mm.) above the top of the embedded pile (15 in. (381 mm.) above its

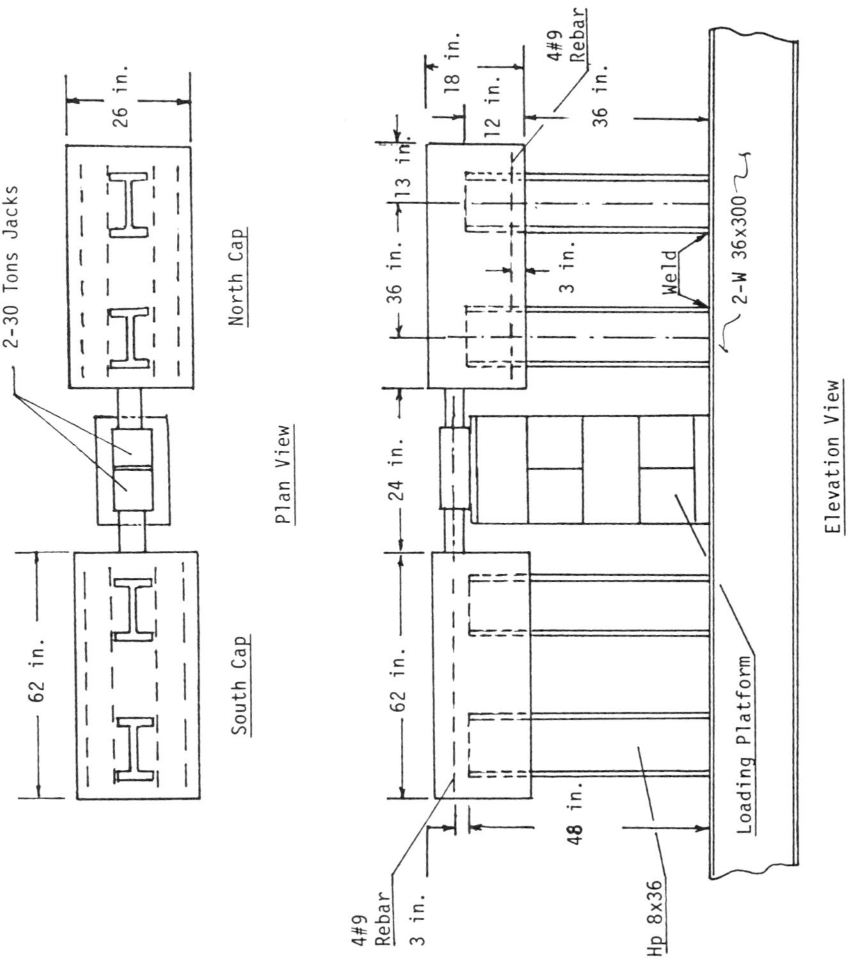

Figure 1 - Full-Scale Test Set-Up (1 in. = 25.4mm)

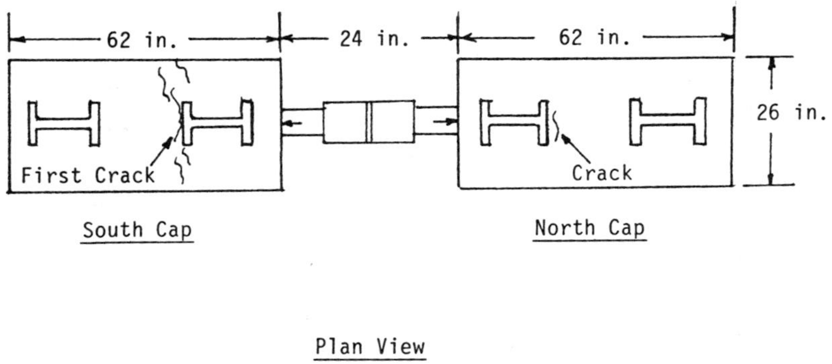

Figure 2 - Crack Pattern at the Bottom of Cap (1 inc = 25.4mm)

base), was gradual with many cracks propagating before the ultimate load was reached. The first crack was observed at a lateral load of 15 kips(06.8 kN)/pile at the junction of the inner pile and the south cap as shown in Fig. 2. The crack widened steadily and consistently as the load was increased. Additional new cracks appeared in the same vicinity, and finally the cap failed at a lateral load of 29 kips (129 kN)/pile. The cap was thoroughly examined and it was found that there was no sign of local crushing of the concrete at its junction.

The north cap, which has reinforcement at its bottom (3 in. (76 mm.) above its base), was intact at the failure of the south cap. A hair line crack at a lateral load of approximately 20 kips (89 kN)/ pile was observed. Thereafter, no new cracks appeared nor did the previous crack widen with an increase in the lateral loads. The experiment was terminated when the south cap failed.

The average values for the horizontal and vertical deflections at the corners of the caps, for the rotation of the caps, and for the stresses in the piles at the base of the caps are presented in Table 1 and also plotted in Figs. 3 through 6. The horizontal deflections were substantially the same until the lateral load reached about 10 kips (44.5 kN)/pile level. As the load was increased, the ratios of horizontal deflections of the south cap to those of the north cap also increased, reaching about 2.0. As stated earlier, the cracks at the base of the south cap first developed at a lateral load of 15 kips (66.8 kN)/pile. Since the crack openings could not be prevented (as no reinforcement was placed at the base of the south cap), it was reasonable for the south cap to deflect more since the piles were more nearly "pin-connected" to the cap. The north cap, however, maintained its original fixity and, therefore, its lateral deflection was about half of that of the south cap. There was a similar trend for both vertical deflections and rotations of the caps, although the south cap deflected twice as much and also rotated twice as much as the north cap did from the beginning of lateral loading.

Stresses in the piles in the north cap were about two and one half times as great as those in the south cap. It should be noted that in the case of the south cap, as the connection (between cap and piles) became pinned-head or nearly so as a result of crack openings at the connection, it was unable to transfer the cap moments fully to the piles below. This behavior of the south cap accounted for lower stresses in the piles, whereas the north cap due to its fixed (or rigid) connection, or nearly so, was able to transfer the cap moments to the piles more efficiently.

Conclusions

The results of this investigation substantially prove that the location of the reinforcement in the pile cap does affect its moment and lateral load capacity. It is evident from the test results that for a given lateral load, the lateral deflection and the rotation of the concrete cap with reinforcement 15 in. (381 mm.) from its base is appoximately twice that of the concrete cap with reinforcement 3 in.

TABLE 1, TEST RESULTS

Lateral Kips/Pile	Pile Cap	Horizontal Deflection in.	Vertical Deflection in.	Rotation 10^{-4} Rad.	Stress ksi
5	North	0.020	0.006	2.00	2.90
	South	0.029	0.015	4.50	1.00
10	North	0.040	0.013	4.05	4.80
	South	0.062	0.028	9.40	2.20
15	North	0.063	0.019	6.60	9.30
	South	0.101	0.043	13.60	3.98
20	North	0.084	0.026	8.80	
	South	0.147	0.057	18.80	
25	North	0.108	0.033	10.50	16.05
	South	0.203	0.072	23.10	6.70

Note: 1 Kip = 4.45 kN.
1 ksi = 6.895 Pa.
1 in. = 25.4 mm.

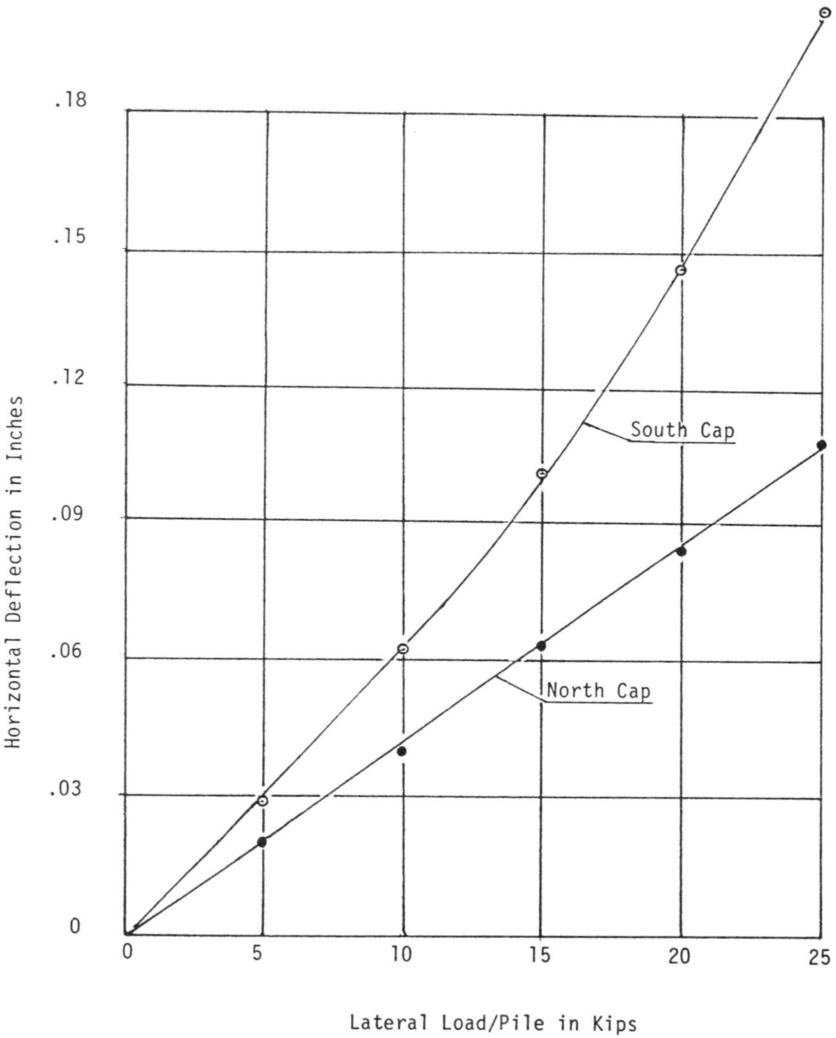

Figure 3 - Load Versus Horizontal Deflection (1 kip = 4.45kN; 1 in. = 25.4mm)

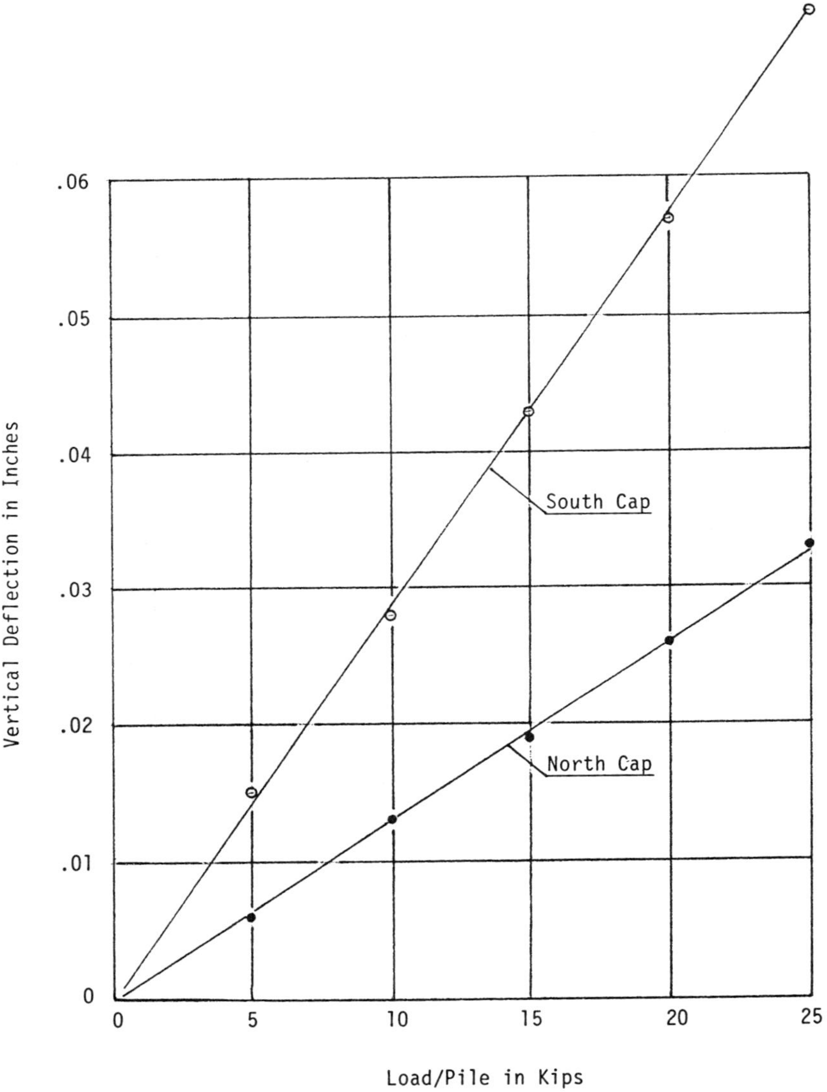

Figure 4 - Load Versus Vertical Deflection (1 kip = 4.45kN; 1 in. = 25.4mm)

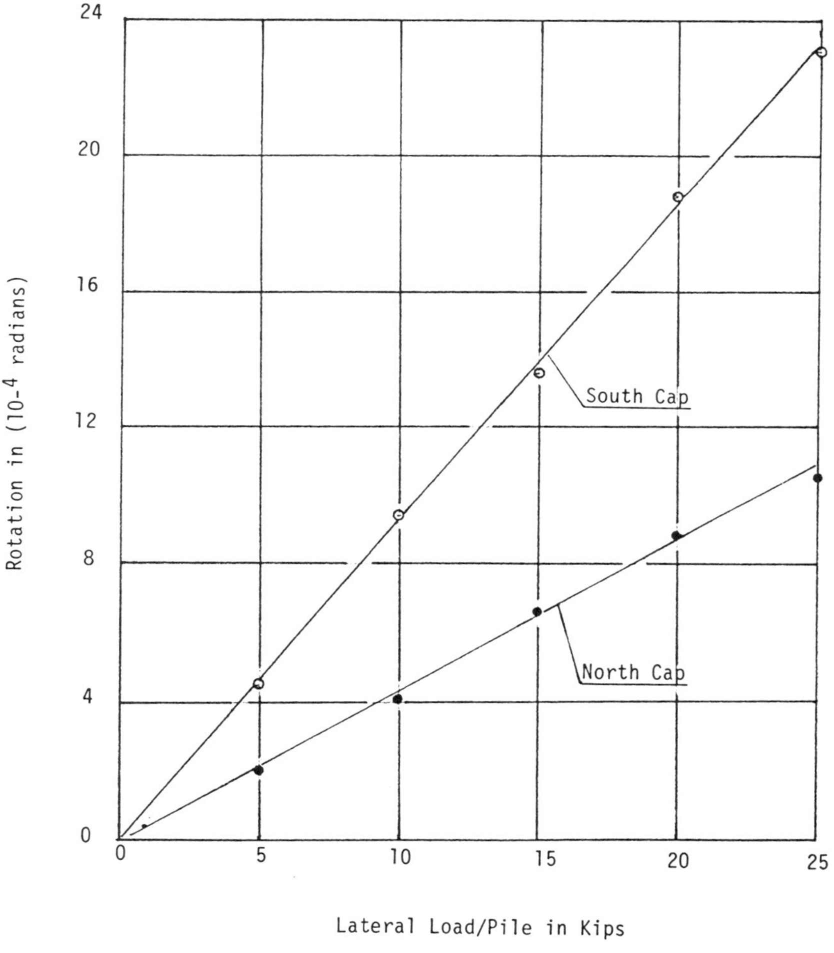

Figure 5 - Load Versus Rotation of Pile Cap (1 kip = 4.45kN)

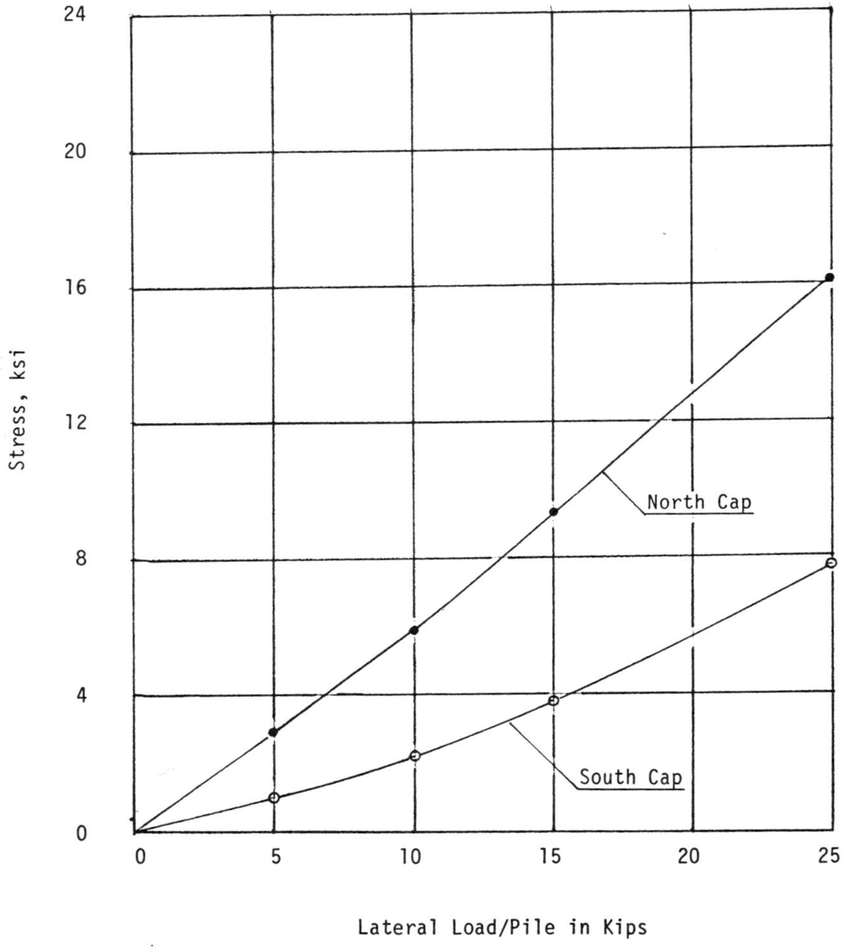

Figure 6 - Load Versus Pile Stress (1 kip = 4.45kN; 1 ksi = 6.895Pa.)

(76 mm.) from its base. The reverse is true for the stress in the piles at the base of the caps. A comparison of values obtained by this investigation with the available data attests their accuracy. Paduana and Yee (8) concluded (from the tests performed on the individual embankment vertical piles) that the fixed head pile resists twice as much lateral load as a free head pile for the same horizontal deflection.

Therefore, based on the above results, it is observed that steel reinforcement placed near the bottom of the pile cap and well below the pile top is advantageous in enhancing its moment and lateral load resisting capacity as compared to steel placed near the top of the cap and above the pile tops and that the design of the pile-pile cap connections is very critical to the performance of pile supported foundations.

Acknowledgment

The author acknowledges the assistance of P. N. Beri, Peter Carrato and Peter Lennon.

Appendix I

(1) Correspondence with various state transportation departments.

(2) "Building Code Requirements for Reinforced Concrete," ACI 318-77, American Concrete Institute, Detroit, MI., 1981

(3) Davis, G. W., "Foundation Pile-Head Bond and Anchorage Tests," Public Roads, Volume 9, No. 9, November 1928, pp. 480-481.

(4) Faber, J. and Mead, F. H., "Reinforced Concrete," The English Language Book Society (Britian), 1965.

(5) "Investigation of the Strength of Connection Between a Concrete Cap and the Embedded End of Steel H-Pile." Research Report No. 1, Department of Highways, State of Ohio, 1947.

(6) Mason, H. G., and Bishop, J. A. "Measurement of Earth Pressure and Deflection Along the Embedded Portion of a 40 Feet Steel Pile," Symposium on Lateral Load Tests on Piles, ASTM STP 154A, 1954.

(7) Mozer, J.D.,"Pile Cap Design Considerations," Presented at the Central Pennsylvania Section, ASCE Pile Seminar, Oct., 1976.

(8) Paduana, Joseph A. and Yee, Wifred S., "Lateral Load Tests on Piles in Bridge Embankment," Transportation Research Record, Transportation Research Board, National Research Council, No. 517, 1974, pp. 77-92.

(9) "Research on Pile Cap Design," Ground Engineering (London), Volume 6, No. 3, May 1973, p. 52.

(10) "Standard Specifications for Highway Bridges", American Association of State Highway and Transportation Officials, 12th Ed., 1977.

(11) Tod, J. D., "The Determination of Tensile Stress-Strain Curve for Concrete Proceedings," The Institute of Civil Engineering (London) Volume 4, pt.1, No. 2, March 1955, pp. 201-211.

(12) Wen, R. K. L., "Model Studies of Laterally Loaded Pile Foundations," Proc. of Highway Research Board, Vol. 34, 1944.

Dynamic Stiffness of Pile Groups

by

Jose M. Roesset, M. ASCE[1]

Introduction

Pile foundations were analyzed for some time neglecting interaction effects between the piles through the soil and enforcing only compatibility of displacements at their heads under the assumption of a rigid mat. The first significant study on the behavior of complete pile foundations, including group effects, under static loads was presented by Poulos in 1971. In his paper Poulos proposed a general methodology to derive the flexibility matrix for the soil surrounding the piles, using Mindlin's equation integrated over a rectangular area. He suggested, in addition, an approximation for a simplified analysis of large pile groups, enforcing compatibility of displacements at the pile heads but using interaction coefficients derived from the study of two piles at a time, and he highlighted the main effects of group behavior: a reduction in the overall stiffness with increasing number of piles and decreasing pile spacing, and an unequal distribution of forces at the pile heads.

The dynamic response of an isolated pile was studied by Novak (1976) using a Winkler type foundation with dynamic springs, complex functions of frequency, to simulate the soil. Blaney (1976) presented a solution for end bearing piles relating forces and displacements along the pile in terms of the mode shapes and wave numbers of the soil deposit, obtained at each frequency by a semi discrete formulation, and expressing the relationship in the form of a boundary matrix. An eigenfunction expansion was also used by Nogami and Novak (1976) and Novak and Nogami (1977) for a uniform, homogeneous, soil deposit. In this case, neglecting the coupling between horizontal and vertical forces and displacements, the mode shapes for the soil stratum can be obtained analytically without the need to discretize the soil into horizontal layers. Similar analytical solutions had been suggested by Tajimi (1969) and used by Kobori, Minai, and Baba (1977) and Kagawa and Kraft (1981). On the other hand, solutions based on a finite element model, discretizing the soil both vertically and radially have

[1]Paul D. and Mary Robertson Meek Centennial Professor. The University of Texas at Austin.

been presented by Kuhlemeyer (1979a, b) and by Angelides (1980), who combined a finite element region, with simulated nonlinear behavior, around the pile and the boundary matrix used by Blaney to reproduce the far field. An extensive set of comparative studies by Sanchez Salinero (1982) indicate that for typical piles all these approaches, from the Winkler assumption with dynamic springs to analytical or semi-analytical expansions, and finite element discretizations, yield results which are very similar. Differences in the results are more noticeable when dealing with very short, stubby piles, very stiff soils, or soil deposits containing thin interspersed layers with very different properties. The dynamic behavior of single piles is thus well understood today and a number of procedures are available to determine pile displacements and internal forces for small amplitude vibrations, when linear soil behavior and perfect bonding between the pile and the soil can be assumed.

Less information was available on the dynamic response of pile groups but this has been an area of considerable interest and research in the last seven years. The first contribution in this field is due to Wolf and Von Arx (1978) who used a finite element model in cylindrical coordinates to compute numerically the influence coefficients (Green's functions) giving the displacements at any point within the soil mass due to a ring load applied at a given depth. The flexibility matrix of the soil can then be obtained at each frequency from the influence coefficients and the dynamic stiffness matrix of the complete system can be assembled from the contributions of the soil and the individual piles. Wolf and Von Arx determined the horizontal and vertical stiffnesses and the seismic response of a large pile foundation in a layered soil stratum of finite depth and pointed out the basic features of the solution: some identical to those reported by Poulos for static loads, others related to dynamic effects, such as a strong dependence of the results on frequency and a larger reduction of the real part of the stiffnesses than the imaginary terms with increasing number of piles or decreasing pile spacing. A somewhat similar approach was followed by Gomez (1982) who determined the horizontal dynamic stiffness of pile groups in a layered soil deposit computing the influence coefficients in terms of the mode shapes of the soil and the boundary matrix used by Blaney, while Nogami (1979) determined the vertical stiffness of pile groups in a uniform soil stratum using the analytic expression for the modes of his previous study (Nogami and Novak. 1976). A more efficient technique to compute the Green's functions and determine the dynamic stiffnesses of pile groups arranged in circular rings was presented by Waas and Hartmann (1981). Finally, a general solution for pile groups in a layered half space, combining a numerical integration for the evaluation of the influence coefficients and an analytical solution for the pile stiffnesses and flexibility matrices, was derived by Kaynia and Kausel (1982), who conducted a number of studies to assess the effect of various approximations.

The extension of the Winkler foundation model to pile groups, determining the dynamic springs through the study of a group of disks in a plane (actually a disk and a set of points), was proposed by Nogami (1980, 1983) and Sheta and Novak (1982). Studies by Sanchez Salinero (1983) indicate that this is still, generally, a reascnable approximation.

All the above solutions assume again a linear problem (linear soil behavior and perfect bonding between the piles and the surrounding soil). It has long been recognized, however, that pile behavior is strongly influenced by nonlinear effects (particularly near the top for an isolated pile) which become significant even at relatively small amplitudes of motion. These effects are incorporated in static analyses through the use of a Winkler foundation with nonlinear springs (P-y or T-z curves) backed by experimental observations (Matlock and Reese 1960, Matlock 1970, Reese, Cox and Koop 1975, Stevens and Audibert 1979). The number of studies conducted to date on the nonlinear dynamic response of piles is limited although a number of approximations are used in practice. Finite element plane stress analyses with a nonlinear soil model and possible separation between the pile (disk) and the soil were conducted by Yegian and Wright (1973) and Thompson (1977), under static loads. These studies showed the importance of the initial state of stresses and separation effects on the soil stiffness. A finite element core region with simulated nonlinear behavior (iterative linear analyses with adjusted soil properties based on the level of strains), but still perfect bonding, and Blaney's boundary matrix to reproduce the far field, were used by Angelides (1980), who was able to reproduce qualitatively the results of analyses with P-y curves but obtained larger stiffnesses due mainly to the perfect bonding. Much simpler approximations are used sometimes in practice, reproducing the soil around a pile by springs and dashpots: the springs are determined directly from the P-y (or T-z) curves, while the dashpot contants are computed from the imaginary part of Novak's dynamic springs, either directly or scaling them proportionally to the springs. These may be reasonable engineering approximations but they need proper justification and their accuracy is questionable. An improved approximation was suggested by Sheta and Novak (1982) computing dynamic springs for a disk surrounded by a ring of soil with reduced properties. This approach can be extended to the analysis of pile groups (or groups of disks) but the basic question is how to determine the size and properties of the rings in practice for a given soil profile and level of excitation.

Much more work remains to be done on the dynamic behavior of complete pile foundations accounting for non-linear effects, both numerically and experimentally. Tests with pile groups are very limited in number (Matlock, Ingram, Kelley, and Bogard 1980, O'Neill 1982) and are needed to substantiate analytical predictions.

The purpose of this paper is to review some of the existing solutions for the dynamic analysis of pile groups, the accuracy of various approximations and the basic features of the available results.

Dynamic Group Effects

The dynamic stiffness of a foundation is defined as the ratio between the applied force and the resulting displacement under a steady state vibration at a given frequency. It is therefore a function of the frequency of excitation. In addition, since the displacement is out of phase with respect to the force, a situation typical of damped systems, the stiffness is defined at each frequency by two constants, one representing its amplitude (the ratio of the amplitude of the force to the amplitude of the displacement), the other representing the phase shift. It is more common, however, to represent a dynamic stiffness as a complex function of frequency. Letting then

$$k(\Omega) = k_r(\Omega) + i\ k_i(\Omega) \tag{1}$$

where k_r represents the real part of the stiffness and k_i the imaginary part, the amplitude would be

$$|k| = k_r^2 + k_i^2 \tag{2a}$$

and the phase $\tan\alpha = k_i/k_r$ (2b)

Alternatively

$$k_r = |k| \cos\alpha \qquad k_i = |k| \sin\alpha \tag{3}$$

The real part of the stiffness is normally associated with the constant of an equivalent spring (at each frequency), whereas the imaginary part, divided by the frequency Ω, can be interpreted as the constant of an equivalent viscous dashpot. To make them more general results are typically expressed in terms of a dimensionless frequency a_o rather than the actual frequency Ω, with

$$a_o = \Omega r/c_s \tag{4}$$

.r is the radius of the foundation (or a characteristic dimension if the shape is not circular) and c_s is the shear wave velocity of the soil for a uniform deposit or the velocity at some point for a variable profile.

Figure 1 shows the real and imaginary parts of the horizontal dynamic stiffness of a single end bearing pile embedded in a soil stratum of finite depth and fixed against rotation at its head. Both terms show oscillations associated with the natural frequencies of the stratum. Moreover because of the assumption of a rigid rock underlying the soil the imaginary part is almost zero below the fundamental frequency of the soil layer (it would be exactly zero if there was no internal soil damping). If there is some internal soil damping, however, both terms can be reasonably approximated by straight lines: a horizontal line for the real part (there is in reality a small decrease of the real part with increasing frequency) and a sloping line for the imaginary part, suggesting that a good approximation can be obtained replacing the pile by an equivalent frequency independent spring and dashpot. It should be noticed that for typical piles the values of the dimensionless frequency a_0 of interest are smaller than 0.2 for wave action and smaller than 0.5 for seismic excitation. Values of a_0 larger than 1 may be, however, of interest where dealing with foundations for vibrating machinery with a high number of rpm's.

As pointed out by Poulos (1971) the two main effects of group action (interaction between the piles through the surrounding soil) for static loads are:

1. a reduction of the foundation stiffnesses over the values computed adding the contributions of the individual piles. This effect is normally represented by a reduction factor α. If k_g is the stiffness of a complete pile foundation with N piles, k_s is the corresponding stiffness of a single pile and x is the distance of each pile to the axis of rotation of the foundation

$$k_{gx} = \frac{Nk_{sx}}{\alpha_x} \quad \text{or} \quad \frac{\Sigma k_{sx}}{\alpha_x} \quad \text{for horizontal forces}$$

$$k_{gz} = \frac{Nk_{sz}}{\alpha_z} \quad \text{or} \quad \frac{\Sigma k_{sz}}{\alpha_z} \quad \text{for vertical forces}$$

$$k_g = \frac{\Sigma x^2 k_{sz}}{\alpha_\phi} \quad \text{for moments} \qquad (5)$$

The reduction factors α_x α_z α_ϕ are functions of the number of piles, their relative stiffness with respect to the soil, their relative spacing and their geometric arrangement. Values of the reduction factor between 2 and 10 can be expected for typical pile foundations under horizontal or vertical forces.

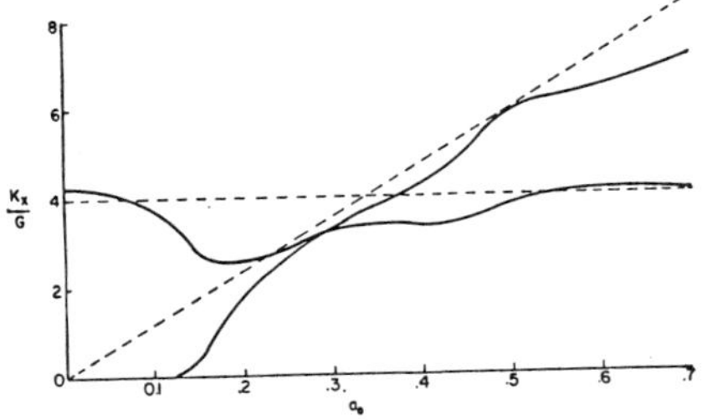

FIG. 1 DYNAMIC STIFFNESS FOR SINGLE PILE

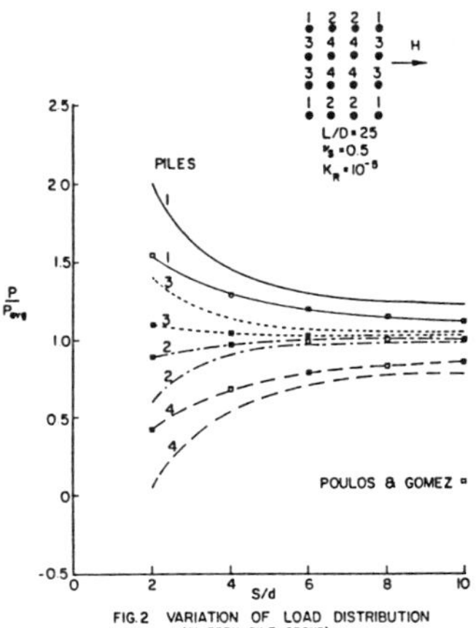

FIG. 2 VARIATION OF LOAD DISTRIBUTION
(SIXTEEN PILE GROUP)

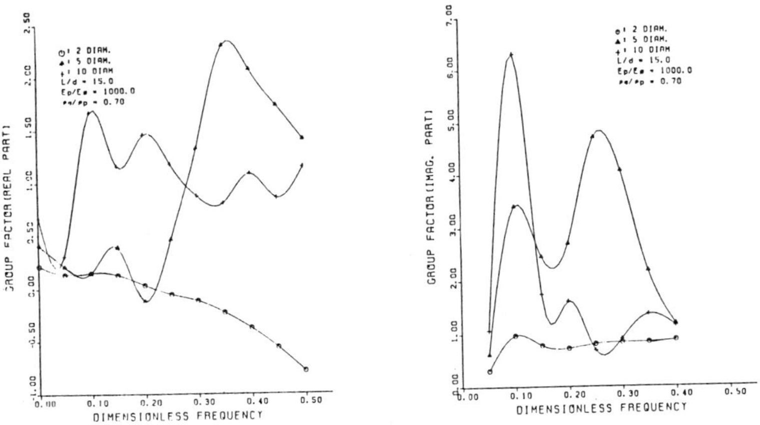

FIG. 3 DYNAMIC GROUP FACTOR FOR 4 X 4 PILE GROUP

2. An unequal distribution of forces among the different piles. Figure 2 shows a typical distribution of horizontal forces among the piles for a 4x4 pile group as a function of the relative pile spacing, s/d, where s is the distance between the axes of the piles and d is their diameter. The piles have a length to diameter ratio of 25, Poisson's ratio for the soil is 0.5 and the stiffness ratio K_r defined by Poulos as E_pI_p/E_sL^4 is 10^{-5}. E_p and E_s are the moduli of elasticity of the pile and the soil, I_p is the bending moment of inertia of the pile and L is the pile length. Shown in the figure are results published by Poulos (1971) and those obtained by Gomez (1982). Both exhibit the same trends, indicating that the corner piles take a larger fraction of the load while the center piles carry the smallest forces. Gomez' results show however, somewhat smaller effects (a less uneven distribution of forces) than those of Poulos, a trend that occurred in all cases studied.

Under dynamic loads one can expect that both effects will depend on frequency. Kaynia and Kausel (1982) expressed their results in the form of a group factor, defined as the ratio of the group stiffness at each frequency to the sum of the static stiffnesses of the individual piles. At zero frequency this ratio is the inverse of the reduction factor defined by (5). Gomez used instead the ratio of the group stiffness at each frequency to the sum of the stiffnesses of the individual piles at the same frequency, the ratio being computed separately for the real and the imaginary terms. While the first definition allows to use the results from the static stiffness of each pile, the computation of the dynamic stiffnesses of a single pile can be performed now very economically with a variety of available programs. More importantly the second definition allows to isolate and illustrate better group effects under dynamic loads. The differences in the results from both definitions are in any case very small for dimensionless frequencies of the order of 0.5 or less.

Figure 3 shows results obtained by Gomez for a 4x4 pile group embedded in a soil stratum underlain by rigid rock. The ratio of the modulus of elasticity of the pile to that of the soil was equal to 1000, the ratio of the mass densities 0.70, and the length to diameter ratio of the piles was 15. Results for this case have also been presented by Kaynia and Kausel assuming the piles embedded in a homogeneous half space (it should be noticed that Kaynia uses a dimensionless frequency which is the double of the usual one with the diameter of the pile instead of the radius as characteristic length). In spite of this basic difference (a finite stratum versus a halfspace) the results of both studies compare reasonably well. It can be seen from these results that the group factors for the real and imaginary terms of the stiffness exhibit a number of oscillations with frequency which become more apparent as the spacing between the pile increases (although the amplitude of the oscillations is larger in some cases for a relative spacing s/d of 5, the number of oscillations over the range of frequencies studied increases always with s/d). The first valley and peak in the curves is associated with the existence of a rigid base (soil layer of finite depth) and do not appear in the half space solution. The other peaks, however, occur at frequencies for which the wavelength is similar to (or a multiple of) the pile spacing. These frequencies increase as the pile spacing decreases, which explains the smoother shape of the curves for s/d = 2.

For closely spaced piles the group factor for the real part of the stiffness decreases steadily with increasing frequency and becomes negative for values of a_0 larger than about 0.25. This type of behavior is associated with inertia effects and suggests the existence of a soil mass vibrating in phase with the piles (a constant mass M would result in a real stiffness of the form $k - M\Omega^2$, with a parabolic variation with frequency). The group factor for the imaginary part remains on the other hand almost constant and close to unity (in other cases the group factor for the imaginary part increases slowly but steadily with frequency for these close spacings). While these results are only valid for the particular case studied the general trends are similar for a number of different foundations analyzed, suggesting that it may be possible to derive simplified formulae to predict the dynamic response of pile foundations with closely spaced piles, at least for dimensionless frequencies of 0.5 or less.

The behavior of pile foundations with larger pile spacings or at higher frequencies is clearly more complicated because of the wave interference phenomena leading to the large oscillations in the group factors. A question remains as to whether these effects would actually occur in practice when accounting for nonlinear soil behavior or separation effects. The results of the tests conducted by Matlock et al (1980) on pile groups indicate much smaller interaction coefficients under static loads for pile spacing of 5 or 10 diameters than would be predicted by elastic solutions.

Figure 4 shows the distribution of forces among the piles as a function of frequency. The forces in piles 2 and 3, on the sides of the foundation, change slightly with frequency but are always close to the average load (total force divided by the number of piles). The effect of frequency is much more dramatic on the corner and center piles. Under static loads the corner piles carry forces which vary from 1.2 to 1.7 times the average load as the pile spacing decreases from 10 to 2 diameters, while the center piles

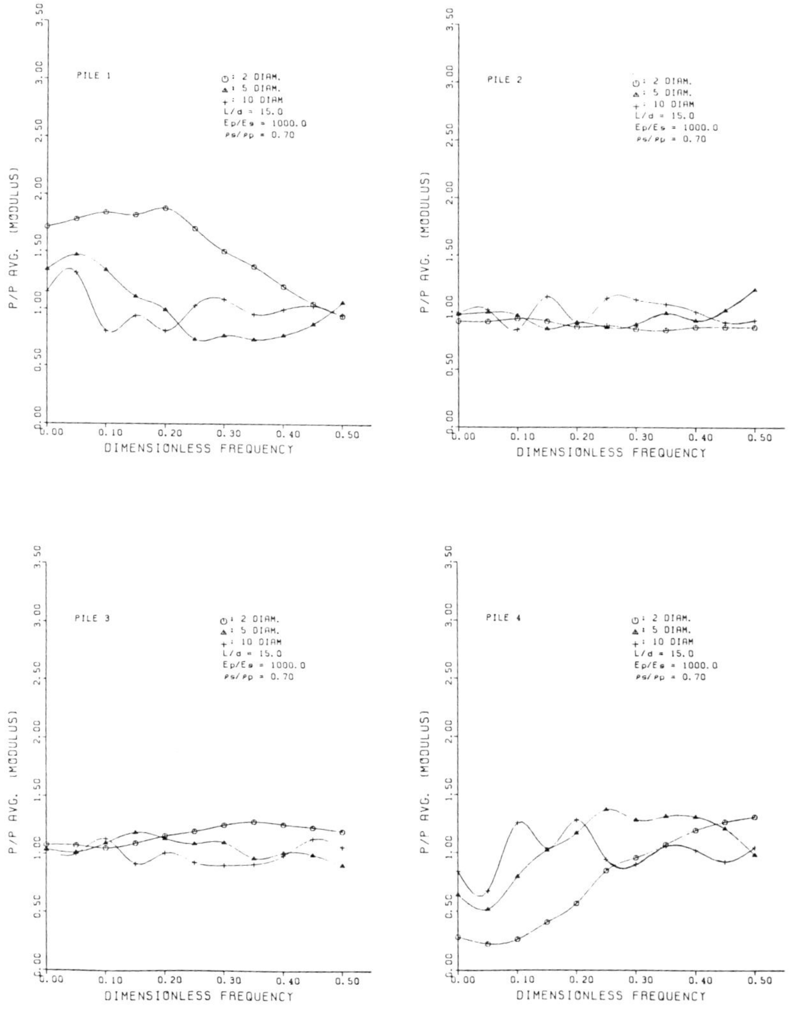

FIG 4 DYNAMIC FORCE DISTRIBUTION FOR 4 X 4 PILE GROUP

carry from 0.8 to 0.3 times the average load respectively. At a dimensionless frequency of 0.5 on the other hand, the center piles carry approximately 1.2 times the average load for a pile spacing of 2 diameters while the corner piles carry about 90 percent. It appears that as the frequency increases the distribution of forces among the piles becomes generally more uniform. At values of the dimensionless frequency between 0 and 0.5 this distribution can vary however substantially, suggesting that a design of the piles using the distribution factors from a static analysis may be unconservative under dynamic conditions (particularly for the center piles).

While the results presented here are for a limited number of cases, a substantial number of pile groups have been studied under horizontal and vertical dynamic loads and their results can be found in the references. All these studies indicate that frequency effects are much more important for pile groups, or complete file foundations, than for single piles.

Effect of Various Approximations

As pointed out in the Introduction a number of approaches have been developed in recent years to compute the dynamic stiffness of pile foundations including group effects. Some of these are based on fully three dimensional formulations considering all the piles simultaneously, while others take advantage of various simplifying assumptions. All of them introduce, however, some approximations to reduce the cost of computation, which is normally considerable. The main approximations used in practice are:

1. When computing the influence coefficients, to determine the flexibility matrix of the soil, forces are applied at various points along an object pile and the displacements are obtained at the locations of the other piles. The dimensions of the cross section of the object pile are taken into account by applying uniformly distributed forces over a rectangular area (Poulos) or a cylindrical surface (Kaynia and Kausel), by considering ring loads (Wolf and Von Arx) or by applying forces along the circumference that will produce a uniform (rigid body) displacement of the cross section (Gomez). The displacements are computed, however, at points along the axes of the other piles and not around their outer surface. This implies that the other piles are replaced by equivalent lines (their axes) neglecting their cross sectional dimensions.

2. In the computation of the influence coefficients (Green's functions) the displacements are only obtained in the direction of the applied forces. When the stiffness matrix of the soil is combined with those of the piles to solve the complete system this implies that compatibility of displacements between the piles and the surrounding soil is only enforced, and soil reaction are only considered, in the direction of the excitation. Clearly if a horizontal (or vertical) force is applied on a pile the others will experience horizontal displacements in two orthogonal directions as well as vertical displacements.

3. For a group of equal piles arranged in a circular ring and a horizontal stratified soil deposit (soil properties uniform in horizontal planes) all piles would experience the same displacements and forces under vertical excitation. This reduces considerably the number of degrees of freedom since only one pile per ring must be considered. An approximation can be introduced for lateral loads or moments (Waas and Hartman, 1981) assuming that piles in a ring have radial and vertical displacements proportional to the cosine of their position angle with respect to the direction of the load, and tangential displacements proportional to the sine (this implies that all piles in a ring have the same horizontal displacements in the direction of the load). An additional approximation can be introduced sometimes for piles arranged in rectangular grids assuming equivalent circular rings.

4. For a group of two piles i,j, an interaction coefficient a_{ij} is computed as the ratio of the displacement u_{ij} at the top of pile i due to a unit force at the head of pile j, to the displacement u_{ii} caused by a unit force acting on the pile i by itself. The total displacement u_i at the head of i is then expressed as the summation of the displacements caused by forces p_j acting at the heads of all the piles.

$$u_i = \sum_j u_{ij}\, p_j = u_{ii} \sum_j a_{ij} p_j = \frac{1}{k_{is}} \sum_j a_{ij}\, p_j \qquad (6)$$

where k_{is} is the appropriate head stiffness of pile i by itself.

Assuming a rigid cap or mat connecting the heads of all the piles and imposing a unique displacement u on the foundation, denoting by A the matrix with general term a_{ij}, and letting B (general term b_{ij}) represent its inverser, the net force at the head of pile i is

$$p_i = u \sum_i b_{ij}\, k_{js} \qquad (7)$$

and the head stiffness of pile i accounting for group effects is

$$k_{ig} = \sum_j b_{ij} k_{js} \tag{8}$$

The total stiffness of the foundation is finally

$$K = \sum_i k_{ig} = \sum_i \sum_j b_{ij} k_{js} \tag{9}$$

This is the approximation that was proposed by Poulos (1971) for the static analysis of pile groups. When all piles are equal the coefficients a_{ij} can be computed first as a function of the relative distance between the two piles for the vertical case, and as a function of this distance and the angle between the line of the piles and the direction of the force for horizontal excitation (the values need only be computed for angles of 0 and 90°). Poulos presented in graphical form the variation of the interaction coefficients with distance for the case of a homogeneous soil deposit and two different ratios $k_r = E_pI_p/E_sL^4$, but in a real application one must derive them for the actual soil profile and characteristics of the piles. For the dynamic case the coefficients a_{ij} must be computed, in addition, at each frequency.

It should be noticed, finally, that while this approach allows to determine the stiffnesses of the complete foundation and the forces at the heads of the individual piles, additional assumptions are necessary to determine internal forces along each pile for design purposes. An assumption which has been used in practice is to study each pile individually, under the force acting at its head, scaling uniformly the soil properties to obtain the desired head displacement.

5. Replacing the soil surrounding the piles by a Winkler foundation, neglecting therefore the interaction between soil reactions and pile displacements at different depths. The Winkler assumption is commonly used in the static analysis of single piles, with nonlinear springs (references 9, 10, 19, 23, 25, 28). Its use for dynamic springs, complex functions of frequency, which can be expressed analytically in terms of Bessel functions, and the extension to pile groups, considering a gruop of plane disks has been proposed by Nogami (1980, 1983) and Sheta and Novak (1982).

Particularly efficient solutions can be obtained when several of the above simplifying assumptions are combined. Considering two disks in a plane, or more conveniently, under assumption 1, a disk and a point, representing the center of the second disk, one can obtain interaction coefficients for the soil reactions, or Winkler springs, similar to those of Poulos' method. The displacements u_{ii} of a single disk by itself due to a unit load, and u_{ij} of a disk due to a unit load applied on the other, can be obtained analytically in terms of Hankel functions and modified Bessel functions. The stiffness of the Winkler springs associated with each pile including group effects is then simply

$$k_g = \frac{1}{u_{ii} + u_{ij}} \tag{10}$$

whereas for a single disk

$$k_s = \frac{1}{u_{ii}} \tag{11}$$

The pile head interaction coefficients of Poulos' approach can then be obtained solving single piles on Winkler foundations with stiffnesses k_s and k_g respectively. An additional approximation can be introduced for a homogeneous soil profile, or for preliminary calculations, assuming that the pile head stiffnesses are proportional to the soil stiffnesses raised to the power 0.75 for horizontal vibration and 0.5 for vertical excitation, or using any of a variety of approximate formulae suggested in the literature (see for instance reference 20). The advantage of this procedure is that the coefficients a_{ij} for the pile heads can be obtained entirely from closed form analytical expressions. Alternatively one can form an interaction matrix A and apply expressions (6) to (9) to the computation of the Winkler springs of each pile (disk) including the effects of all the others. In this way each pile can be studied individually to determine not only its contribution to the stiffness of the complete foundation but also the internal forces.

To assess the effect of the first assumption Sanchez Salinero (1983) determined the dynamic soil reactions for two disks, using the boundary integral equation (boundary element) method, and for a disk and a point, using the appropriate analytical expressions. He used then the results to compute the group factors (as previously defined) for two piles considered as beams on a Winkler foundation with the spring constants obtained by both approaches. Figure 5 shows typical results for pile spacings of 2 and 5 diameters and vertical (axial) excitation while figures 6 and 7 show the corresponding results under lateral loads when the piles are at 0 and 90° with respect to the

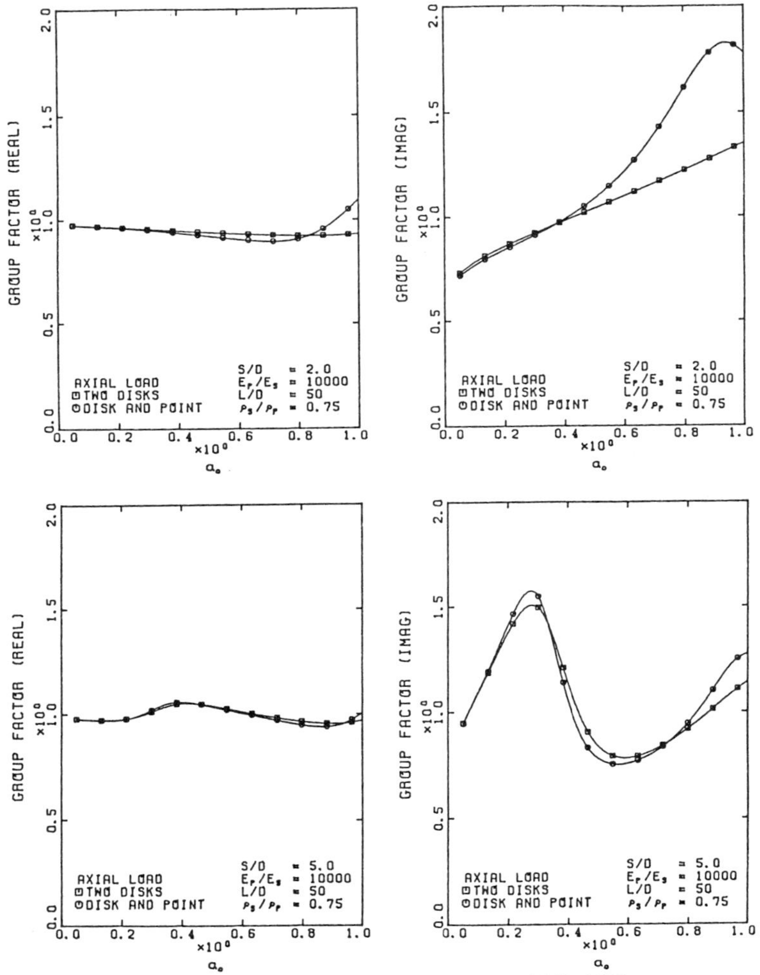

FIG.5 EFFECT OF PILE SIZE - AXIAL LOAD

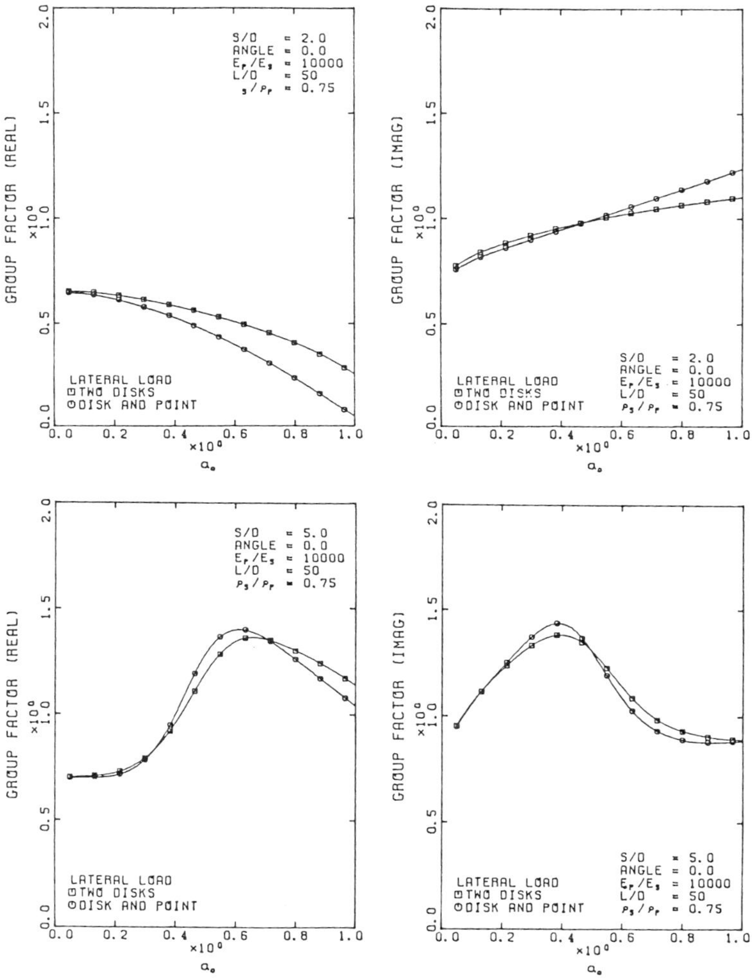

FIG.6 EFFECT OF PILE SIZE - LATERAL LOAD 0°

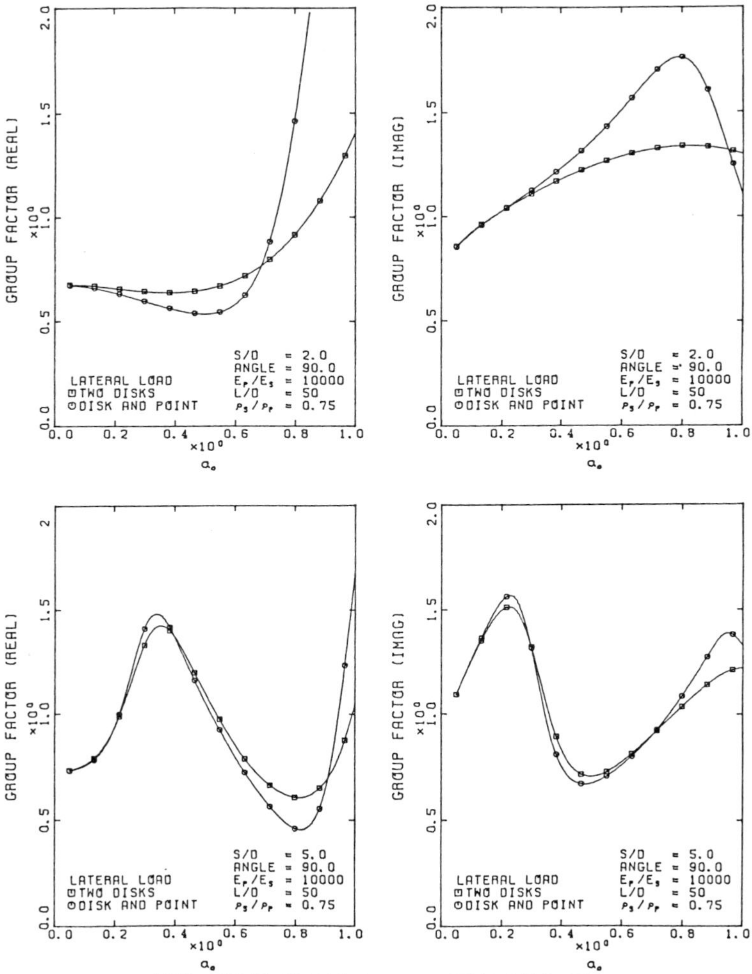

FIG. 7 EFFECT OF PILE SIZE - LATERAL LOAD 90°

direction of the force. The piles had a length to diameter ratio of 50, the ratio of the moduli of elasticity was 10,000 and the ratio of densities 0.75. Poisson's ratio for the soil was 0.25. It can be seen that the effect of replacing the second disk by a point (or the second pile by a line) is small for a spacing of 5 diameters (it is even smaller over the complete range of frequencies studied for spacings of 10 diameters). The agreement between the results is still good for spacings of 2 diameters in the low frequency range (values of the dimensionless frequency a_o of up to 0.3 or even 0.5). It deteriorates, however, for larger frequencies and particularly, under lateral forces, when the line of the piles is at 90° with respect to the direction of the load. The differences for this case tend to increase for larger Poisson's ratios. It should also be noticed that accounting for the size of the disks leads to smoother results with smaller oscillations.

Table 1 shows the effect of the second assumption on the reduction factor for a group of 4x4 piles under a static horizontal load as a function of the pile spacing. The effect on the distribution of forces at the pile heads is shown in Table 2. Shown in the tables are also the results combining assumptions 2 and 4 (Poulos' approach). The length to diameter ratio of the piles was 25 and Poisson's ratio for the soil was 0.25. Assumption 1 was used in all cases. The results obtained using Poulos' approach are those computed by Gomez (1982), which are somewhat different from those reported in Poulos' paper (see figure 2). It can be seen that the results obtained with Poulos' approach are always in good agreement with those of a complete solution when compatibility is only enforced in the direction of the load (x direction). The differences increase, however, as the ratio k_r increases (a stiffer pile or softer soil), and for smaller spacings. The effect of enforcing full compatibility between the piles and the soil is small for $k_r = 0.1$ but increases for $k_r = 10^{-5}$ (very flexible piles or very stiff soil) and small pile spacings. The largest effects are caused by enforcing compatibility in the other horizontal direction, the additional changes resulting from imposing vertical compatibility being small. It is interesting to notice that the more accurate solutions exhibit in all cases smaller interaction effects than predicted by approximate formulations. Similar conclusions are reached from dynamic analyses at various frequencies.

Figures 8 and 9 show the group factors (real and imaginary parts) for two piles at 0 and 90° with respect to the direction of the force, obtained from a three dimensional solution (with simplifying assumptions 1 and 2) and an equivalent Winkler foundation (assumptions 1, 2 and 5). It can be seen that for spacings of 5 diameters the results are very similar although the Winkler assumption produces slightly higher oscillations (the agreement improves with increasing spacing between the piles). For a spacing

Table 1

Reduction Factor for a 4x4 Pile Group

k_r	s/d	Poulos	only x	x and y	x, y, and z
	2	4.01	3.92	3.20	3.18
10^{-5}	6	2.17	2.16	1.98	1.98
	10	1.71	1.71	1.64	1.64
	2	7.17	6.73	6.71	6.67
10^{-1}	6	4.85	4.66	4.65	4.60
	10	3.86	3.78	3.76	3.73

Table 2

Horizontal Load Distribution P/Pav.

k_r	s/d	Poulos	only x	x and y	x, y, and z	Pile
	2	1.55	1.53	1.30	1.28	1
10^{-5}		0.92	0.91	1.00	1.02	2
		1.10	1.10	1.00	0.98	3
		0.43	0.46	0.70	0.72	4
	6	1.20	1.20	1.15	1.15	1
		1.00	1.00	1.01	1.01	2
		1.02	1.02	1.00	1.00	3
		0.78	0.78	0.84	0.84	4
	10	1.12	1.12	1.10	1.10	1
		1.00	1.00	1.01	1.01	2
		1.00	1.00	1.00	1.00	3
		0.88	0.88	0.89	0.89	4

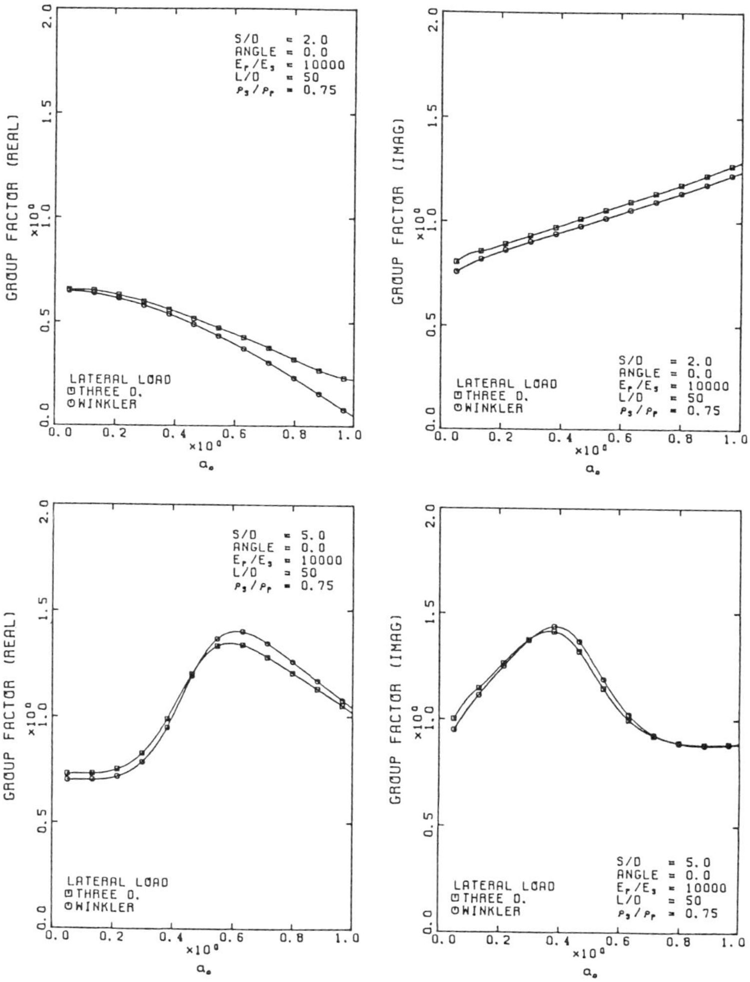

FIG. 8 EFFECT OF WINKLER ASSUMPTION - LATERAL LOAD 0°

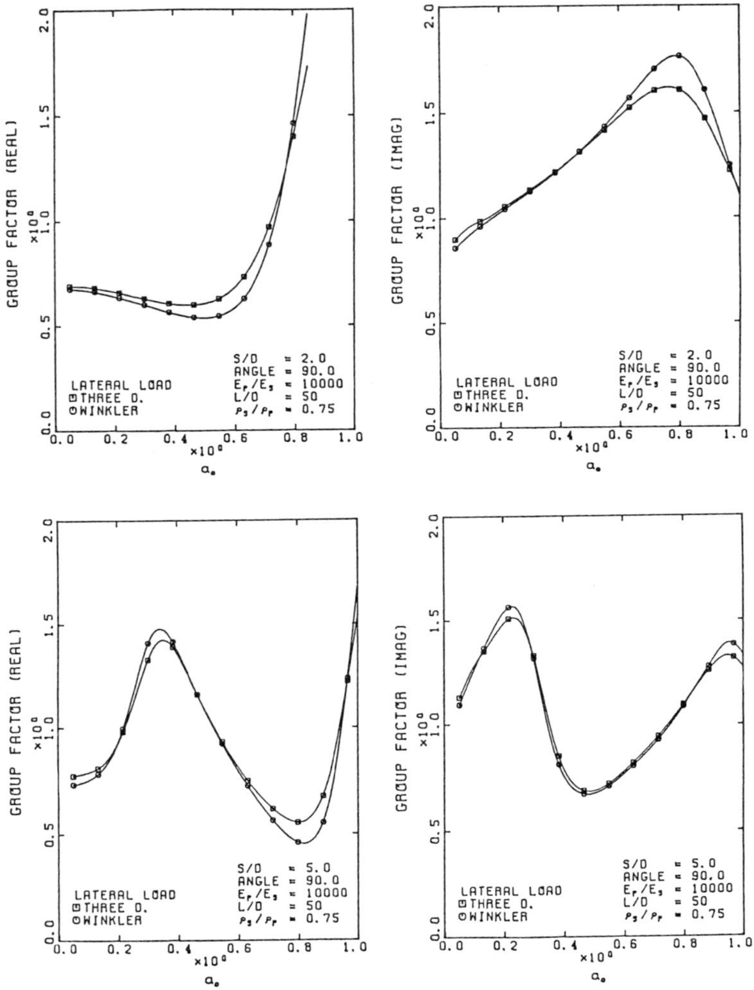

FIG.9 EFFECT OF WINKLER ASSUMPTION LATERAL LOAD 90°

of 2 diameters the results are still close for dimensionless frequencies smaller than 0.5 but the agreement deteriorates somewhat with increasing frequencies. It appears, however, that in this range the Winkler assumption introduces errors which are smaller than those associated with other normal assumptions (particularly neglecting the size of the other pile).

Conclusions

A number of procedures have been proposed in recent years for the dynamic analysis of complete pile foundations including group effects. Most of these procedures introduce some sort of approximation to reduce the cost of computation. Dynamic group effects depend highly on frequency, particularly for large pile spacings. When the pile spacing is of the order of five diameters or more it appears that a number of simplifying assumptions can be used with very little error. These simplified models are, however, less accurate when dealing with closely spaced piles and dimensionless frequencies larger than 0.5. Significant questions remain as to the validity in practice of linear elastic solutions which assume perfect bonding between the piles and the soil and linear soil behavior. More experimental work is badly needed.

References

1. Angelides D.C. and Roesset J.M. (1980). "Nonlinear Dynamic Stiffness of Piles." Research Report R80-13. Department of Civil Engineering. M.I.T. April.

2. Blaney G.W., Kausel E. and Roesset J.M. (1975). "Dynamic Stiffness of Piles." Proc. of the 2nd International Conference on Numerical Methods in Geomechanics. Blacksburg, Virginia.

3. Gomez R. and Roesset J.M. (1982). "Dynamic Stiffness of Pile Groups." Internal Report Department of Civil Engineering. The University of Texas at Austin.

4. Kagawa T. and Kraft L.M. Jr. (1981). "Dynamic Characteristics of Lateral Load-Deflection Relationships of Flexible Piles." International Journal of Earthquake Engineering and Structural Dynamics. Vol. 9.

5. Kaynia A.M. and Kausel E. (1982) "Dynamic Stiffness and Seismic Response of Pile Groups." Research Report R82-03. Department of Civil Engineering. M.I.T.

6. Kobori, T, Minai R. and Baba K (1977) "Dynamic Behavior of a Laterally Loaded Pile." Proc. of the 9th International Conference on Soil Mechanics and Foundation Engineering - Specialty Session 10 - Tokyo, Japan.

7. Kuhlemeyer R. L. (1979a) "Vertical Vibration of Piles". Journal of the Geotechnical Engineering Division. ASCE. Vol. 105. No. GT2.

8. Kuhlemeyer R. L. (1979b). "Static and Dynamic Laterally Loaded Floating Piles." Journal of the Geotechnical Engineering Division. ASCE. Vol. 105. No. GT2.

9. Matlock H. and Reese L.C. (1960). "Generalized Solutions for Laterally Loaded Piles". Journal of the Soil Mechanics and Foundations Division. ASCE. Vol. 86. SM5.

10. Matlock H. (1970) "Correlations for Design of Laterally Loaded Piles in Soft Clay." Proceedings of the 2nd Offshore Technology Conference. Paper OTC1204.

11. Matlock H., Ingram W.B., Kelley A.E. and Bogard D. (1980). "Field Tests of the Lateral Load Behavior of Pile Groups in Soft Clay." Proceedings of the 12th Offshore Technology Conference - paper OTC3871.

12. Nogami T. and Novak M. (1976). "Soil-Pile Interaction in Vertical Vibration" International Journal of Earthquake Engineering and Structural Dynamics - Vol. 4.

13. Nogami T. (1979) "Dynamic Group Effect of Multiple Piles under Vertical Vibration." Proceedings of the Third ASCE Engineering Mechanics Specialty Conference. Austin, Texas.

14. Nogami T. (1980) "Dynamic Stiffness and Damping of Pile Groups in Inhomogeneous Soil." ASCE. Special Technical Publication on Dynamic Response of Pile Foundations.

15. Nogami T. and Hsiao Liang (1983) "Behavior of Pile Groups Subjected to Dynamic Loading" Proceedings Fourth Canadian Conference on Earthquake Engineering. Vancouver.

16. Novak M. and Nogami T. (1977) "Soil Pile Interaction in Horizontal Vibration." International Journal of Earthquake Engineering and Structural Dynamics, Vol. 5.

17. O'Neill M., Hawkins R.A. and Mahor L.J. (1982) "Load Transfer Mechanisms in Piles and Pile Groups." Journal of the Geotechnical Engineering Division. ASCE. Vol. 108, no. GT12.

18. Poulos H.G. (1971) "Behavior of Laterally Loaded Piles: II-Pile Groups." Journal of Soil Mechanics and Foundations Division. ASCE. Vol 97. No. SM5.

19. Reese L.C., Cox W.R. and Koop F.D. (1975) "Field Testing and Analysis of Laterally Loaded Piles in Stiff Clay." Proceedings of the 7th Offshore Technology Conference. Paper No. OTC2312.

20. Sanchez Salinero I. (1982) "Static and Dynamic Stiffnesses of Single Piles." Research Report GR82-31. Department of Civil Engineering. The University of Texas at Austin.

21. Sanchez Salinero I. (1983). "Dynamic Stiffness of Pile Groups: Approximate Solutions." Research Report GR83-5. Department of Civil Engineering. The University of Texas at Austin.

22. Sheta M. and Novak M. (1982) "Vertical Vibration of Pile Gruops." Journal of the Geotechnical Engineering Division. ASCE. Vol. 108. No. GT4.

23. Stevens J.B. and Audibert J.M.E. (1979) "Re-examination of P-y curve Formulations." Proceedings of the 11th Offshore Technology Conference paper No. OTC3402.

24. Tajimi H. (1969) "Dynamic Analysis of a Structure Embedded in an Elastic Stratum." Proceedings of the 4th World Conference in Earthquake Engineering. Santiago, Chile, Vol. 3.

25. Thompson G.R. (1977). "Application of the Finite Element Method to the Development of P-y curves for Saturated Clays." M.S. Thesis. Department of Civil Engineering. The University of Texas at Austin.

26. Waas G. and Hartmann H.G. (1981) "Analysis of Pile Foundations under Dynamic Loads." SMIRT Conference-Paris.

27. Wolf J.P. and Von Arx G.A. (1978). "Impedance Function of a Group of Vertical Piles." Proceedings of the Specialty Conference on Soil Dynamics and Earthquake Engineering. ASCE - Pasadena, California - Vol. 2.

28. Yegian M. and Wright S.G. (1973). "Lateral Soil Resistance-Displacement Relationships for Pile Foundations in Soft Clays." Proceedings of the 5th Offshore Technology Conference. Paper No. OTC1893.

WINKLER SOIL MODEL FOR AXIAL RESPONSE ANALYSIS
OF PILE GROUPS

by

Toyoaki Nogami[1], M. ASCE and Stephen K. Paulson[2]

ABSTRACT

An efficient and convenient approach for analysis of axially loaded pile group response is presented in this paper. In the approach, the soil is idealized as a nonlinear "Winkler model for a pile group" attached to a group of piles. The model is conceptually identical to the commonly used Winkler model for single piles but can account for pile-soil-pile interaction. This idealization of the soil medium enables use of an efficient method for calculating pile group response. Using the approach presented, pile group responses are computed and compared with the responses obtained by those obtained idealizing the soil as an elastic continous medium. The calculated pile group responses are also compared with those observed in a full-scale field load test on a pile group. It is confirmed from these comparisons that the approach presented can favorably predict the behavior of axially loaded pile groups.

INTRODUCTION

The behavior of loaded pile groups is affected by various factors and is extremely complex. Various approaches have been proposed for computation of axially loaded pile group response. They are characterized by the type of soil model used in the analysis.

A finite element soil model (Ottaviani, 1975), while providing the most rational representation of the soil-pile system, is expensive to use, requires empirical input, and is considered to be a research tool.

A Mindlin's solution, obtained for an elastic continuous medium, was used in the analysis of nonlinear pile groups. Nonlinear behavior was taken into account by assigning an elasto-plastic behavior at the soil-pile interface and adjusting the Young's modulus of the soil (Poulos, 1980; and Poulos and Randolph, 1983).

---

1. Assoc. Prof. of Civil Engineering, University of Houston, Houston, Texas 77004

2. Senior Project Engineer, Gulf Oil Exploration and Production Company, Houston, Texas 77236

A Winkler soil model, characterized by the force-displacement relationship (load transfer curve), is often used for a single pile response analysis. In order to account for group effects, O'Neill (1977 and 1979) used a Mindlin's solution to modify the load transfer curves due to interaction effects. Focht and Koch (1973) approximated the group effects by modifying the pile-head displacement computed for a single pile. In their analysis, load transfer curves were used for a single pile response analysis and modification to account for group effects was made using Mindlin's solution. Randolph developed expressions for the pile-head response of a single pile using an elastic Winkler model. (Randolph, 1980; Poulos and Randolph, 1983; and Randolph and Worth 1979). In order to account for group effects, the single pile response was modified using a simplified interaction factor expression found from finite element analyses.

The Winkler model concept has also been directly applied in modeling the soil in a pile group response analysis (Nogami, 1980, 1983 and 1984). The model is convenient and efficient in modeling soil behavior in pile group response computations. However, this model is only applicable to low load level elastic conditions and is not well suited for pile groups subjected to large load levels.

NONLINEAR WINKLER MODEL FOR PILE GROUP

The soil medium around a group of piles in a Winkler model can be idealized as a set of horizontal springs interconnecting the pile shafts and fixed boundary points as shown in Fig. 1. Under this idealization, the soil reactions against the pile shafts are related to the pile shaft displacements only at the level where the soil reactions are considered. This idealization is conceptually identical to a traditional Winkler model and is called a "Winkler model for a pile group" (Nogami, 1980, 1983 and 1984).

When N piles are embedded in the soil medium idealized as a Winkler model for a pile group, the soil reaction forces against the piles at the depth z are expressed by

$$\{p(z)\} = [k_s]\,\{w(z)\} + \{c\} \tag{1}$$

where $\{p(z)\}$ and $\{w(z)\}$ = vectors containing the soil reaction forces against the N piles at depth z and the displacements of the N piles at depth z, respectively; $[k_s]$ = NxN soil stiffness matrix at the depth z; and $\{c\}$ = vector containing a component of the soil reaction which is independent of the pile shaft displacement.

Nonlinearity in the soil-pile system is induced by the nonelastic behavior of the soil and slippage at the soil-pile interface. It has been assumed that a strong nonlinearity develops within the soil region Δr around each pile and the soil response at $r>\Delta r$ is affected very little by this induced non-

Fig. 1 Winkler model for pile group

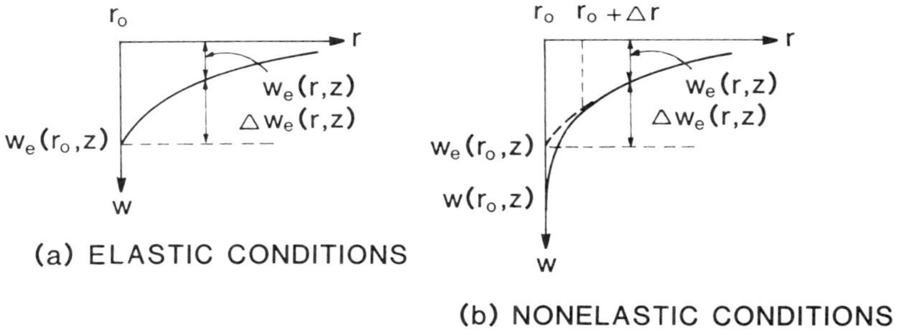

Fig. 2 Variation of soil displacement with distance r

linearity. Referring to Fig. 2, the vertical soil displacement at $r(>\Delta r)$ induced by the vertical displacement of a single pile is expressed by

$$w(r,z) = w_e(r_o,z) - \Delta w_e(r,z) \tag{2}$$

where r_o = radius of a pile; and the subscript, e, indicates the linear elastic condition. When the relationship between the soil reaction and pile shaft displacement (the so called load transfer curve or p-w curve) is given as shown in Fig. 3, the pile shaft displacement under linear and nonlinear conditions can be obtained from the following relationships, respectively

$$w_e(r_o,z) = f_o\, p(z) \qquad \text{and}$$

$$w(r_o,z) = f_o'\, p(z) + d \qquad p_i \leq p(z) \leq p_{i+1} \tag{3}$$

where f_o, f_o' and d = constants to define a linearized load transfer curve. The difference in the displacements between r_o and r, Δw_e, is calculated from a plane strain solution given by,

$$\Delta w_e\,(r,\, z) = \frac{\ell n\, r/r_o}{2\pi\, G_s(z)}\, p(z) = \Delta f\, p(z) \tag{4}$$

where G_s = shear modulus of the soil.

Equations 2, 3 and 4 lead to the following force-displacement relationship for a Winkler model for a pile group:

$$\{w(z)\} = [f_s]\, \{p(z)\} + \{d\} \tag{5}$$

The term located at (i,j) in $[f_s]$ is

$$f_s(i,j) = f_o - \Delta f(r_{ij}) \qquad \text{for } i \neq j$$

$$f_s(i,j) = f_o' \qquad \text{for } i = j \tag{6}$$

where r_{ij} = spacing distance between the ith and j^{th} piles. "f_o'" located at at the position (i, i) in $[f_s]$ and "d" located at the position i in $\{d\}$ are determined from the load transfer curve corresponding to the ith pile. Once $[f_s]$ and $\{d\}$ are known, $[k_s]$ and $\{c\}$ in Eq. 1 can be calculated from

$$[k_s] = [f_s]^{-1}$$

$$\{c\} = -\,[f_s]^{-1}\,\{d\} = -\,[k_s]\,\{d\} \tag{7}$$

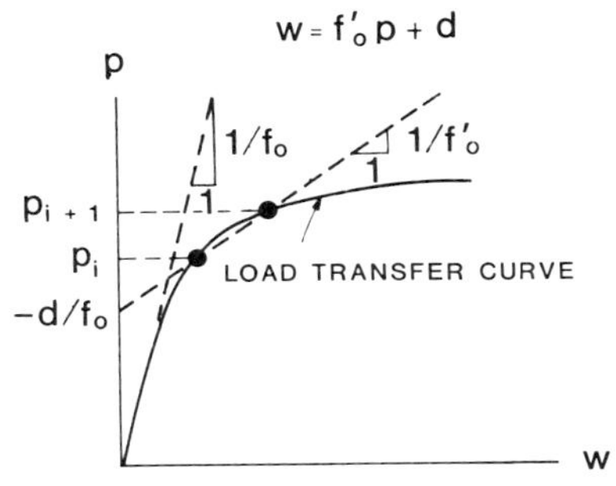

Fig. 3 Linearized load transfer curve for the load level between p_i and P_{i+1}

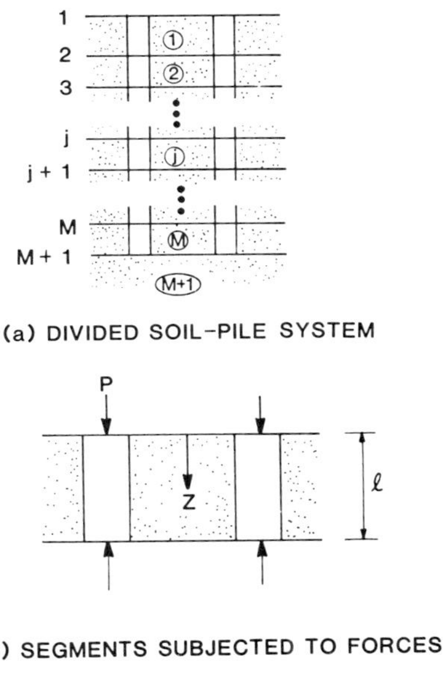

Fig. 4 Divided soil-pile system and segments subjected to forces

FORMULATION AND COMPUTATION OF PILE GROUP RESPONSE

The soil-pile system made of N piles is divided into M homogeneous pile segments and horizontal soil layers as shown in Fig. 4. When the soil medium around the pile shafts in a group is idealized as a Winkler model for a pile group, the soil reactions against the pile shafts induced by the vertical pile displacements are expressed by Eq. 1. Thus the governing equation of the axial response of the pile group can be written by

$$[EA]\left\{\frac{d^2w}{dz^2}\right\} = -[k_s]\{w\} - \{c\} \tag{8}$$

where $[EA]$ = NxN diagonal matrix containing the axial stiffnesses of N piles.

Solving Eq. 8 for the boundary conditions at both ends of the pile segments as shown in Fig. 4, one can find the following relationship (see Appendix II for derivation):

$$\{Z\}_i^B = [t_i]\{Z\}_i^A + \{q\}_i \qquad i = 1,2,3, ---, M \tag{9}$$

where the expressions of $[t_i]$ and $\{q\}_i$ are given in Appendix II; $\{Z\} = \{\underline{w}, \underline{P}\}^T$ = vectors containing the axial displacements and forces of N piles; and $\{Z\}_i^A$ and $\{Z\}_i^B$ = $\{Z\}$ at the upper and lower ends of the ith segment, respectively.

When the soil displacements at the pile tips are expressed by

$$\{w_{base}\} = [f_{base}]\{p_{base}\} + \{q_{base}\}, \tag{10}$$

the vector at the base of the pile tip, $\{Z\}_{M+1}^A$, is

$$\{Z\}_{M+1}^A = \begin{Bmatrix} \underline{f}_{base}\ \underline{p}_{base} + \underline{q}_{base} \\ \underline{p}_{base} \end{Bmatrix} \tag{11}$$

where $\underline{p}_{base} = \{p_{base}\}$ = vector containing the soil reaction forces at the bases of N piles; $\underline{q}_{base} = \{q_{base}\}$; and $\underline{f}_{base} = [f_{base}]$ = soil flexibility matrix at the bases of pile tips.

The compatibility and equilibrium conditions at the boundary between the adjacent segments require

$$\{Z\}_i^B = \{Z\}_{i+1}^A \qquad i = 1,2,3, ---, M \tag{12}$$

Successive applications of Eqs. 9 and 12 for i=1 through M lead to

$$\{Z\}_{i+1}^A = [T]_i\{Z\}_1^A + \{Q\}_i \qquad i = 1,2,3, ---, M \tag{13}$$

where,

$$[T_i] = [t_i]\,[t_{i-1}] \text{ --- } [t_2]\,[t_1]$$

$$\{Q\}_1 = \{q\}_1 \tag{14}$$

$$\{Q\}_i = \sum_{n=1}^{i-1} [t_i]\,[t_{i-1}] \text{ --- } [t_{n+1}]\,\{q\}_n + \{q\}_i \quad \text{for } i \geq 2$$

Substituting Eq. 11 into $\{Z\}_{M+1}^{A}$ in Eq. 13 and rearranging it, the pile-head forces of N piles are expressed by

$$\{P\} = [K_p]\,\{w\} + \{R\} \tag{15}$$

where $\{w\}$ and $\{P\}$ = vectors containing axial displacements and forces of N piles at their heads; and

$$[K_p] = [\underline{f}_{base}\ \underline{T}_{11} - \underline{T}_{22}]^{-1}\,[\underline{T}_{21} - \underline{f}_{base}\ \underline{T}_{11}]$$

$$\{R\} = [\underline{f}_{base}\ \underline{T}_{11} - \underline{T}_{22}]^{-1}\,\{\underline{Q}_p - \underline{f}_{base}\ \underline{Q}_w\} - \{q_{base}\} \tag{16}$$

Subvectors $\underline{Q}_p$ and $\underline{Q}_w$ and submatrices $\underline{T}_{11}$, $\underline{T}_{12}$, $\underline{T}_{21}$, and $\underline{T}_{22}$, in Eq. 16 are located in the following locations, respectively

$$\{Q\}_{M+1} = \begin{Bmatrix} \underline{Q}_w \\ \underline{Q}_p \end{Bmatrix} \quad \text{and} \quad [T_{M+1}] = \begin{bmatrix} \underline{T}_{11} & \underline{T}_{12} \\ \underline{T}_{21} & \underline{T}_{22} \end{bmatrix} \tag{17}$$

When the piles are attached to a flexible common pile cap, the response of pile-heads attached to the pile cap can be obtained by solving the following equation:

$$\{P\} = [K]\,\{w\} + \{R\} \tag{18}$$

where $[K] = [K_p] + [K_c]$; $[K_c]$ = stiffness matrix of the pile cap; and $\{P\}$ = vector containing the applied forces at the cap.

Based on the idealization of the soil medium and formulations of the pile group response above described, a computer program "GROUP8" has been developed for the computation of both axial and flexural pile group responses. The following are computation procedures used in GROUP8:

1. Assuming the range between p_i and p_{i+1}, determine f_o' and d for all segments of pile shafts (see Fig. 3).

2. Construct $\lfloor k_s \rfloor$ and $\{c\}$ for the Winkler model for pile group defined in Eq. 1 for all soil layers.

3. For all segments, compute $[T_i]$ and $\{Q\}_i$ defined in Eq. 13 after first computing $\lfloor t_i \rfloor$ and $\{q\}_i$ defined in Eq. 9.

4. Compute pile-head stiffness, $[K_p]$, and $\{R\}$ from Eq. 16.

5. Construct $[K\rfloor$ defined in Eq. 18, when piles are attached to a common cap.

6. Compute the pile-head displacements by solving Eq. 18.

7. Compute the pile-head forces from Eq. 15 with the pile-head displacements computed at Step 6.

8. Compute the axial displacements and forces in the piles at the connections between the segments from Eq. 13 with $\{Z\}_1^A$ computed in the previous steps.

9. Compute the soil reaction force at the mid-point of each segment from Eq. 1 where $\{w\}$ is the average displacement between those at the upper and lower ends of the segment.

10. If the computed soil reaction force p is outside of the range between p_i and p_{i+1}, redefine a linear relationship (f_o' and d) by changing p_i and p_{i+1}.

11. Repeat the above procedures from Step 2 through 10 until the computed soil reaction force p is within the range between p_i and p_{i+1} assumed.

COMPARISON WITH RESPONSES COMPUTED BY USING AN ELASTIC CONTINUOUS SOIL MODEL

One of the most critical assumptions used in the present approach is idealizing the soil medium around the pile group as a Winkler model for a pile group. In order to examine this assumption, the elastic responses of pile groups computed using the present approach are compared with those computed by other approaches based on idealizing the soil as an elastic continuous medium.

Under elastic conditions, f_o' and d in Eq. 3 are equal to f_o and zero, respectively. Thus, according to Eqs. 5, 6 and 7, $\{c\}$ is equal to zero and the only input parameter to construct $\lfloor k_s \rfloor$ is the value f_o. Because of the definition of f_o given in Eq. 3, the value f_o can be conveniently determined by finding the stiffness of an equivalent Winkler model for a single pile (k_o) and using the relationship $f_o = k_o^{-1}$: an equivalent Winkler model for a single pile is defined as a Winker model which produces the pile-head displacement of a single pile attached to a Winkler model identical to that embedded in a continuous soil medium. The stiffness of an equivalent Winkler model for a single pile is assumed to be expressed by

$$k_o\ (z) = G_s(z)\ \bar{k}_o \tag{19}$$

where $\bar{k}_o$ is assumed to be independent of depth z and uniform along the entire length of the pile. Thus, the equivalent Winkler model is defined in terms of a single value $\bar{k}_o$ for a given soil profile and pile. In finding $\bar{k}_o$, displacements of single piles computed by Mattes and Poulos (1969) and Poulos (1979) were used as the displacements of single piles embedded in an elastic continuous soil medium.

Considering two identical piles subjected to identical loads at the both heads, the interaction factor, α, is defined by

$$\alpha = \frac{\text{additional displacement due to adjacent pile}}{\text{single pile displacement}}$$

Figs. 5 and 6 show the comparisons of the interaction factors computed by the Winkler model approach (Nogami, 1984) with those computed by Poulos (1974 and 1979) and Banerjee and Davies (1977). The soil was treated as a three dimensional continuous medium in the latter analyses. It is seen in the figures that a Winkler model for a pile group can be effectively used for modelling a continuous soil medium in pile group response analysis.

COMPARISON WITH FIELD TEST RESULTS

Field Test

A series of full-scale axial static load tests on groups of vertical piles were reported by O'Neill et al (1981). The test piles were steel pipes, 10.75 inches in outside diameter with a wall thickness of 0.365 inches, and driven to 40 feet below the ground surface. Nine of the piles formed a pile group with nominally three diameters between centers and were attached to a common reinforced concrete cap with 4.25 foot thickness (Fig. 7). A three foot space was left between the bottom of the pile cap and the soil surface. Two other piles were also tested to get information on the behavior of single isolated piles. They were identical to the group piles except that they were not capped and extended to 3.3 feet above the ground surface.

A detailed soil property and characterization description of the test site is found in Refs. 3 and 10. Figs. 8, 9 and 10, reproduced from the references, show the stratigraphy, undrained shear strength and Young's modulus of the soil at the test site, respectively. In general, the soil is a very stiff, saturated clay preconsolidated by desiccation. The layers depicted in Fig. 8 become generally less plastic, less compressible, and sandier with depth. Stratum C is silty and represents a zone of weathering on the prehistoric surface of a Pleistocene terrace. The soils above this

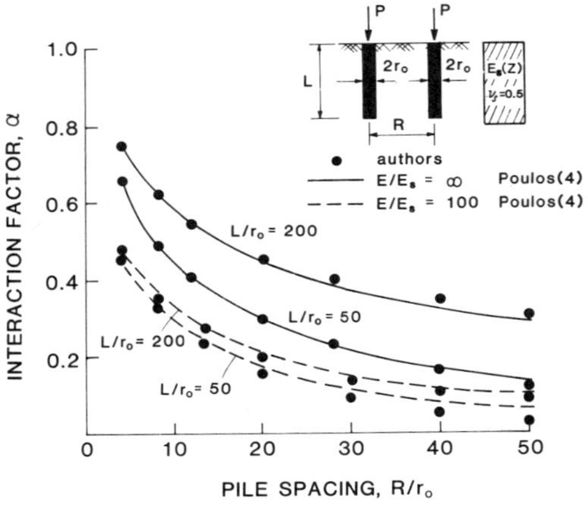

Fig. 5 Group factors for homogeneous soil profiles computed using the Winkler model for a pile group and continuous model (after Nogami, 1984)

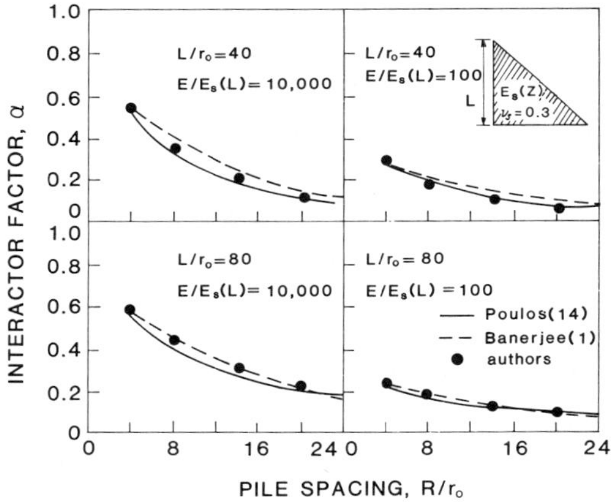

Fig. 6 Group factors computed, using Winkler model for a pile group and continuous model, for nonhomogeneous soil profiles (after Nogami, 1984)

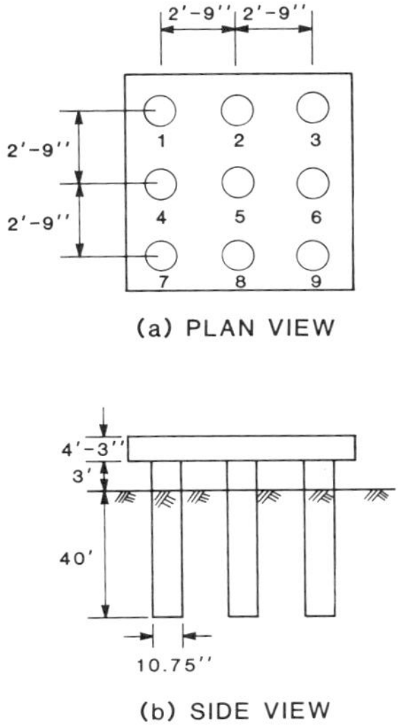

Fig. 7 3x3 pile group used for the field load test

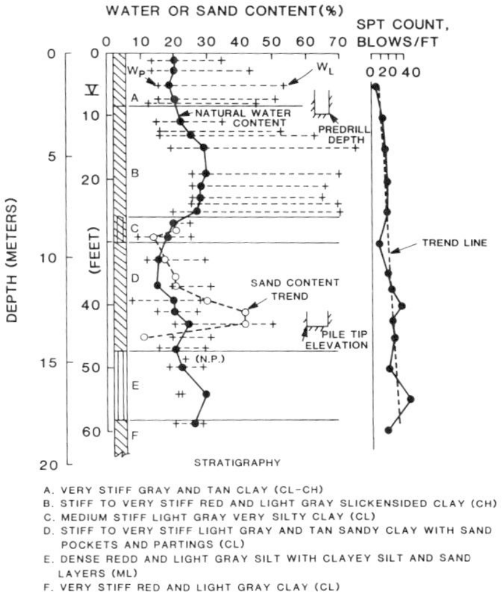

Fig. 8 Stratigraphy of soil at the test site (after O'Neill, 1981)

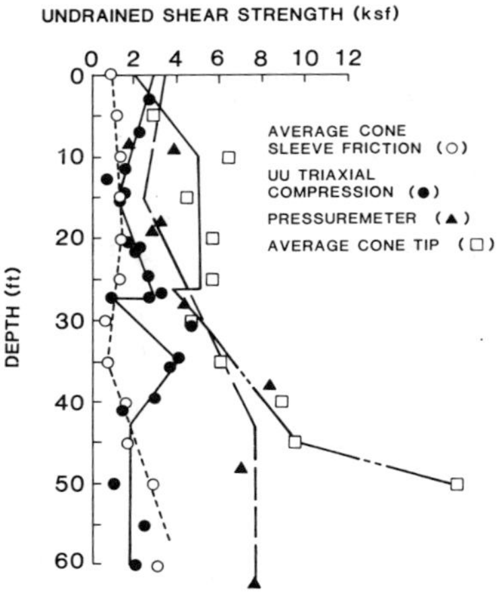

Fig. 9 Distribution of undrained shear strength of the soil at the test site (after O'Neill, 1981)

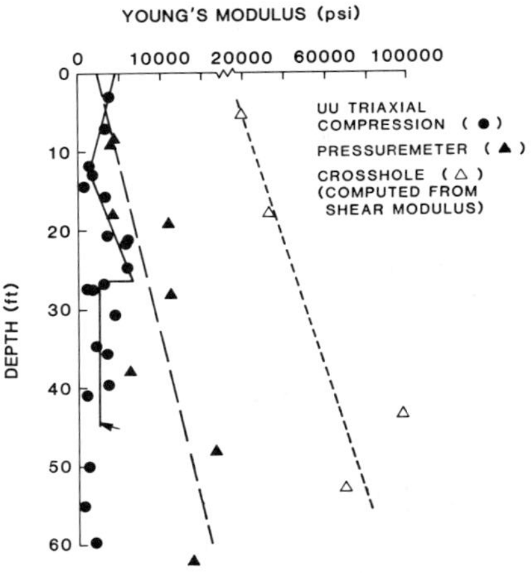

Fig. 10 Distribution of Young's modulus of the soil at test site (after O'Neill, 1981)

stratum are slickensided, and the soils below that depth contain numerous sand seams.

Determination of Input Parameters For the Pile Group Response Analysis

In order to determine f_o (= k_o^{-1}) in terms $\bar{f}_o$ (= $\bar{k}_o^{-1}$), the elastic constants of the soil must be known (see Eq. 19). Four different Young's modulus distributions were assumed as shown in Fig. 11, where the distributions A and D correspond to the values determined from the pressuremeter and cross-hole seismic survey, respectively. The value $\bar{f}_o$ for each of these four soil profiles were determined under the elastic behavior as shown in Fig. 12. Using four sets of soil profiles and their $\bar{f}_o$ values, the displacements of the pile group were analyzed within an elastic range of behavior and the best-fit soil profile was determined as shown in Fig. 13. The best-fit soil profile obtained is close to the one obtained from a cross-hole seismic survey, as shown in Fig. 11, where the broken curve corresponds to the best-fit soil profile.

Unit load transfer curves were constructed from the following relationship modified from the one originally proposed by Vijayvergiya (1977):

$$p = p_{max}\,[2(\beta\, w/w_c)^{0.5} - (\beta\, w/w_c)] \tag{20}$$

where p = unit side resistance; w = pile displacement; p_{max} = 0.5 c_u; c_u = undrained shear strength; β = scaling factor; and w_c = pile deflection corresponding to p_{max}. The undrained shear strength c_u used for the computation of p_{max} is given in Fig. 14. The value w_c in the above relationship was computed from the following expression given by Vijayvergiya (1977):

$$w_c = \frac{p_{max}\, r_o}{G_s}\left\{0.67 + \ell n\, \frac{2\rho\, L(1-\nu_s)}{r_o}\right\} \tag{21}$$

where L = pile penetration depth; ρ = G_s at mid-depth of the pile divided by G_s at pile tip; and ν_s = Poisson's ratio of soil (assumed to be 0.5). The expression of the unit load transfer curve originally given by Vijayvergiya corresponds to the expression in Eq. 20 with β = 1.0. β = 0.3 was used in the present analysis. Furthermore, since f_o was defined independently from the analysis of a single isolated pile for an elastic behavior, the load transfer curve expressed by Eq. 20 was modified as shown in Fig. 15, to make the load transfer curve for small displacement levels compatible with f_o.

Comparison with Field Test Results

Using the inputs defined above, the response of a reference pile (single isolated pile) was computed for the loads applied in the field test, and the computed values were compared with field observed values in Figs. 16 and 17.

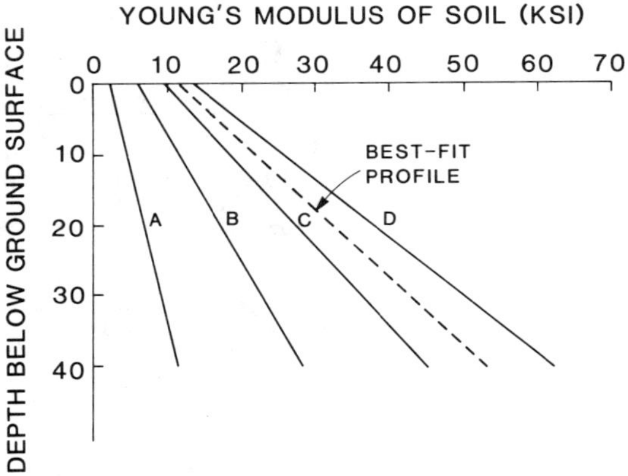

Fig. 11 Assumed distribution of the soil Young's modulus to find the best-fit profile

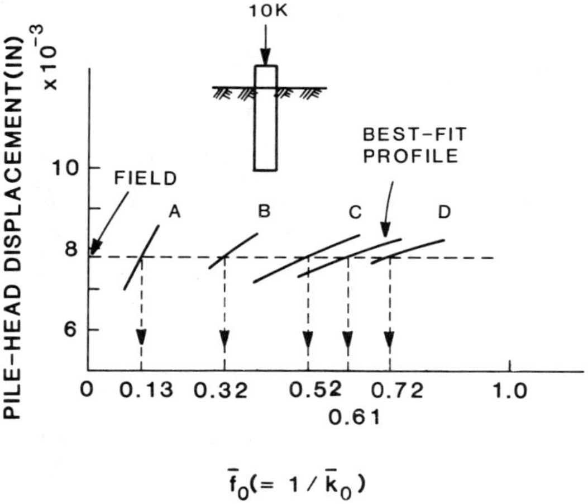

Fig. 12 Determination of $\bar{f}_0$ for assumed Young's modulus distribution

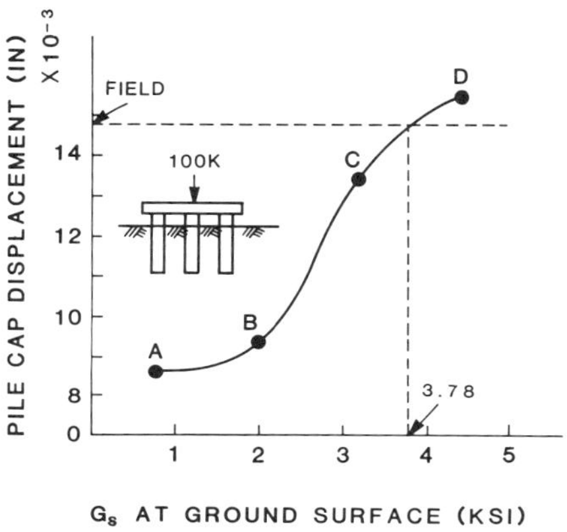

Fig. 13 Determination of best-fit shear modulus distribution of test site

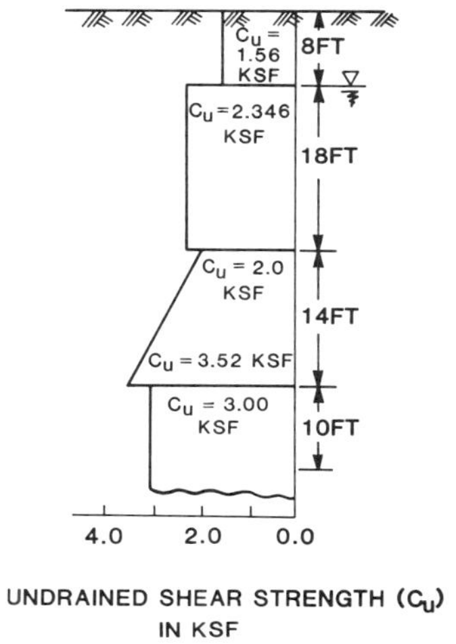

Fig. 14 Undrained shear strength of soil used for computations

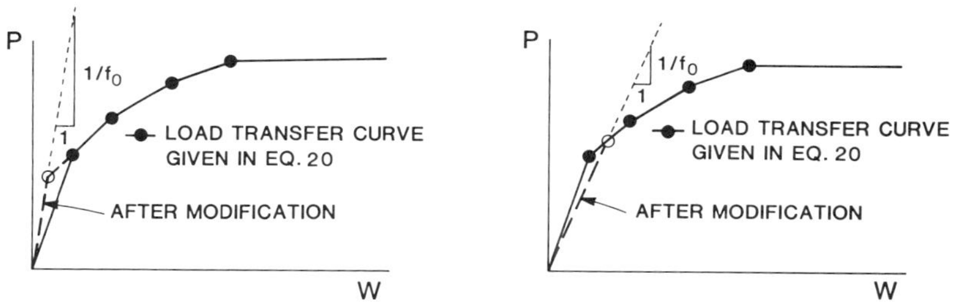

Fig. 15 Correction of empirical load transfer curve

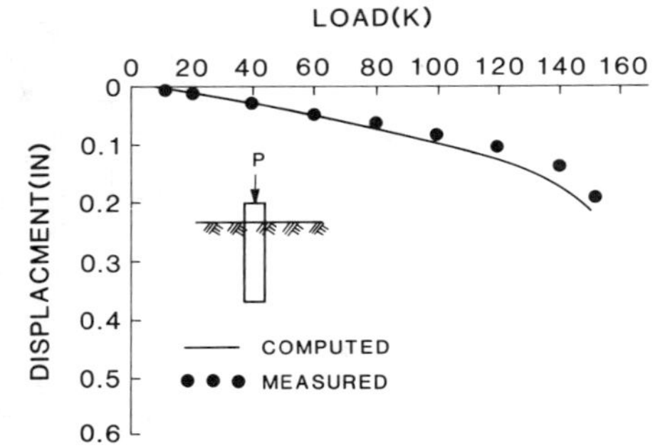

Fig. 16 Computed and field observed pile-head displacements of a single pile

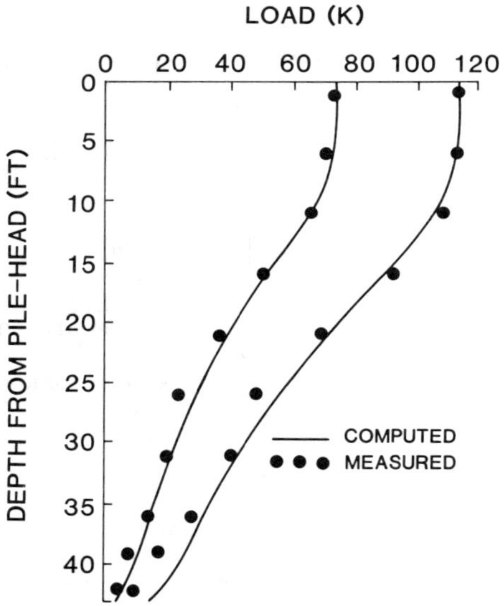

Fig. 17 Computed and field observed axial forces of a single pile along the pile shaft

The input used in the computation can reproduce the response of a single isolated pile adequately to those observed in the field.

Using the above input, the response of 3 x 3 pile group was analyzed for the loads applied in the field test. Figs. 18 and 19 show the comparison of the computed pile group response with those observed in the field test. The computed and field observed pile-head forces are given in Table 1. In general, a good agreement between the computed and field observed values can be seen. Thus, if the input parameters are selected from the behavior of a single isolated pile, the present approach can predict the nonlinear behavior of pile group reasonably well.

CONCLUSIONS

This paper presents a nonlinear "Winkler model for a pile group" and an efficient computational method for pile group response with this model. The input for defining the Winkler model for a pile group are f_0 (or k_0), load transfer curves, and the shear modulus of the soil. The shear modulus corresponding to a low strain level of undisturbed soil is recommended for the input shear modulus of soil. This soil modulus is conveniently obtained from a cross-hole seismic survey. The value f_0 (or k_0) and load transfer curves are either determined from the behavior of a single pile observed in a field load test at the site or selected so that they can adequately produce the behavior of a single pile at the site.

ACKNOWLEDGEMENTS

This paper describes some of the results obtained in the early portion of a series of research projects on group pile foundations initiated in 1979. The projects were supported by Gulf Oil Corporation. Graduate students Francisco Mora, of the Rensselaer Polytechnic Institute, and Peter Tsiatsios and H.L. Chen, of the University of Houston, assisted during the studies of thse projects. The authors express their appreciation to these dedicated graduate students.

APPENDIX I: REFERENCES

1. Banerjee, P. K. and Davies, T. G., "Analysis of Pile Groups Embedded in Gibson Soil," Proceedings, Ninth International Conference on Soil Mechanics and Foundation Engineering, Vol. I, 1977, pp. 381-386.
2. Focht, J. A., Jr. and K. J. Koch, " Rational Analysis of Lateral Performance of Pile Groups," OTC 1896, 5th OTC, II, May, 1973, pp. 701-708.
3. Mahar, L. J., and O'Neill, M. W., "Geotechnical Characterization of Desiccated Clay," Journal of Geotechnical Engineering Division, ASCE, Vol. 109, No. GT1, 1983, pp. 56-71.

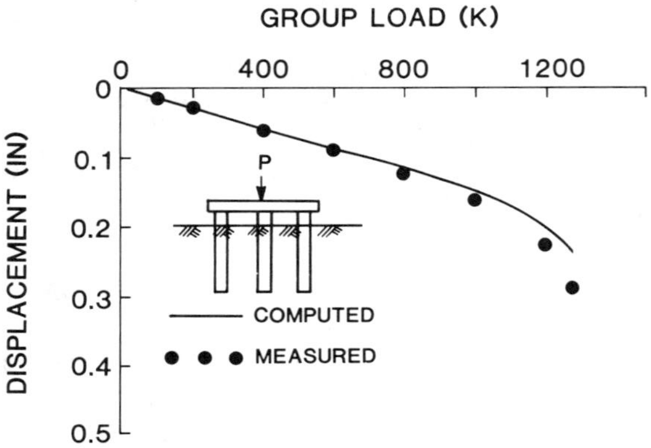

Fig. 18 Computed and field observed pile-head displacement of a pile group

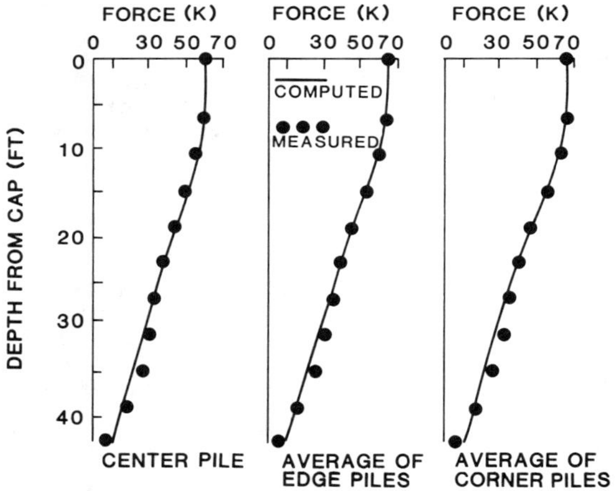

Fig. 19a Computed and field observed axial forces of grouped piles along the pile shafts (applied load = 581.5 K)

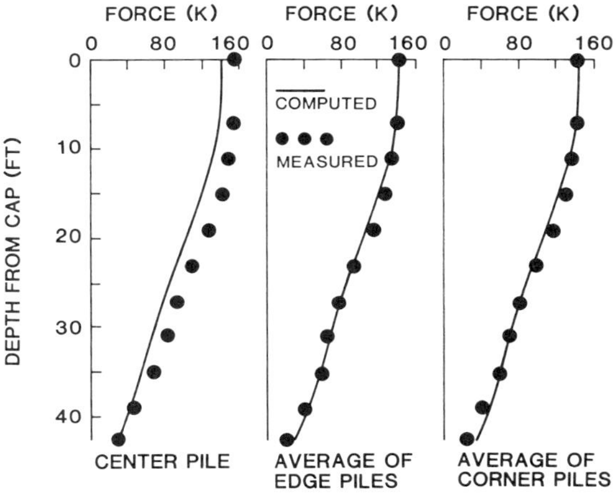

Fig. 19b Computed and field observed axial forces of grouped piles along the pile shafts (applied load = 1274.7 K)

TABLE 1

COMPARISON BETWEEN MEASURED AND COMPUTED PILE-HEAD FORCES

Nominal Group Load (Kips)	Pile	Measured Load (Kips)	Computed Load (Kips)
	1	68.0	66.5
581.4	2	64.1	63.7
	3	67.5	66.5
	4	68.6	63.4
	5	59.9	60.4
	6	61.0	63.7
	7	65.7	66.5
	8	62.2	63.7
	9	64.3	66.5
	1	137.8	130.6
1166.1	2	135.8	129.1
	3	131.8	130.6
	4	124.6	129.1
	5	128.8	126.9
	6	116.6	129.1
	7	132.8	130.6
	8	127.3	129.1
	9	130.5	130.6
	1	150.9	143.0
1274.7	2	152.2	141.0
	3	131.0	143.0
	4	135.1	141.0
	5	154.4	138.3
	6	126.0	141.0
	7	140.9	143.0
	8	139.4	141.0
	9	144.9	143.0

4. Mattes, N. S. and Poulos, H. G., "Settlement of Single Compressible Piles," Journal of Soil Mechanics and Foundations Division, ASCE, Vol. 95, No. SM1, 1969, pp. 189-209.
5. Nogami, T., "Dynamic Stiffness and Damping of Pile Groups in Inhomogeneous Soil," Dynamic Response of Pile Foundations: Analytical Aspects, ASCE, Oct., 1980, pp. 31-52.
6. Nogami, T., "Dynamic Group Effect in Axial Responses of Grouped Piles," Journal of Geotechnical Engineering Division, ASCE, Vol. 109, No. GT2, 1983, pp. 228-243.
7. Nogami, T. and Chen, H. L., "Simplified Approach for Axial Pile Group Response Analysis," Journal of Geotechnical Engineering Division, ASCE, Vol. 110, No. GT9, 1984.
8. O'Neill, M. W., Ghazzaly, O. I., and Ha, H. B., "Analysis of Three-Dimensional Pile Group with Nonlinear Soil Response and Pile-Soil-Pile Interaction," Proc., 9th OTC, II, OTC 2838, May, 1977, pp. 245-256.
9. O'Neill, M. W., Ghazzaly, O. I., and Ha, H. B., "Assessment of Hybrid Model for Piel Groups," TRR 733, T.R.B., Washington, D.C., 1979, pp. 36-43.
10. O'Neill, M. W., Hawkins, R. A., and Mahar, L. J., "Field Study of Pile Group Action: Final Report," Report Nos. FHWA/RD-81/002-FHWA/RD-81/008, Federal Highway Administration, U.S. D.O.T., March, 1981.
11. Ottaviani, M., "Three-Dimensional Finite Element Analysis of Vertically Loaded Pile Groups," Geotechnique, 25, No. 2, June, 1975, pp. 159-174.
12. Poulos, H. G. and Davis, E. H., "Elastic Solution for Soil and Rock Mechanics," John Wiley & Son, New York, 1974.
13. Poulos, H. G., "Group Factors for Pile Deflection Estimation," Journal of Geotechnical Engineering Division, ASCE, Vol. 105, No. GT12, 1979, pp. 1489-1509.
14. Poulos, H. G., "Settlement of Single Piles in Nonhomogeneous Soil," Journal of Geotechnical Engineering Division, ASCE, Vol. 105, No. GT5, 1979, pp. 627-641.
15. Poulos, H. G., "An Approach for the Analysis of Offshore Pile Groups," Numerical Methods in Offshore Piling, I.C.E., London, 1980, pp. 119-126.
16. Poulos, H. G. and Randolph, M. K., "Pile Group Analysis: A Study of Two Methods," Journal of Geotechnical Engineering Division, ASCE, Vol. 109, No. GT3, March, 1983, pp. 355-372.
17. Randolph, M. F. and Worth, C. P., "An Analysis of Vertical Deformation of Pile Groups," Geotechnique, Vol. XXIV, December, 1979, pp. 217-229.
18. Randolph, M. F., "PILGT: A Computer Program for the Analysis and Design of Pile Groups Under General Loading Conditions," Soil Report TR91 CUED/D, Cambridge University, 1980.
19. Vijayvergiya, V. N., "Load-Movement Characteristics of Piles," Proc., Ports '77, ASCE, II, 1977, pp. 269-286.

APPENDIX II. DERIVATION AND EXPRESSION OF MATRIX [t]

The behavior of N-pile group is governed by Eq. 8. It is assumed that the soil displacement at the pile shafts within a layer is expresssed by

$$\{w\} = \sum_n z^n \{a_n\} \qquad n = 0,1,2, --- \tag{22}$$

where $\{a_n\}$ = unknown constants; and coordinate z is defined in Fig. 4.

Solving Eq. 8 for $\{w\}$ after substituting Eq. 22 and setting the obtained $\{w\}$ equal to $\{w\}$ in Eq. 22, the displacement of the pile is expressed by

$$\{w\} = \{A\} + z\,\{B\} + \sum_{n=1} \frac{z^{2n}}{2n!}[\delta]^n\{A\} + \sum_{n=1} \frac{z^{2n+1}}{(2n+1)!}[\delta]^n\{B\}$$
$$+ \frac{z}{2}\,\{c\} + \sum_{n=1} \frac{z^{2n+2}}{(2n+2)!}[\delta]^n\{c\} \qquad n = 1,2,3,\ --- \qquad (23)$$

Where $\underline{A}$ and $\underline{B}$ = vectors containing unknown constants.

Since the expression of the axial force in the pile shaft can be obtained from P = EA dw/dz, the following expression is obtained in a matrix form,

$$\begin{Bmatrix} \underline{w}(z) \\ \underline{p}(z) \end{Bmatrix} = \begin{bmatrix} \underline{I} & \underline{0} \\ \underline{0} & \underline{EA} \end{bmatrix} \left[\begin{bmatrix} \underline{I} & z\underline{I} \\ \underline{0} & \underline{I} \end{bmatrix} + \sum_n \begin{bmatrix} \frac{z^{2n}}{2n!}\underline{\delta}^n & \frac{z^{2n+1}}{(2n+1)!}\underline{\delta}^n \\ \frac{z^{2n-1}}{(2n-1)!}\underline{\delta}^n & \frac{z^{2n}}{2n!}\underline{\delta}^n \end{bmatrix} \right] \begin{Bmatrix} \underline{A} \\ \underline{B} \end{Bmatrix}$$
$$+ \begin{bmatrix} \frac{z}{2}\underline{I} + \sum_n \frac{z^{2n+2}}{(2n+2)!}\underline{\delta}^n \\ \frac{1}{2}\underline{I} \quad \sum_n \frac{z^{2n+1}}{(2n+1)!}\underline{\delta}^n \end{bmatrix} \begin{Bmatrix} \underline{c} \\ \underline{c} \end{Bmatrix} \qquad n = 1,2,3,\ --- \qquad (24)$$

where $\underline{w}$ and $\underline{p}$ = vectors containing the axial displacements and forces of N piles in a group, respectively; $\underline{I}$ = $[I]$ = identity matrix; $\underline{EA}$ = $[EA]$; $\underline{\delta}$ = $[EA]^{-1}$ $[k_s]$; and $\underline{c}$ = $\{c\}$.

The axial displacements and forces of the piles at the upper and lower ends of the segments are denoted as, respectively

$$\begin{Bmatrix} \underline{w}(0) \\ \underline{P}(0) \end{Bmatrix} = \{Z\}^A \quad \text{and} \quad \begin{Bmatrix} \underline{w}(\ell) \\ \underline{P}(\ell) \end{Bmatrix} = \{Z\}^B \qquad (25)$$

Substituting Eq. 25 into Eq. 24 and eliminating the unknown constants, the following relationship is obtained:

$$\{Z\}^B = [t]\{Z\}^A + \{q\} \qquad (26)$$

where

$$\lfloor t \rfloor = \begin{bmatrix} \underline{I} + \sum_{n} \frac{\ell^{2n}}{2n!}\,\underline{\delta}^{n} & \ell\underline{I} + \sum_{n} \frac{\ell^{2n+1}}{(2n+1)!}\,\underline{\delta}^{n} \\ \underline{EA} \sum_{n} \frac{\ell^{2n-1}}{(2n-1)!}\,\underline{\delta}^{n} & \underline{EA} + \underline{EA} \sum_{n} \frac{\ell^{2n}}{2n!}\,\underline{\delta}^{n} \end{bmatrix}$$

$$\{q\} = \begin{bmatrix} \frac{1}{2}\,\underline{I} + \sum_{n} \frac{\ell^{2n-2}}{(2n+2)!}\,\underline{\delta}^{n} \\ \frac{1}{2}\,\underline{EA} + \underline{EA} \sum_{n} \frac{\ell^{2n+1}}{(2n+1)!}\,\underline{\delta}^{n} \end{bmatrix} \left\{ \begin{matrix} \underline{c} \\ \underline{c} \end{matrix} \right\} \qquad (27)$$

In the computations, the maximum number of terms used in the series expressions in $\lfloor t \rfloor$ and $\{q\}$ was three ($n = 3$).

THE SIMPLE-PILE APPROACH TO PILE INSTALLATION IN CLAYS

Mohsen M. Baligh, M. ASCE*

ABSTRACT

The Simple Pile Approach provides a simple yet realistic method for elucidating and predicting pile installation effects on the surrounding soils. This article describes the fundamentals of the approach and presents solutions for the effects of quasi-static deep undrained penetration of a simple solid pile on the deformations and strains everywhere in saturated isotropic clays and the deviatoric stresses in the outer soil located at some distance from the pile. Extensions of these solutions including stresses and pore pressures to more realistic conditions (e.g., anisotropic clays and drained penetration in sands) by means of the Strain Path Method are outlined. Evaluations of cavity expansion methods indicate that spherical expansion solutions have no direct application to pile installation studies and cylindrical cavity expansion solutions have the potential of providing reasonable predictions only in the outer "elastic" soil located at distances in excess of four to five radii from the pile axis and far behind the pile tip (about 10 radii).

INTRODUCTION

Pile installation causes severe disturbances and leads to significant changes in stresses, pore water pressures and properties of the soil around the pile. Since installation represents the first step in the life of a pile, the sound understanding and formulation and ultimately the realistic prediction of pile behavior hinges on the successful analysis of pile installation effects.

The development of reliable predictive methods for installation effects is complicated by the following factors:

a) the singularities and high gradients in these <u>two-dimensional problems</u> where the field variables depend on the radial and vertical locations;
b) the large deformations and strains in the soil as observed in laboratory and in situ situations;
c) the very complicated behavior of soils including nonlinearities, inelasticity and anisotropy and, possibly, time-dependent (visco-elastic) and frictional response;
d) the presence of water (and possibly gas) in the pores requiring the treatment of the soil as a multi-phase medium; and
e) the characteristics of the soil-pile interface that cannot be simulated by linear modelling e.g. Coulomb friction.

* Professor, Dept. of Civil Engineering, Massachusetts Institute of Technology, Cambridge, Mass. 02139

Therefore, the rational and realistic analysis of pile installation effects represents a most challenging and important geotechnical problem. This article presents a new approach to the installation process: the Simple-Pile Approach that on one hand provides a more realistic tool for investigating pile installation effects as compared to existing methods and, on the other, involves a reasonable degree of complexity to make it useful to the practicing engineer. For the sake of clarity, the article is limited to undrained deep pile installation in saturated (incompressible) isotropic homogeneous clays.

CAVITY EXPANSION APPROACH

Modern Cavity Expansion Methods (CEM) appear to have been first developed for applications to metal indentation (Bishop et al., 1945) and were then generalized and utilized in geotechnical applications primarily for the interpretation of the pressuremeter test and for estimating the bearing capacity of deep foundations and disturbances due to pile installation: Chadwick (1959); Gibson and Anderson (1961); Ladanyi (1963); Butterfield and Bannerjee (1970); Baguelin et al. (1972); Ladanyi (1972); Palmer (1972); Vesic (1972); Prevost and Hoeg (1975); Carter et al. (1978); Randolph et al. (1979). At present, CEM are the most widely utilized approach for estimating pile installation effects on the surrounding soil. CEM assume conditions of radial symmetry and hence restrict the dependence of field variables (i.e., displacements, strains, stresses and pore pressures) to the radial coordinate only.

1. Advantages of CEM

The advantages of CEM in the analysis of pile installation effects stem from the simplicity of the approach. The assumed condition of radial symmetry (cylindrical or spherical) reduces the problem to a one-dimensional situation and enables solutions to be obtained with a modest level of effort. The following advantages of CEM are worth noting:

(a) The number of field variables (unknowns) is drastically reduced and hence the generation and understanding of solutions is greatly facilitated. For example, instead of considering two deformation components and four strain components in the actually two-dimensional axisymmetric installation problem, cylindrical expansion considers only one deformation and two strains in the soil.

(b) The field equations reduce from a complicated set of partial differential equations to a single ordinary differential equation that can be directly integrated (at least numerically) for virtually any comprehensive set of constitutive relations of the soil. Levadoux and Baligh (1980) describe the soil behavior aspects needed to obtain complete solutions by CEM.

(c) In view of the simplicity of the solution methods, many of the complicated aspects of penetration can be incorporated in CEM, e.g., large strains, high gradients, multiphased media, etc.

2. Disadvantages of CEM

Simplified approaches often spearheaded development and opened new doors but, unfortunately, have also caused major retrogressions in the history of science and engineering principally due to hasty generalizations and unjustified hypotheses. As a simplified approach, CEM inherently possess a significant retrogressive potential because of the following reasons:

(a) The applicability of one-dimensional Cavity Expansion solutions to the two-dimensional pile installation problem is by no means clear since it involves the "fitting" of a known solution to a problem of larger dimensions. The process can thus be confusing and controversial since it cannot be achieved uniquely, must neglect some aspects of the problem and inherently assumes that the neglected aspects are not important.
(b) The limitations and errors involved in utilizing the simplified CEM have not been clearly identified and/or evaluated.

THE SIMPLE SOLID PILE

In an attempt to investigate the applicability of CEM to pile installation and develop a better understanding of the mechanisms and limitations involved, Baligh (1975) obtained the solid simple pile solution to simulate the deep steady (two-dimensional) quasi-static undrained penetration of an axisymmetric solid pile in a saturated (incompressible) homogeneous isotropic clay initially subjected to an isotropic state of stress. These soil conditions are exactly the same as required in spherical CEM. Uniqueness of deformations and strains are discussed by Baligh (1984).

1. Background: Spherical Cavity Expansion

Consider a spherical cavity expanded in an infinite incompressible, isotropic and homogeneous soil initially subjected to an isotropic state of stress. Using a spherical coordinate system, Fig. 1a, the cavity has a radius R(t) at time t and starts from a zero radius, R(0)=0 at t=0. Conditions of spherical symmetry and soil incompressibility require that, at any time t, the radial location of a soil element, $\rho(t)$, be related to its initial location ρ_o ($=\rho(0)$) by the expression

$$\rho(t) = [\rho_o{}^3 + R^3(t)]^{1/3} \qquad (1)$$

Equation (1) means that, at any time t, if the cavity radius R(t) is known, soil deformations are totally described and hence soil strains are fully determined without the need to know the shearing behavior of the soil. This result has very important consequences. In particular, it means that the problem is totally strain-controlled and, for any given constitutive relation, stresses and pore pressures can be simply computed to satisfy equilibrium requirements. This is a major attractive feature of CEM because the computed strains apply to all material properties provided the conditions required to keep spherical symmetry are maintained (e.g., isotropy and homogeneity).

Figure 1a shows the distortion of a square grid at any time t, during spherical cavity expansion. With time, the radius R(t) increases but the distorted grid remains geometrically similar to that in Fig. 1a. Therefore, cavity expansion fails to capture two crucial aspects of vertical penetration: 1) the dependence of soil deformations on the vertical coordinate expected during vertical penetration and, 2) the steady state.

An exactly equivalent method of simulating spherical cavity expansion is to consider a single spherical source located at $\rho=0$ and discharging (emitting) an incompressible material at a rate (of volume) V per unit time. Assuming spherical symmetry, this means that, at any time t, the material occupies a spherical cavity of radius R(t) given by

$$R(t) = \left[\frac{3}{4\pi} V \cdot t\right]^{1/3} \tag{2}$$

For a given value of R(t), the soil deformations are described by Eq. (1) and soil strains are exactly the same as in the spherical cavity expansion solution and the deformed grid is also given by Fig. 1a. However, the advantage of utilizing the single source formulation instead of the conventional cavity expansion approach is to develop the capability of treating steady penetration situations of interest and problems involving multiple sources, Baligh (1984).

In a spherical coordinate system, the radial velocity of a soil element located at any radius ρ is given by $\upsilon_\rho = \frac{V}{4\pi} \cdot \frac{1}{\rho^2}$ and other velocity components vanish. On the other hand, in a cylindrical coordinate system, Fig. 1a, the non zero velocity components are given by:

$$\upsilon_r^\circ = \frac{V}{4\pi}\frac{\sin\phi}{\rho^2}; \quad \upsilon_z^\circ = \frac{V}{4\pi}\frac{\cos\phi}{\rho^2}; \quad \rho^2 = r^2 + z^2; \quad \phi = \arctan\frac{r}{z} \tag{3}$$

where the superscript o refers to the single spherical source situation.

2. The Simple Pile Solution

Considering a single spherical source in a uniform flow with velocity U in the z-direction, the soil particle velocity components υ_r and υ_z in a cylindrical coordinate system are given by:

$$\upsilon_r = \upsilon_r^\circ \; ; \; \upsilon_z = U + \upsilon_z^\circ \tag{4}$$

where υ_r° and υ_z° correspond to the point source solution given by Eq.(3). Clearly, the addition of a uniform flow changes only the vertical velocity by a uniform amount U. This apparently small deviation from the spherical cavity solution causes drastic changes in soil deformations and strains and, in fact, gives rise to the simple

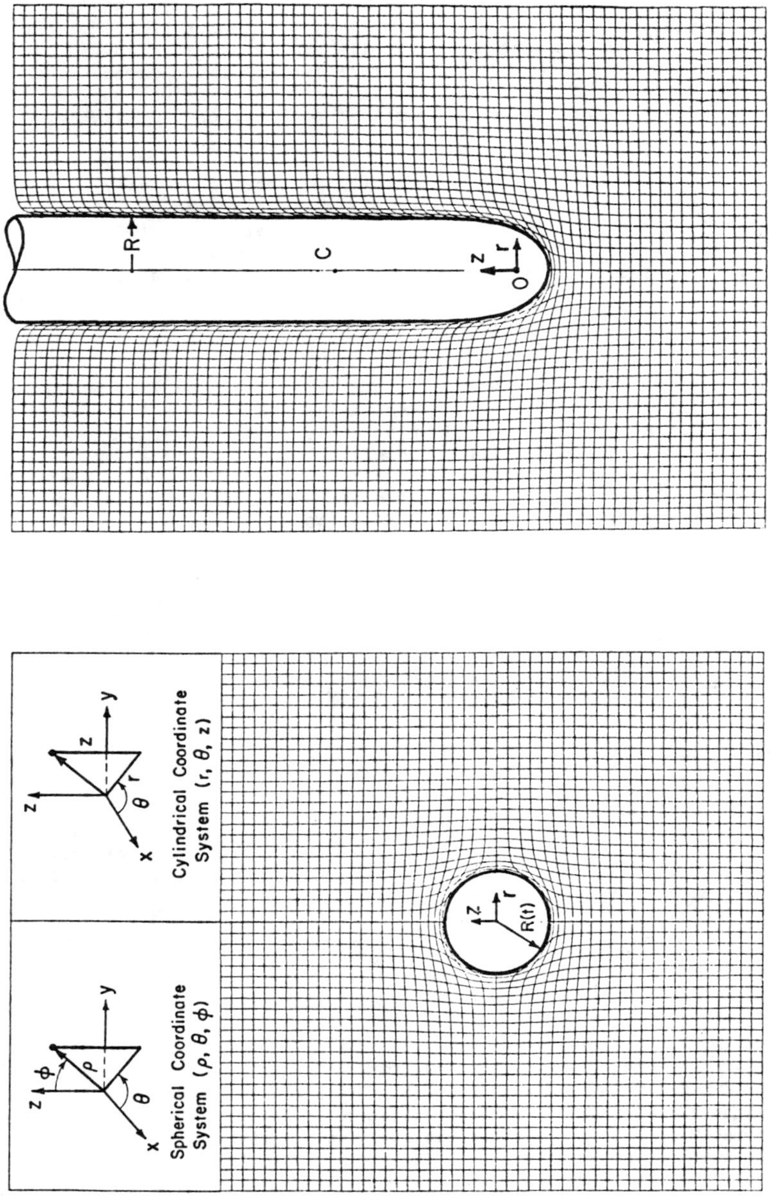

Fig. 1 Deformation of a Square Grid in Saturated Clays

(a) During deep spherical cavity expansion (or single source)
(b) During penetration of a "simple pile" (Baligh, 1975)

pile in Fig. 1b where the essential features of penetration now start to emerge.

Fig. 1b presents soil deformations obtained on the basis of the expressions of particle velocities in Eqs. (4) and (3) using the following straightforward procedure of integration:
(a) Starting at a sufficient distance ahead of the pile (i.e., at large negative values of z in Fig. 1b), select a particle located at a radial distance $r=r_o$, say.
(b) From Eqs. (4) and (3) compute the particle velocities and hence determine its location after a sufficiently small time increment Δt.
(c) Perform step (b) repeatedly to trace the stream line or the particle path as it moves past the pile.
(d) Repeat step (a) for different values of r_o to obtain soil deformations at various locations.

Comparing the results in Figs. 1a and 1b, the following effects of adding a uniform velocity field, U, to a single spherical source (or spherical cavity) solution on deformations are identified:

1. When the radii of the pile and sphere are approximately the same, as in Fig. 1, the distortions observed ahead of the pile exhibit some similarity with those corresponding to spherical cavity solutions because the scale of Fig. 1 was selected in order to enhance the similarity between spherical cavity expansion and simple pile penetration. In actual fact, the two solutions are quite different especially in the far field where cavity expansion significantly underpredicts deformations (and strains).

2. Soil deformations behind the tip of the pile, Fig. 1b, bear absolutely no resemblance to the spherical cavity solution in Fig. 1a. In fact, deformations around the pile shaft are also different from cylindrical cavity expansion predictions as discussed subsequently.

3. Results

3.1 Simple Pile Geometry

From Eqs. 4 and 3, it can be shown that the simple pile is a blunt penetrometer having a radius, r, at any vertical distance, z, given by the equation

$$\frac{r}{R} = [(1 + \cos\phi)/2]^{1/2}; \quad \phi = \arctan\frac{r}{z} \tag{5}$$

where the pile shaft radius at $z \to \infty$ (or $\phi \to 0$), R, is given by

$$R = \left[\frac{V}{\pi U}\right]^{1/2} \tag{6}$$

and the pile tip (r=0) is located at z= -R/2. Practically, the simple pile can be considered to have a uniform shaft radius R from point C in Fig. 1b located 4R behind the tip. Although seemingly complicated and probably not very practical, the simple pile geometry was selected in order to simplify the generation and presentation of solutions.

3.2 Classification of Soil Zones

Figure 2a illustrates the geometry of the simple solid pile and divides the soil around the pile into three regions I, II and III as well as three segments A, B, and C such that the soil can be classified into nine categories (I-A, I-B, I-C, II-A, II-B,...III-C). This zoning is not really needed since solutions (deformations, strains and stresses) are continuous and smooth but will hopefully facilitate the discussion of the results, illustrate important mechanisms and, in the future, enable a more focused attention on the difficulties of soil behavior.

The regions I, II, and III describe the severity of soil disturbances caused by installation. Based on the octahedral straining levels in the soil, the proposed boundaries in Fig. 2 between various regions are given by the expression

$$r/R = \varepsilon^{-1/2} \left[\frac{(1 + \cos\phi)}{24} (1 + \cos\phi + \sin^2\phi)\right] \tag{7}$$

where ε is the strain level considered; ε=0.1 for boundary I-II and ε=0.02 for boundary II-III. Region I includes the inner soil that is heavily strained by pile installation (more than 10%, say) and extends from the pile surface to a distance approximately equal to the pile radius R around and ahead of the pile. Region III contains the outer soil in the far field where strains are relatively small and hence possibly allow consideration of the soil as an elastic material. For the sake of illustration, a yielding strain level of 2% is selected to plot the boundary of region III (for the isotropic clays initially subjected to isotropic stresses considered herein) and hence the outer soil in Fig. 2 is located beyond a radius r=4.5R around the shaft and a depth in excess of 2.5R ahead of the tip. Finally, in the intermediate inner region II, the soil is in a plastic (yielded) state but is sheared to moderate strain levels (between 2 and 10%).

The three segments A, B, and C reflect the mode of shearing of the soil. Segment A enclosed in a 60° cone ahead of the pile ($150° < \phi < 180°$) includes soil that is principally sheared in a triaxial compression mode. Segment C enclosed in a 60° cone behind the origin ($0 < \phi < 30°$) contains the soil where the shearing strains are similar to an expanded cylindrical cavity. Finally, Segment B ($30° < \phi < 150°$) is the transition zone where most of the changes in soil shearing takes place.

3.3 Strain Components

Penetration of an axisymmetric pile develops four strain components in the soil: the vertical strain, ε_{zz}, the radial strain, ε_{rr}, the tangential (or hoop) strain, $\varepsilon_{\theta\theta}$, and the shear strain ε_{rz}. In order to calculate strains, consider a soil element at some location such as A in Fig. 2b (corresponding to ϕ=140°). After a time interval Δt (= $\Delta s/|\upsilon|$ where Δs is the distance traveled along the particle streamline and $|\upsilon|$ is the particle speed) the soil element moves to point B (ϕ=130°) and each strain component such as ε_{zz} increases by an amount $\Delta\varepsilon_{zz} = \dot{\varepsilon}_{zz} \times \Delta t$

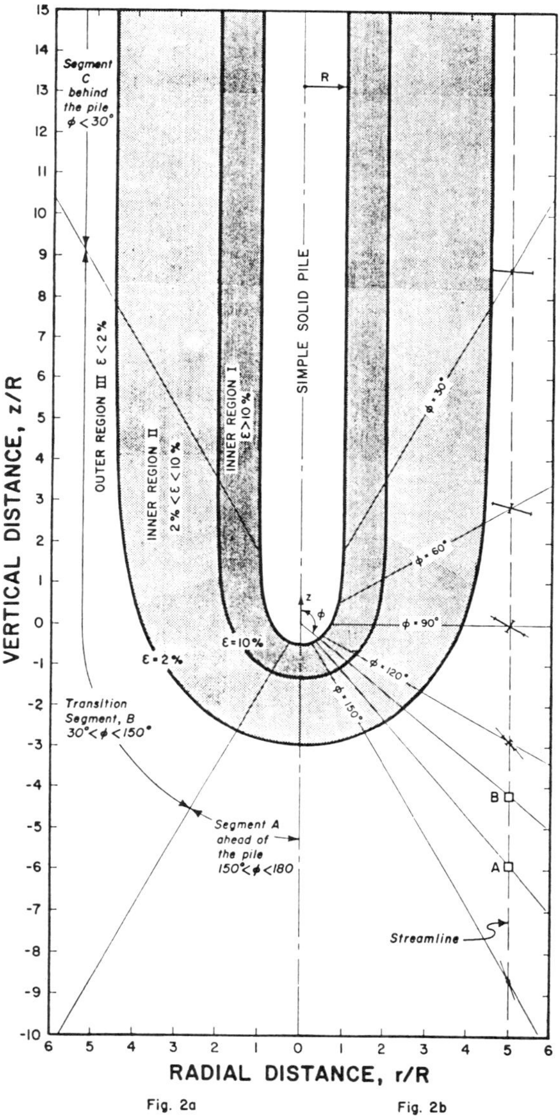

Fig. 2 Simple Solid Pile Penetration

a - Soil Zones

b - Major and Intermediate Principal Strain Directions and Magnitudes in the Outer Soil

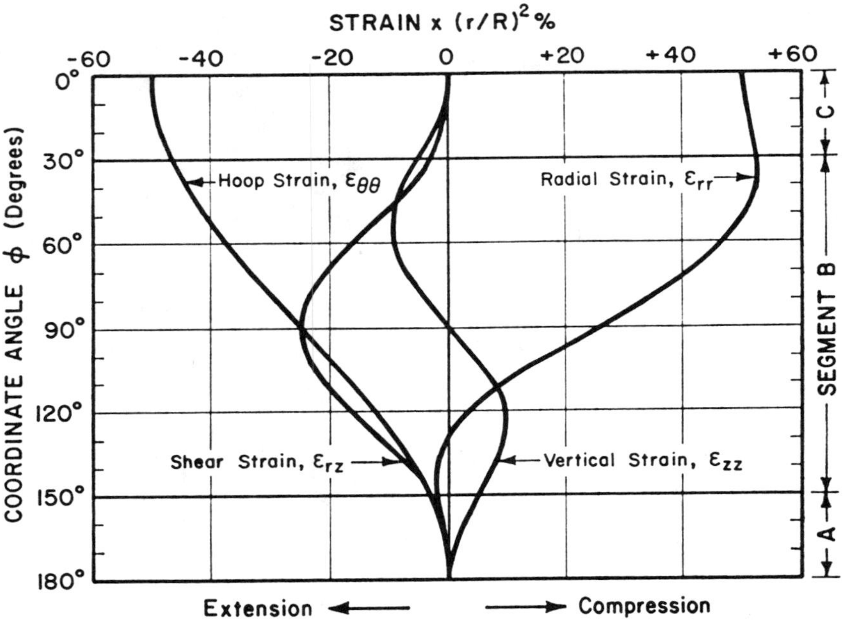

Fig. 3 Strain components in the outer soil due to simple pile penetration

where $\dot{\varepsilon}_{zz}$ is the strain rate. Therefore, knowing the velocity of the soil particles (Eqs. (4) and (3)) streamlines can be traced and the strains computed by simple summation provided the strain rate components are determined. Appendix A presents closed form expressions for the strain rates that are obtained from the velocities by differentiation.

When the strains are reasonably small, the changes in soil geometry are negligible and the integration can be carried out analytically because soil particles move along essentially vertical streamlines involving minor horizontal deviations. Appendix A presents closed form expressions for the strains in the far field (or the outer region III in Fig. 2a).

For soil elements located in the outer region III, Fig. 2b illustrates the major and intermediate principal strain directions and figure 3 presents the strain components during simple pile penetration and indicate the following important features:

1. For any given angle ϕ ($0<\phi<180°$), all strain components decay in proportion to $(R/r)^2$. Noting that, according to Eq. 1, spherical cavity expansion predicts that far field strains are proportional to $(R/r)^3$, it is thus clear that, for any shearing behavior of the clay, spherical cavity solutions will grossly underpredict the strains (and hence displacements and stresses) in the outer soil and therefore should be avoided in the modelling of pile installation.

2. The tangential strain $\varepsilon_{\theta\theta}$ is the minor principal strain everywhere in the soil.

3. In <u>Segment A</u> ahead of the pile, $150°<\phi<180°$, the shearing strains ε_{rz} are relatively small such that, as a first approximation, the major principal strain can possibly be assumed to act in the vertical direction and that
$\varepsilon_{zz} \simeq \frac{1}{4}(R/z)^2$; $\varepsilon_{rr} \simeq \varepsilon_{\theta\theta} \simeq -\varepsilon_{zz}/2$. These are the conditions imposed during undrained triaxial compression shearing.

4. <u>In Segment C</u> behind the pile, $0<\phi<30°$, the shearing strains are also small and the major principal strain is very close to the horizontal direction: $\varepsilon_{rr} \simeq (1/2)(R/r)^2$; $\varepsilon_{\theta\theta} \simeq -\varepsilon_{rr}$ and $\varepsilon_{zz} \simeq 0$. These are the conditions assumed by cylindrical cavity expansion.

5. In the intermediate <u>Segment B</u>, $30°<\phi<150°$, most of the changes in soil strains take place and the vertical strain ε_{zz} reverses from compression, when $\phi>90°$, to extension, when $\phi<90°$. Furthermore, when $\phi=90°$ (or $z=0$) the shear strain ε_{rz} reaches a maximum value and the magnitudes of ε_{rr} and $\varepsilon_{\theta\theta}$ are the same (but with different signs) and equal one half their final values at $\phi \to 0$.

It is also important to note that if the outer region III can be assumed to behave elastically because the strains are sufficiently small, the isotropy of the clay enables direct application of the above remarks to the stresses in the outer soil by simply substituting the word stresses instead of strains.

3.4 Strain Paths

In view of the assumed incompressibility of the clay, three strain components are sufficient to fully describe the state of strain during undrained axi symmetric penetration. The three deviatoric strains E_i (= E_1, E_2, E_3) defined in Fig. 4 are selected to study the strain paths of soil elements. The E_i-strains provide a clear picture of the shearing modes of the soil because, under idealized testing conditions, conventional triaxial tests impose E_1 shearing strain, pressuremeter tests (or cylindrical cavity expansion) apply E_2 and simple shear tests impose an E_3 mode.

Based on the closed form solutions presented in Appendix A, Figure 4 shows the strain path projections in the E_i-space in the far field (outer region III) during penetration of a simple pile. Several features are indicated by Figure 4:

1) The strains caused by penetration are not monotonically increasing and that significant reversals of the E_1 and E_3 strains (i.e., ε_{zz} and ε_{rz} strains) take place. Reversals of E_2, the pressuremeter or cylindrical cavity mode of shearing, does not exist in Fig. 4 because of the particular geometry of the simple pile. However, results obtained by Levadoux and Baligh, 1980, indicate that reversals of E_2 take place during penetration of the more practically interesting cylindrical piles with conical tips.
2) Perfect symmetry and anti-symmetry exist in the strain paths of E_3 and E_1, respectively, when plotted vs. E_2.
3) When the soil element is at an elevation R/2 behind the tip ($z=0$ or $\phi=90°$) E_1 vanishes, E_3 reaches a maximum and E_2 equals half its maximum value.
4) Far behind the tip (i.e., $z\to\infty$ or $\phi\to 0$), both E_1 and E_3 vanish and E_2 reaches a maximum value corresponding to cylindrical cavity expansion.

In describing the deviatoric strain paths of three elements located in the inner soil (Regions I and II) close to the pile, Fig. 5 indicates the following:

1. The strain levels in the inner soil are much greater than normally imposed in common laboratory and pressuremeter tests and illustrated by the shaded zones in Fig. 5. Therefore, the post peak behavior of the clay should be expected to have an important effect on stresses and pore pressures especially in Region I very close to the pile.
2. Initial failure of the soil located near the axis ahead of the pile (i.e., in Segment A, Fig. 2) takes place essentially due to E_1 straining (vertical compression) with possibly some contribution of E_3 straining well before E_2 (cylindrical cavity expansion) is felt.
3. The strain paths for soil elements in the inner soil exhibit the same reversals discussed earlier for the outer soil in region III. However, in view of the large strains in the inner soil, strain reversals have very important consequences on stresses and pore pressures in the vicinity of the pile as discussed below.

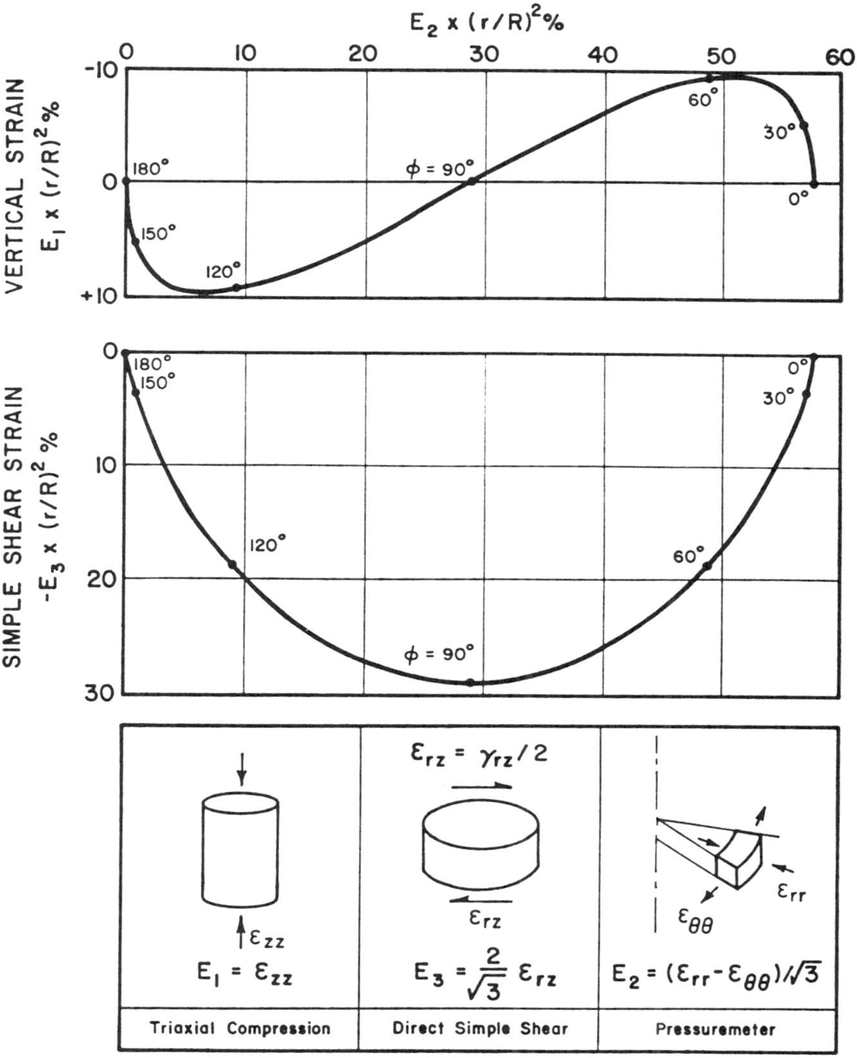

Fig. 4 Deviatoric strain paths in the outer soil during simple pile penetration

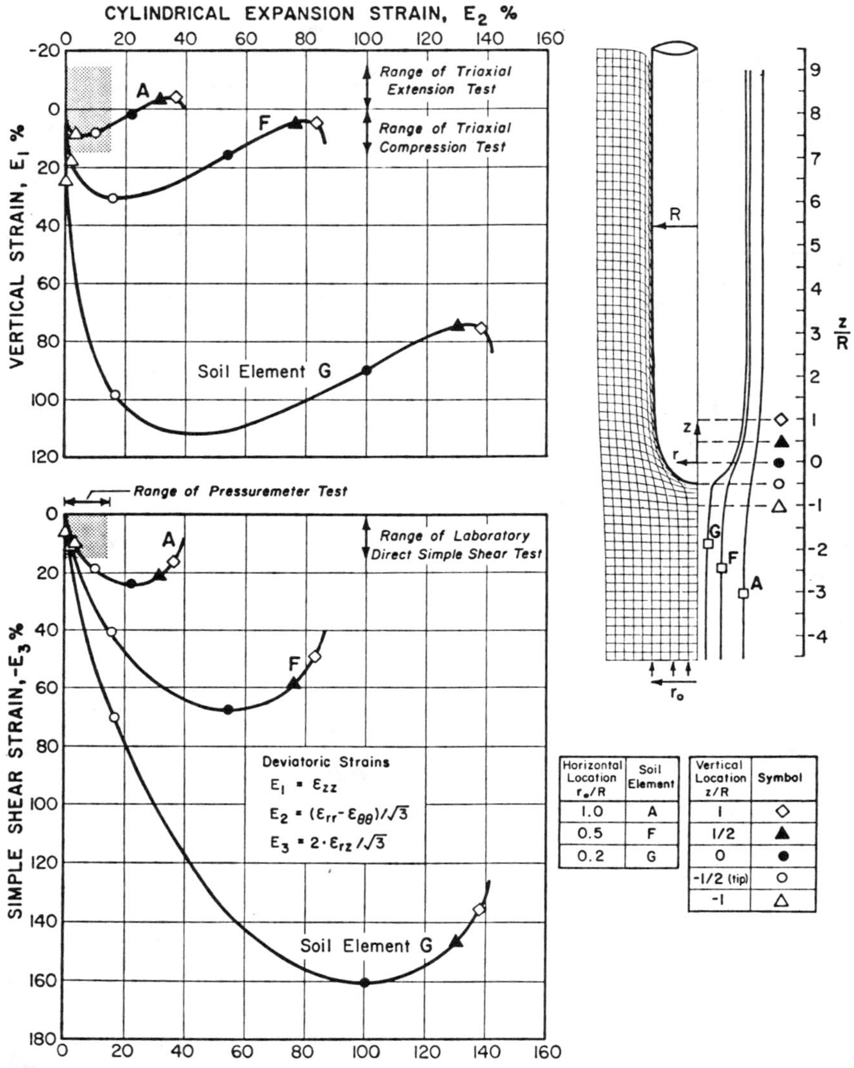

Horizontal Location r_o/R	Soil Element
1.0	A
0.5	F
0.2	G

Vertical Location z/R	Symbol
1	◇
1/2	▲
0	●
-1/2 (tip)	○
-1	△

Fig. 5 Deviatoric strain paths in the inner soil during simple pile penetration

IMPORTANCE OF STRAIN PATHS

The analysis of pile installation effects on the soil surrounding pile shafts by means of cylindrical Cavity Expansion Methods neglects the strain path (or history) of soil elements and hence inherently assumes that the final state of strain after shearing is sufficient to provide a good indication of the important shearing modes of the soil and, most importantly, can lead to adequate predictions of stresses and pore pressures. From a theoretical standpoint, these assumptions would be justified if and when the soil can be considered to behave elastically. Earlier discussions indicate that, during pile installation, elastic behavior (if any) cannot be assumed for the inner soil located within at least four to five radii from the pile. From a pragmatic standpoint, implementation of these assumptions to the inner soil where yielding takes place and where predictions are of real interest, can lead to serious errors and misleading results because of two basic reasons:

1. Pile installation causes large straining of the soil with significant strain reversals and rotations of the principal directions (see Fig. 5).
2. Fundamentals of plasticity theory indicate that, in non-monotonic strain paths and/or strain paths involving rotations of the principal directions, if the last strain increment is of significant magnitude (exceeding 1%, say) it is likely to have more influence on the stresses than the magnitude of the final strains after shearing.

In order to illustrate the effects of strain reversals at large strain levels (as encountered during pile installation) the results of an undrained Direct Simple Shear (DSS) test are shown in Fig. 6. The soil consists of a plastic clay (w_L=75%, w_P=25%) with a natural water content w_N=45% from Empire, Louisiana. The sample was K_o-consolidated to a vertical stress in excess of three times the estimated in situ maximum past pressure and then rebounded to a vertical effective consolidation stress $\bar{\sigma}_{vc}$= 3.5 Kg/cm^2 corresponding to an overconsolidation ratio of two. The sample was then sheared to an engineering strain γ_{xz}=25% and the shearing direction reversed to reduce γ_{xz} to almost zero. Results of the sample shearing presented in Fig. 6 indicate that, once high levels of straining have been reached, the following general aspects of clay behavior of major significance in pile installation processes become evident:

1. During monotonic loading (without reversals), the exact level of straining is not very important in determining the soil stresses and pore pressures. In Fig. 6, even though the clay exhibits a relatively high degree of strain softening, stresses at γ_{xz}=15% are not very different from γ_{xz}=20%, say, during the first shearing stage (loading).
2. After the shearing direction is reversed, major changes in stresses take place even at very small levels of strain increments (less than 1%).
3. Given the magnitude of the final strain after shearing without the

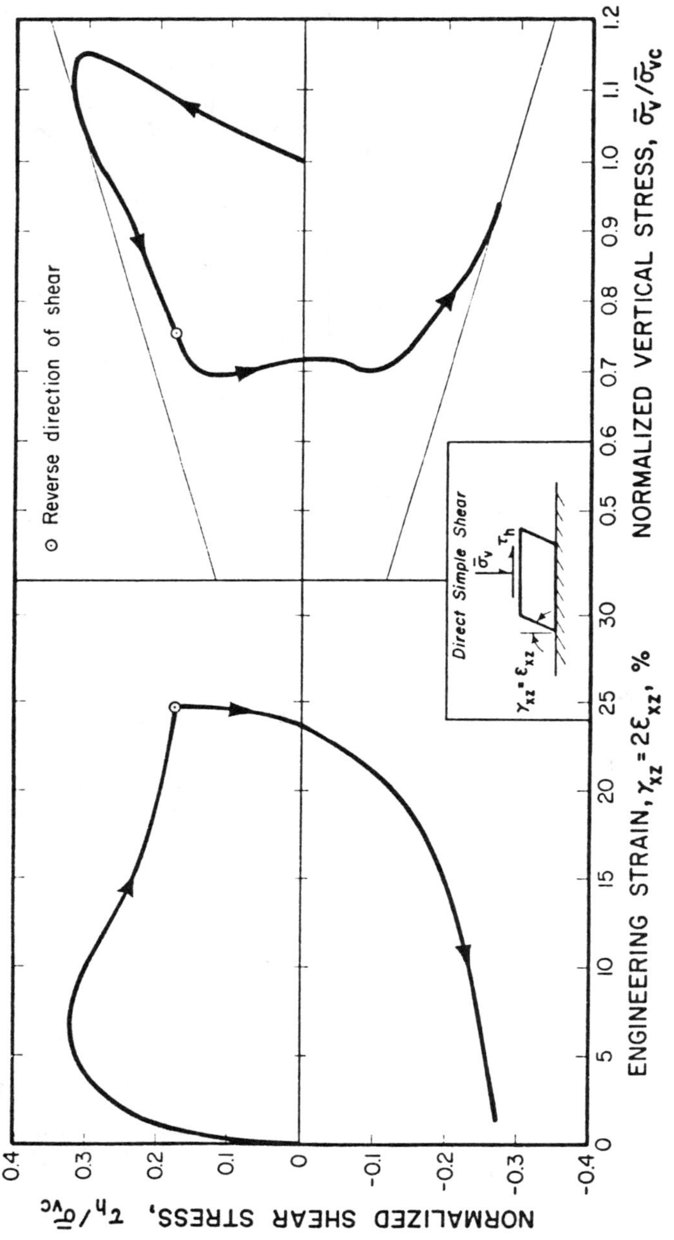

Fig. 6 Effects of Strain Reversals during Undrained Simple Shear Testing

detailed strain history of the soil, it is impossible to predict stresses with any degree of confidence. For example, at γ_{xz}=20% (say), Fig. 6 indicates that the shear stress can either be +0.2 $\bar{\sigma}_{vc}$ or -0.125 $\bar{\sigma}_{vc}$ and hence even the direction of the shear stresses can be erroneously predicted.

DISCUSSION OF THE SIMPLE PILE SOLUTION

1. Advantages of the simple pile approach

1. The simple pile approach introduces a relatively simple method to determine the effects of pile installation on soil deformations and strains in a comprehensive manner and hence enables the development of a better understanding of pile installation mechanisms.
2. The approach is more realistic than the widely used cavity expansion methods (CEM) in simulating pile installation because it considers the two-dimensional nature of penetration.
3. Results of simple pile solutions provide the necessary tools to evaluate the applicability and accuracy of CEM.
4. The simple pile approach offers a rational and systematic method of dividing the soil around the pile into different zones and characterizing the difficulties associated with the prediction of stresses and pore pressures in each zone.

2. Disadvantages of the simple pile solution

1. From a practical standpoint, the assumed geometry of the simple pile is not particularly attractive because actual piles have different geometries.
2. The simple pile solutions apply to idealized and hence restrictive soil characteristics (incompressible isotropic clays initially subjected to isotropic stresses) and installation conditions (steady quasi-static).
3. Compared to cavity expansion methods, the simple pile approach cannot utilize simple one-dimensional techniques to determine the stresses and pore pressures in the inner soil close to the pile but requires more comprehensive (and hence complicated) soil models in order to conduct a two-dimensional incremental stress analysis.

3. Extensions

Most of the limitations and disadvantages of the simple pile solutions can be eliminated by the Strain Path Method (Baligh, 1984). Extensions of the simple pile approach have already been carried out and solutions including stresses and pore water pressures everywhere in the soil, have been obtained for:

1. more realistic pile geometries including solid cylindrical piles with conical tips (Levadoux and Baligh, 1980) and open ended piles (Chin and Baligh, 1984);

2. more realistic nonlinear path-dependent soil response including anisotropic strain-softening clays initially subjected to anisotropic stresses (Levadoux and Baligh, 1980 and Kavvadas and Baligh, 1982).

Results of these analyses and the associated extensive experimental verification program established the validity of the simple pile solutions presented herein in describing the important aspects of pile installation. At present, work at the Massachusetts Institute of Technology is underway to utilize the Strain Path Method in extending simple pile solutions to compressible soils in order to investigate the drained penetration in sands, and the penetration of dynamically driven rough piles.

CONCLUSIONS

1. The simple pile approach provides a simple yet reasonably realistic method for elucidating the important mechanisms controlling pile penetration and for estimating the effects of pile installation.

2. Results presented herein apply to incompressible homogeneous isotropic clays initially subjected to isotropic stresses and include deformations and strains everywhere in the soil as well as the shear (deviatoric) stresses in the outer "elastic" soil. Methods and extensions to achieve more realistic and complete solutions are outlined.

3. The inner soil within a radius of 4 to 5 pile radii from the axis (or possibly more) is sheared to sufficiently large strain levels to cause failure and is subjected to significant strain reversals.

4. The large strain reversals that take place in the inner soil impose serious limitations on the validity of cavity expansion solutions that assume monotonic shearing especially in the immediate vicinity of the pile (region I within one radius from the shaft).

5. Although instrumental in developing the simple pile solutions, spherical cavity expansion solutions are nowhere applicable during pile penetration and should therefore be abandoned as a tool for estimating pile installation effects.

6. When carefully used, cylindrical cavity expansion solutions can provide reasonable estimates of the soil conditions at a sufficiently large distance behind the tip ($0<\phi<30$, say) and only in the far field ($r \geq 4.5R$, say) where the outer soil can be assumed elastic and where stresses and pore pressures are therefore not strongly influenced by the actual strain paths of the soil.

APPENDIX A

STRAIN RATES AND FAR FIELD STRAINS

A-1 Strain Rates

In a cylindrical coordinate system, the non vanishing components of the strain rates during axisymmetric penetration in an infinite medium consist of three normal strain rates (considered positive when compressive)

$$\dot{\varepsilon}_{rr} = -\frac{\partial \upsilon_r}{\partial r} ; \quad \dot{\varepsilon}_{zz} = -\frac{\partial \upsilon_z}{\partial z} ; \quad \dot{\varepsilon}_{\theta\theta} = -\frac{\upsilon_r}{r} \qquad \text{(A-1)}$$

and one shearing strain rate $\dot{\varepsilon}_{rz} = -\frac{1}{2}\left(\frac{\partial \upsilon_r}{\partial z} + \frac{\partial \upsilon_z}{\partial r}\right)$ (A-2)

where the velocity components υ_r and υ_z in the positive r and z directions, respectively.

From eqs. (4), (3), (A-1) and (A-2), the strain rates during simple pile penetration in an incompressible soil are the same as the single spherical source and given by:

$$\dot{\varepsilon}_{rr} = -\frac{V}{4\pi\rho^3} A_{rr}(\phi); \quad \dot{\varepsilon}_{zz} = -\frac{V}{4\pi\rho^3} A_{zz}(\phi); \quad \dot{\varepsilon}_{\theta\theta} = -\frac{V}{4\pi\rho^3}; \quad \dot{\varepsilon}_{rz} = -\frac{V}{4\pi\rho^3} A_{rz}(\phi) \qquad \text{(A-3)}$$

where $A_{rr}(\phi) = \cos^2\phi - 2\sin^2\phi$; $A_{zz}(\phi) = \sin^2\phi - 2\cos^2\phi$; $A_{rz}(\phi) = -\frac{3}{2}\sin(2\phi)$ and ϕ = arc tan(r/z) as in Fig. 1.

A-2 Far Field Strains

At some distance from the pile, changes in soil geometry may be neglected, streamlines become approximately vertical and hence any strain rate component such as $\dot{\varepsilon}_{zz}$, is related to the corresponding strain ε_{zz} by the expression:

$$\dot{\varepsilon}_{zz} \simeq -\frac{U}{r}\sin^2\phi \frac{d\varepsilon_{zz}}{d\phi} \qquad \text{(A-4)}$$

where ϕ=arc tan(r/z). This is simply because $\dot{\varepsilon}_{zz} = \frac{d\varepsilon_{zz}}{dt} \simeq U\frac{d\varepsilon_{zz}}{dz}$ and $\frac{d\varepsilon_{zz}}{dz} = -\frac{\sin^2\phi}{r}\frac{d\varepsilon_{zz}}{d\phi}$. Bearing in mind that $R^2 = V/\pi U$, Eqs. A-3 and A-4

$$\varepsilon_{rr} = \left(\frac{R}{2r}\right)^2 B_{rr}(\phi); \quad \varepsilon_{zz} = \left(\frac{R}{2r}\right)^2 B_{zz}(\phi); \quad \varepsilon_{\theta\theta} = \left(\frac{R}{2r}\right)^2 B_{\theta\theta}(\phi); \quad \varepsilon_{rz} = \left(\frac{R}{2r}\right)^2 B_{rz}(\phi)$$

where $B_{rr}(\phi) = 1 + \cos\phi\,(1 + \sin^2\phi)$; $B_{zz}(\phi) = -\cos\phi\sin^2\phi$ (A-5)

$B_{\theta\theta}(\phi) = -(1 + \cos\phi)$; $B_{rz}(\phi) = -\sin^3\phi$

When $\phi \to 0$, $B_{rr} \to 2$; $B_{zz} \to 0$; $B_{\theta\theta} \to -2$ and $B_{rz} \to 0$ (A-6)

When $\phi \to \pi$, $B_{rr} \to -\frac{1}{2}\left(\frac{r}{z}\right)^2$; $B_{zz} \to \left(\frac{r}{z}\right)^2$, $B_{\theta\theta} \to -\frac{1}{2}\left(\frac{r}{z}\right)^2$ and $B_{rz} \to 0$ (A-7)

APPENDIX B

NOTATION

The following symbols are used in this paper:

CEM = Cavity Expansion Methods
E_i = deviatoric strain components;
R = radius of spherical cavity or shaft radius of a simple pile;
r = radial coordinate in a cylindrical system;
t = time;
U = uniform vertical velocity;
V = rate of volume emitted by a point source;
v_i = velocities;
z = vertical coordinate in a cylindrical system;
ε = octahedral strain = $[E_1^2 + E_2^2 + E_3^2]^{1/2}/\sqrt{2}$;
ε_{ij} = tensorial strain components;
$\dot{\varepsilon}_{ij}$ = tensorial strain rate components;
γ_{ij} = engineering strains = 2x tensorial strains;
θ = coordinate angle in a cylindrical (or spherical) system;
ρ = radial coordinate in a spherical system;
ϕ = coordinate angle in a spherical coordinate system;
Ψ = stream function.

APPENDIX C

REFERENCES

Baligh, M.M. (1975) "Theory of Deep Site Static Cone Penetration Resistance," Research Report No. R75-56, Order No. 517, Department of Civil Engineering, MIT, Cambridge, Mass., Sept., 133 p.

Baligh, M.M. (1984) "The Strain Path Method" submitted for publication, Journal of the Geotechnical Engineering Division, ASCE, January 1984.

Bishop, R.F., Hill, R. and Mott, N.F. (1945) "Theory of Indentation and Hardness Tests," Proceedings, Physical Society of London, Vol. 57, Part 3, No. 321, pp. 147-159.

Baguelin, F., et al. (1972) "Expansion of Cylindrical Probes in Cohesive Soils," Journal of the Soil Mechanics and Foundation Div., ASCE, Vol. 98, SM11, pp. 1129-1142.

Butterfield, R., and Bannerjee, P.K. (1970) "The Effect of Pore-water Pressures on the Ultimate Bearing Capacity of Driven Piles," Proceedings, 2nd Southeast Asian Conference on Soil Engineering in Singapore, June, pp. 385-394.

Carter, J.P., Randolph, M.F. and Wroth, C.P. (1978) "Stress and Pore Pressure Changes in Clay During and After the Expansion of a Cylindrical Cavity," Report No. TR51, Dept. of Civil Eng., University of Cambridge, England.

Chadwick, P., (1959) "The Quasi-Static Expansion of Spherical Cavity in Metals and Ideal Soils," Quarterly Journal of Mechanics and Applied Mathematics, Vol. 12, Part 1, pp. 52-71.

Chin, C.T. and Baligh, M.M. (1983) "Deformations and Strains due to Open-Ended Pile Installation in Saturated Clays," Research Report R83-17, Order No. 757, Dept. of Civil Eng., MIT, Cambridge, Mass.

Gibson, R.E. and Anderson, W.F. (1961) "In-Situ Measurement of Soil Properties with the Pressuremeter," Civil Engineering and Public Works Review, Vol. 56, No. 658, pp. 615-618.

Kavvadas, M. and Baligh, M.M. (1982) "Nonlinear Consolidation Analyses around Pile Shafts," 3rd International Conf. on Behavior of Offshore Structures, M.I.T.

Ladanyi, B. (1963) "Expansion of a Cavity in a Saturated Clay Medium," Journal of the Soil Mechanics and Foundations Div., ASCE, Vol. 89, SM4, July, pp. 127-161.

Ladanyi, B. (1972) "In-situ Determination of Undrained Stress-Strain Behavior of Sensitive Clay with the Pressuremeter," Canadian Geotechnical Journal, Vol. 9, No. 3, Aug., pp. 313-319.

Levadoux, J.N. and Baligh, M.M. (1980) "Pore Pressure During Cone Penetration," Research Report R80-15, Order No. 666, Department of Civil Eng., M.I.T., Cambridge, Mass., 310 p.

Palmer, A.C. (1972) "Undrained Plane-Strain Expansion of a Cylindrical Cavity in Clay: a Simple Interpretation of the Pressuremeter Test," Geotechnique, London, England, Vol. 22, No.3, pp. 451-457.

Prevost, J.H. and Hoeg, K. (1975) "Analysis of Pressuremeter in Strain-Softening Soil," Journal of the Geotechnical Eng. Div., ASCE, Vol. 101, GT8, August, pp. 717-732.

Randolph, M.F., Steenfelt, J.S. and Wroth, C.P. (1979) "The Effect of Pile Type of Design Parameters for Driven Piles," Proceedings, 7th European Conference on Soil Mechanics and Foundation Eng., Brighton, England.

Vesic, A.S. (1972) "Expansion of Cavities in Infinite Soil Mass," Journal of the Soil Mechanics and Foundations Div., ASCE, Vol. 98, SM3, March, pp. 265-290.

PILE FOUNDATION MOVEMENTS DURING CONSTRUCTION

By Roy A. Bell[1], Larry R. Taylor[2], and Edward E. Rinne[3]

INTRODUCTION

It is widely recognized that ground heave and accompanying pile heave may occur when a high density of displacement piles is driven. Lateral movement of piles caused by installation of subsequent piles has also been observed (1, 3, 4, 5, 6) under certain conditions of site geometry, driving sequence, and soil types, although the sparsity of documented case histories and general experience within the consulting profession indicate that such cases are rare. When lateral pile movements are significant, considerable time and expense may result from structurally accommodating the misaligned piles into the structure.

The mechanism causing lateral movement is poorly understood although in the worst case (4, 6) lateral movements have implied incompressible soil behavior with the volume of displaced soil equal to the volume of the piles. This condition has been noted in saturated, insensitive fine-grained soils with an apparent very high undrained modulus.

Two recent case histories summarized in this paper provide further data on the nature of lateral soil and pile movements due to pile driving. In one case, the effect of predrilling to reduce the movements is also discussed. Inclinometer pore pressure and surface movement data presented in this paper may be helpful in developing better predictive techniques to anticipate lateral movements and plan accordingly prior to construction.

CASE A. SHIPYARD WHARF AND WORKBAY
LOS ANGELES HARBOR, CALIFORNIA

Project Considerations--The project consisted of a large ship-lift and 260-foot (79 m) wide pile-supported side transfer and workbay platforms adjacent to and inboard from an existing timber wharf. A simplified site plan of the facility is shown on Fig. 1.

1 Partner, Dames & Moore, San Francisco, CA.

2 Staff Engineer, Dames & Moore, Los Angeles, CA.

3 Engineering Manager, J.H. Kleinfelder & Associates, Walnut Creek, CA.

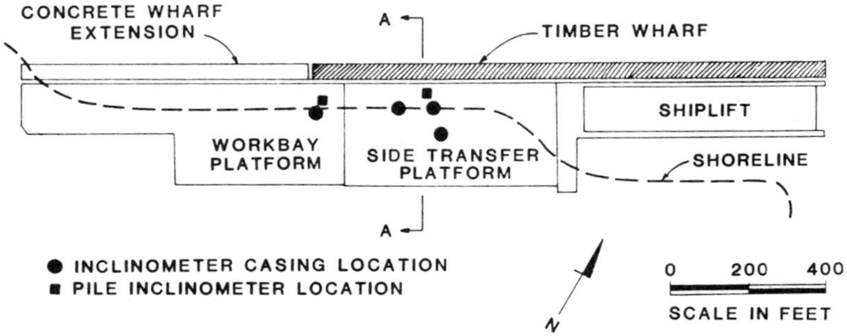

Figure 1: Site Plan

Over most of the site 18-and 24-inch (0.45- and 0.6-m) octagonal prestressed concrete piles with lengths ranging from 85 to 110 feet (26 to 34 m) were driven. Piles were typically driven in rows parallel to the shoreline slope with piles spaced 7 feet (2.1 m) on centers and rows spaced 22 feet (6.7 m) apart. Approximately 4000 piles were driven at the site.

An existing 40-foot (12 m) wide timber wharf was linearly extended with a concrete wharf structure as an early priority in the project. This necessitated driving most of the offshore piles before piles were driven onshore. The driving sequence for the remainder of the nearshore and onshore piles generally progressed from the shoreline inboard. An access corridor to the timber wharf through the side transfer area was driven last. At that stage, grade beams and concrete decks had been poured over previously driven piles in the adjacent area.

Site Conditions--The subsurface profile shown in Fig. 2 depicts a loose to medium dense sandy hydraulic fill approximately 25 feet (7.5 m) thick over soft to medium stiff clay and silt and a very dense fine to medium sand bearing stratum at approximately 78 feet (24 m) below MLLW. The fine-grained deposits are generally of low plasticity and have a water content close to the liquid limit. Zones of medium plastic clay (w_L = 50 to 60) and thin layers of silty sand are also present. The clay/silt appears normally consolidated beneath the fill with an undrained shear strength ratio (S_u/σ_v) of approximately 1/3. The shear strength of the fine-grained soils found offshore roughly equals that of onshore soils at the same elevations, indicating substantial removal of overlying deposits either geologically or through harbor dredging.

The submerged mudline profile beneath the timber wharf shown on Fig. 2 is based on soundings taken during the later stages of construction and is shown in contrast to the flatter mudline slope assumed during design. Also indicated on Figure 2 is the access corridor fill which was removed following driving of the outer rows of piles in that area.

Field Instrumentation--Lateral pile movements were first suspected approximately four months after pile driving began. Preliminary survey data indicated that unrestrained pile tops had been displaced laterally on the order of 24 inches (600 mm) in some locations.

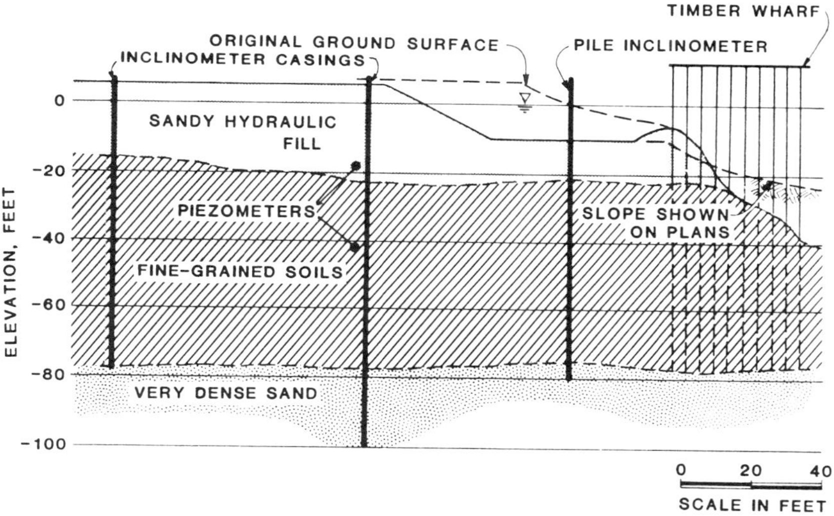

Figure 2: Section A-A

After the suspected pile movements had been verified by surveying, a field instrumentation program was established to complement survey data and more thoroughly evaluate the nature and cause of the soil and pile movements. The objectives of the field monitoring program were to observe:

1. The magnitude, direction, and areal extent of lateral soil and pile movements;

2. The influence of both pile location and driving sequence on lateral movements; and

3. Pore pressure changes in the underlying soils caused by pile installation.

Field instrumentation included four PVC slope indicator casings located near the crest of the shoreline slope with tip elevations ranging from 76 to 100 feet (23 to 30 m) below MLLW. Two casings were cast into 18- and 24- inch (0.45 to 0.61-m) octagonal concrete piles which were driven to pile tip elevations of approximately 80 feet (24 m) below MLLW. One pile with casing was driven at approximately the mid-height of the shoreline slope and the other was driven approximately 25 feet (7.6 m) north of the toe of the shoreline slope. The locations of the inclinometer casings are indicated on Fig. 1 and 2.

Two pneumatic piezometers were installed at elevations of 18 feet (5.5 m) and 43 feet (13 m) below MLLW near the crest of the slope and monitored during installation of adjacent piles. The shallower piezometer was situated in the more granular fill soil and the deeper piezometer was in the fine-grained natural deposits underlying the fill. The piezometer locations are indicated on Fig 2.

Pile tops were selectively surveyed immediately after driving and periodically thereafter. Several monitoring points were also established on the timber wharf in the vicinity of the access corridor and repeatedly surveyed during pile driving operations.

Evaluation of Lateral Soil and Pile Movements--Maximum soil or pile lateral movements measured with inclinometers are summarized in Table 1. Where piles were installed without preaugering, measured lateral soil/pile movements in the vicinity of the shoreline slope ranged from approximately 7.5 to 17 inches (190 to 430 mm). Survey data indicated pile top movements in these areas ranged from 6 to 24 inches (150 to 600 mm).

TABLE 1

SUMMARY OF INCLINOMETER DEFLECTION DATA

Inclinometer Number	Maximum Deplacement (mm)[a] North	East or West	Total	Location
1	190	40	200	Crest of slope
2	240	130	275	Crest of slope
3 (18" pile)	420	115	435	Mid-height of slope
4 (24" pile)	130	20	130	Toe of slope
5	140	0	140	Crest of slope
6[b]	0	15	15	25m inland from slope

[a] Measured at 3m below top of casing for Inclinometers 1, 2, 5, and 6.

[b] Lateral translation of casing suspected after 12-21-83.

Where piles were preaugered in the access corridor, maximum lateral soil/pile movements on the order of 5 inches (130 mm) were measured. Maximum lateral movement of the timber wharf in the access corridor area was on the order of 11 inches (280 mm) and resulted in significant misalignment of crane rails along a 100 foot (30 m) section of the wharf. This movement occurred in an area where soundings showed a steeply sloping mudline beneath the wharf (see Fig. 2).

Fig. 3 presents typical inclinometer deflection profiles for areas where piles were driven both without (Figure 3a) and with (Figure 3b) predrilling. The data indicate that predrilling (to approximately 50 feet (15 m) below MLLW) effectively reduced lateral displacements by roughly 50 percent. The reduction was probably even greater (on the order of 75 percent) since it is estimated that approximately 20 percent of the movement indicated on Figure 3b resulted from driving two nearby piles without predrilling.

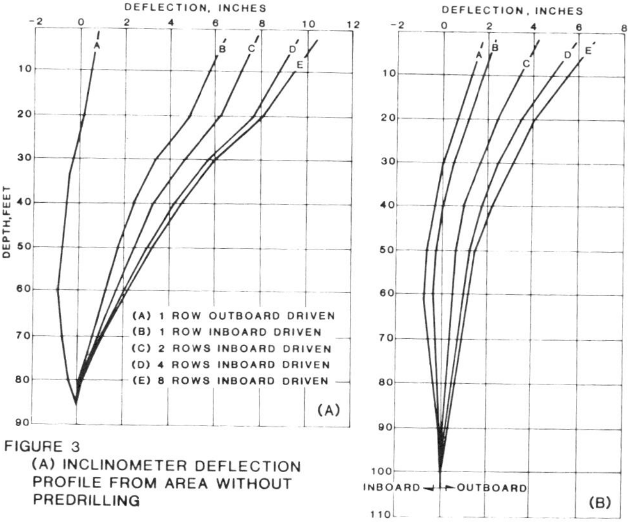

FIGURE 3
(A) INCLINOMETER DEFLECTION PROFILE FROM AREA WITHOUT PREDRILLING

(B) INCLINOMETER DEFLECTION PROFILES FROM AREA WITH PREDRILLING

Figure 3: Incinometer Deflection Profiles

Resolution of the data indicate that the maximum lateral soil and pile movements occurred in a direction away from and roughly perpendicular to the shoreline slope. For piles located in or near the crest of the slope, the displacements induced by installation of subsequent piles in rows parallel to the slope and more than 10 to 15 pile diameters in a landward direction were generally much larger than those induced by installation of piles in the same row a distance of 2-1/2 to 3 pile diameters away.

Inclinometer deflection profiles indicate that the movements occurred in a more or less rotational mode with little or no relative displacement at the tip of casings or piles, as shown on Fig. 3 and 4.

All of the available data indicate that the largest soil and pile lateral displacements occurred within and near the shoreline slope and the magnitude of the movements decreased as the distance from the slope increased. Roughly 90 percent of the total lateral movement experienced by piles in the vicinity of the slope resulted from pile installation within a zone extending approximately 50 feet (15 m) inland from the crest of the slope. The deflection profiles for a pile near the timber wharf in the access corridor area are presented in Fig. 4a in relation to the driving sequence and location shown on Fig. 4b. Fig. 5 presents the relationship between the amount of lateral pile movement and the horizontal distance between the monitored and subsequently driven piles. To account for differences in the number and size of piles installed between measurement intervals, the total lateral movement is divided by the volume of the piles driven during the interval.

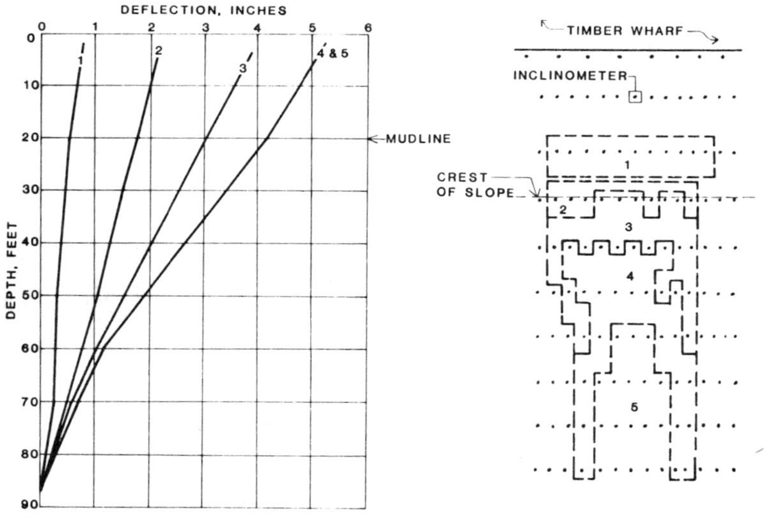

(a) PILE INCLINOMETER DEFLECTION PROFILES

(b) PILE LAYOUT AND DRAWING SEQUENCE IN ACCESS CORRIDOR

Figure 4:
Influence of Driving Sequence and Location of Pile Movement

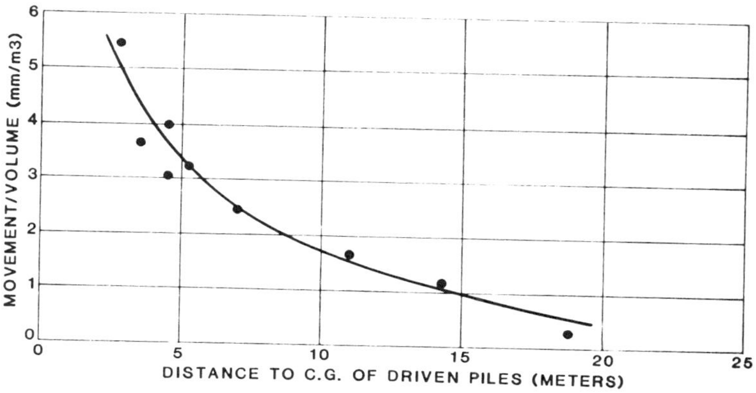

Figure 5:
Influence of Pile Driving Location

Fig. 6 presents a typical time history of pore pressure changes in the fine-grained soils during pile driving in the access corridor. The data indicate that initially large excess pore pressures generated during pile installation are partially dissipated very quickly but there is a gradual accumulation of excess pore pressure over a sustained period of pile driving. The accumulated pore pressure reached a maximum value of about 1½ times the initial conditions and began to dissipate after pile driving operations had progressed over 40 feet (12m) away from the piezometer location. Forty days after the peak value was reached, the pore pressure was about 20 percent above the initial value. Fig. 7 shows the relationship between the immediate pore pressure increase and the horizontal distance to pile driving locations. These data indicate that the excess pore pressures generated during pile installation are somewhat localized with the major zone of influence within approxi-mately 5 pile diameters of the location of the driven pile.

It is possible that the cumulative increase in excess pore pressures observed during pile driving influenced lateral movements to some extent. Reduction of the undrained shear strength of clayey soils shortly after pile installation within pile groups on the order of 10 to 30 percent has been noted in literature (1, 2, 3, 5). Since increasing pore pressures in the vicinity of recently installed piles were probably reducing available shear strength in a zone where passive soil restraint was minimized by the sloping ground surface, the preference for lateral soil displacement in that direction was increased. Similarly, the increased pore pressures may have had the effect of increasing the area in which pile installation induced significant outboard lateral movements.

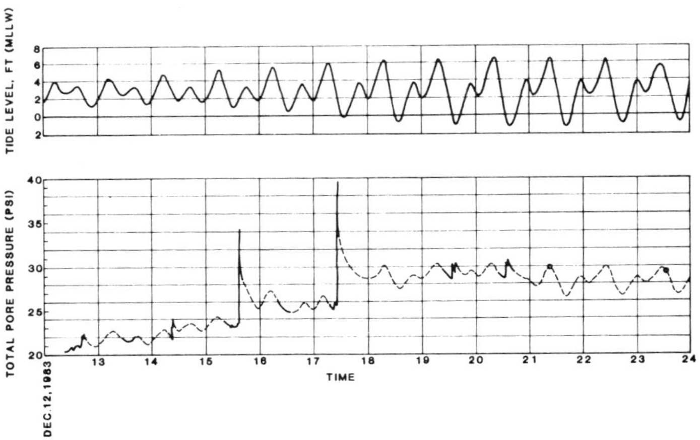

Figure 6: Tide Level and Pore Pressures Monitored in Clay Soils During Pile Driving, December 12–24, 1983

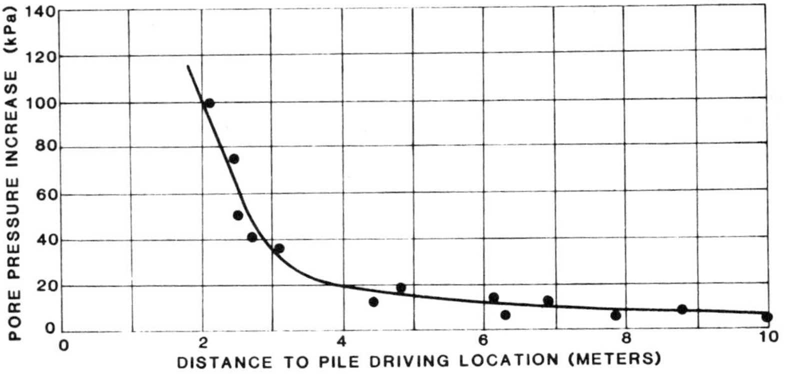

Figure 7:
Influence of Pile Driving Location on Pore Pressure

While only a limited amount of vertical pile heave data was obtained, the available data indicate that piles were subjected to only minor amounts of heave. The data generally suggest that piles near the shoreline were heaved less than 1/4-inch (6 mm) and the amount of heave increased at pile locations further inland to values on the order of 1 to 1-1/4 inches (25 to 30 mm).

Although vertical ground heave was not monitored, visibly detectable amounts of ground heave were not observed around any piles. The absence of significant vertical heave suggests that piles installed beyond the zone of notable lateral displacements may have caused considerable densification of the soil mass; particularly in the upper loose fill soils. In fact, there was some evidence of possible minor subsidence beneath a portion of the workbay deck on the southern side of the workbay platform.

Summary and Conclusions: Case A--The data show that ground displacements caused by pile installation laterally displaced previously installed piles up to 20 inches (0.5 m). The deflected shapes of inclinometers and piles suggest that volumetric strains were distributed fairly linearly with depth and unrestrained piles were not subjected to significant bending. Predrilling was effective in reducing lateral movements by about 50 percent.

Maximum lateral movements were experienced by piles in or near the shoreline slope and occurred in an out-of-slope direction. The lower lateral soil restraint in the out-of-slope direction caused lateral soil displacements to occur in a preferential direction. A temporary reduction of undrained shear strength caused by increased pore pressure in the vicinity of piles recently driven in the slope may also have contributed to lateral movements.

Further inland from the slope the greater degree of lateral soil restraint apparently set up higher lateral pressures around the piles resulting in densification and undrained volume changes without large amounts of lateral soil displacement or vertical ground heave.

CASE B. OIL REFINERY EXPANSION
SUMATRA, INDONESIA

Project Considerations. This project consisted of a large addition to an existing oil refinery in Sumatra, Indonesia.

All new process units and offplot tanks and pipeways were pile supported. Three types of driven piles were used:

1. 18-, 24-, and 30-inch (0.5, 0.6, and 0.8m) diameter steel pipe piles, typically 100 to 130 feet (30 to 40m) long, driven open end.

2. 12-, 16-, and 20-inch (0.3, 0.4, and 0.5m) square reinforced concrete piles cast on-site in segments 35 to 66 feet (10 to 20m) long with steel end plates, spliced by welding the peri-

meter of the end plates; typically driven 100 to 130 feet (30 to 40m) below grade.

3. 16-, 18-, and 24-inch (0.4, 0.5, and 0.6m) diameter prestressed concrete cylinder piles furnished in segments 35 to 66 feet (10 to 20m) long, spliced by welding the steel end plates, bottom segment with conical tip, typically driven 130 to 165 feet (40 to 50m) below grade.

The project included the installation of about 18,000 driven piles, two-thirds of which had been installed at the time of the author's first involvement in the project.

Site Conditions--The site was a former rain forest adjacent to the Rupat Strait. Trees and brush were cleared from the area; the root zone comprising the top 3 to 6 feet (1 to 2m) of the soil profile was left in place. The essentially level site had a natural ground surface about equal to high tide. Test borings drilled for the foundation investigation showed soft clay deposits extended to depths of about 25 to 35 feet (7.5 to 10m) in the process area and up to about 100 feet (30m) below site grade in some of the offplot areas. These soft, slightly overconsolidated clays were underlain by stiff highly overconsolidated clays with some sand lenses. Below these soils was weathered bedrock which was classified as hard sandy clay. The soil profile and properties in the process plant area are summarized in Table 2.

A layer of fine to medium sand, 3 to 6 feet (1 to 2m) thick was placed by hydraulic dredging methods over the site and formed the surface from which the pile foundations were installed. Haul roads were surfaced with imported sandy clays from nearby weathered rock outcrops.

Pile layout and a subsurface section from one typical process plant area that was studied are presented in Figs. 8 and 9. Fig. 8 shows a plan of the pipeway piles, pipeway pile cap configurations and locations of pile cap excavations to the west of the pipeway that were open during driving of some of the piles to the east of the pipeway. The legend in Fig. 8 gives the pile sizes and sequence of construction. A subsurface profile through the zone of maximum movements is shown in Fig. 9. Soil profiles from two nearby test borings are shown at the east and west sides of the section. This site was approximately 3300 feet (1000m) from the Rupat Strait shoreline and therefore not subjected to slope movements or other external loading conditions such as material stockpiles, channels or deep excavations.

Construction Sequence -- The construction sequence was as follows:

1. Hydraulic fill was placed approximately one year before the movement surveys.

2. Pipe piles for support of pipeways were installed soon after fill placement, followed by excavation for pipeway pile caps, pile cap construction, including grade beams, pedestals and anchor bolts for each column, and backfilling to approximate

TABLE 2. SOIL PROPERTIES IN PROCESS PLANT AREA (Haskoning, August 1980)

Soil Layer (1)	Elevation (M) (2)	Total Unit Weight (t/m^3) (3)	Natural Moisture Content (%) (4)	Atterberg Limits LL (5)	Atterberg Limits PI (6)	Average Shear Strength (kg/cm^2) (7)	Triaxial Test Results 0 (Degrees) (8)	Triaxial Test Results C (kg/cm^2) (9)
Hydraulic fill	+5.5 to +4.8	1.85+	30+	--	--	--	--	--
Soft to medium stiff silty clay	+4.45 to +3.30	1.36	95	98	41	0.22	4	0.10
	+3.30 to -0.65	1.37	91	99	44	0.14	5	0.18
	-0.65 to -4.70	1.39	84	97	41	0.10	5	0.16
	-4.70 to -8.30	1.50	72	89	41	0.25	4	0.21
	-8.30 to -10.8	1.60	54	81	37	0.20	7	0.20
Silty sand	-10.8 to -13.5	1.75	29	--	--	--	--	
Medium stiff silty clay with layers of fine sand	-13.5 to -15.7	1.58	58	85	39	0.81	9	0.25
Very stiff silty clay with layers of fine sand	-15.7 to +35.0	1.64 (average)	49 (average)	44-79	35-45	1.46	10.5	0.75
Dense clayey sand	-35.0 to -39.0	1.6	35	--	--	--	--	--
Very stiff to hard silty clay	-39.0 to -41.8	1.6	35	42-47	14	1.7 to 1.8	--	--
Highly weathered bedrock	-41.8	--	--	--	--	--	--	--

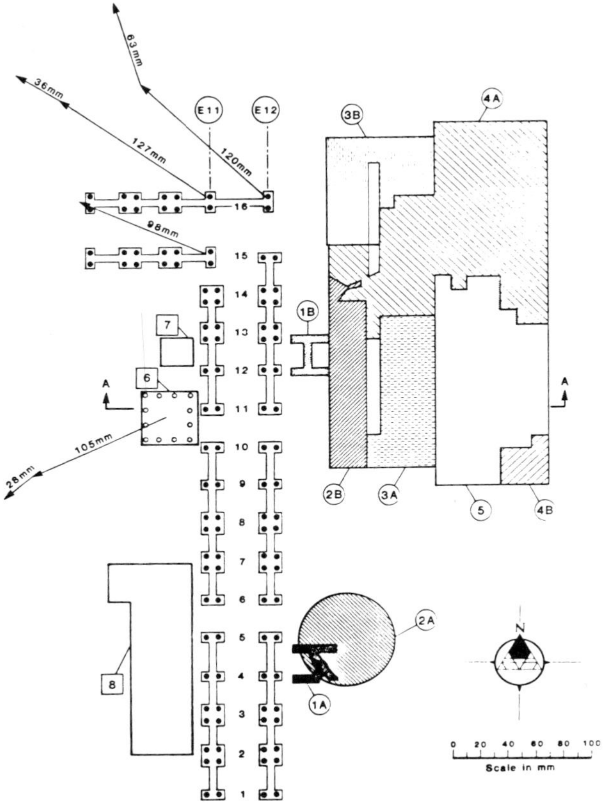

GROUP	DATE DRIVEN	NUMBER & TYPE OF PILES	EXCAVATION
1A	7/4/82 TO 8/23/82	6 - 0.5m x 0.5m x 37.5m 1 - 0.6m DIA. x 40m	6 7
1B	7/4/82 TO 8/23/82	6 - 0.5m x 0.5m x 37.5m 1 - 0.6m DIA. x 40m	
2A	8/23/82 TO 9/7/82	36 - 0.6m DIA. x 40m	
2B	8/23/82 TO 9/7/82	52 - 0.6m DIA. x 40m	
3A	9/7/82 TO 9/14, 15/82	57 - 0.6m DIA. x 40m	8
3B	9/7/82 TO 9/14,15/82	59 - 0.6m DIA. x 40m	
4A	9/14,15/82 TO 10/13/82	89 - 0.5m x 0.5m x 37.5m 64 - 0.6m DIA. x 40m	
4B	9/14,15/82 TO 10/13/82	10 - 0.5m x 0.5m x 37.5m	
5	10/13/82 TO 11/5/82	83 - 0.5m x 0.5m x 37.5m	

Figure 8: Foundation Plan

origin fill grade. The pedestals were probably built within a few millimeters of planned location.

3. Installation of 2 test pile setups, each with 6 reaction piles at locations 1A and 1B shown in Fig. 8. (No excavation was connected with the load tests.)

4. Installation of piles for foundations adjacent to the pipeway, followed by excavation for their pile cap installation, cut off excess portions of piles above grade, pour caps, and backfill. The location of pile caps that were constructed for foundations close to the pipeway during periods between movement surveys are noted in Fig. 8.

5. During construction of pile caps, the area between the pipeway supports was subject to heavy surface loads from the excavation equipment, earth moving trucks, cranes and concrete trucks.

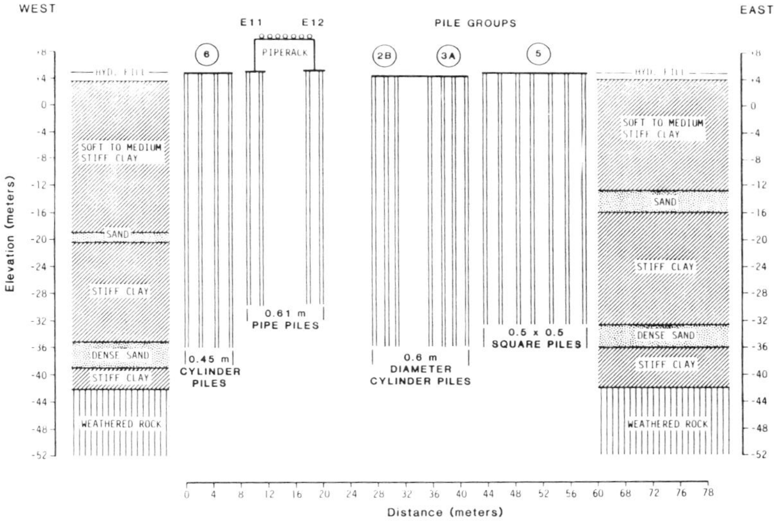

Figure 9: Subsurface Section A-A

Each step in Items 3 through 5 caused lateral movements of the pipeway pedestals. Vertical movements were relatively small and easily corrected by shimming, but lateral movements caused considerable problems during remaining construction and also caused concern for potential damage if movements continued to take place during plant operations.

Field Monitoring--Lateral and vertical movements of piles and pile caps were monitored along pipeway supports and adjacent foundations in four process areas. Individual pile movement monitoring was also performed during and following pile driving within a 100-foot (33m) diameter circular pile group containing 300 prestressed concrete cylinder piles, 24 inch (0.6m) diameter approximately 170 feet (52m) long, with average center-to-center pile spacing about 3 diameters. Lateral pile movements were also monitored at frequent time intervals within 2 large rectangular pile groups during excavation for the pile caps which extended between 6.5 and 10 feet (2 and 3m) below the site fill grade.

Within the area illustrated in Fig. 8, surveys of pile cap locations were performed 6 times over a period of 125 days. As illustrated in Fig. 10, lateral movements of up to 17 inches (440 mm) were observed at the pipeway pedestals. As shown in Fig. 8, vectors of movement were obtained for a 16 pile foundation that moved about 5 inches (130 mm) due to installing groups 2A, 2B, 3A, and 3B. Vectors of movement were also evaluated at pipeway pedestals near a 90-degree bend in the pipeway where cumulative movements were from about 4 to 7 inches (95 to 180 mm) at the times of the two surveys in September 1982. Some surveys of out-of-planned position piles were used to estimate amounts of movement that may have occurred after pile installation. Such surveys were not entirely conclusive due to movement of the set out pins caused by ground displacements that occurred between pin placement and driving of other nearby piles.

Evaluation of Movements--The cumulative amounts of lateral movements that took place are shown in Fig. 10 for each of the surveys. The key numbers of groups of driven piles and pile cap excavations that were open during periods between consecutive surveys from Fig. 8 are shown between the movement survey lines for the indicated period. These data show that lateral movements of the pipeway pedestals were caused primarily by ground displacements from the adjacent pile group installations. However, the 8/23/82 survey shows more movement of column line E11 than occurred at column line E12. This can be attributed to excavations 6 and 7 that were closer to E11 than to E12. If pile displacement was the only cause of movement, E12 would have moved more than E11 during this survey interval. This was further evidenced in several other pipeway/plant areas where there were similar sequences of construction.

The pattern of pipeway pedestal movements appears to be significantly influenced by the presence or absence of grade beams between the pedestals. Visual inspection of the grade beams by excavation alongside subsequent to completion of pile driving indicated the grade beams were not broken. This is not surprising since the cumulative displacement plots in Fig. 10 show the continuous segments of interconnected pedestals remained closely aligned. Large rotation occurred only in the spaces where there were no grade beams.

The lateral pile cap movements are plotted in Fig. 11 in terms of horizontal distances between the pile cap or foundation being monitored for movement and the piles driven during the time interval between movement surveys, as was done in Fig. 5 for the previous case history. To

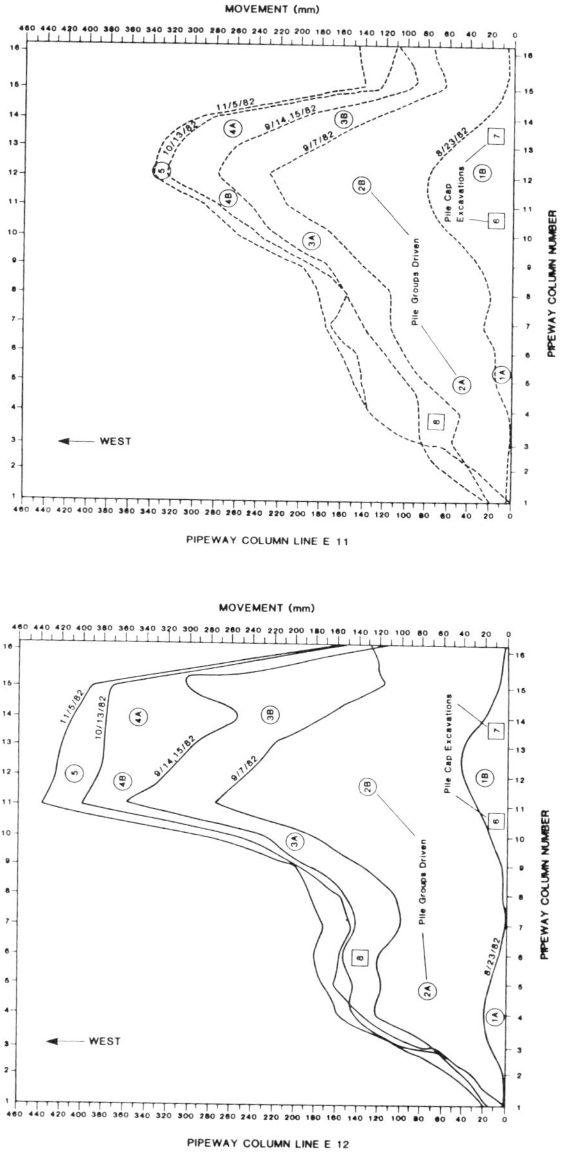

Figure 10: Lateral Movement Survey Measurements

account for different sizes and lengths of piles, the measured movement has been divided by the volume of piling driven below grade during the interval between surveys. The movement data for most of the pipeway pedestals are the components at right angles from the pipeway grade beams, whereas vectorial movement was used for three of the corner pedestals and for the 16-pile group foundation.

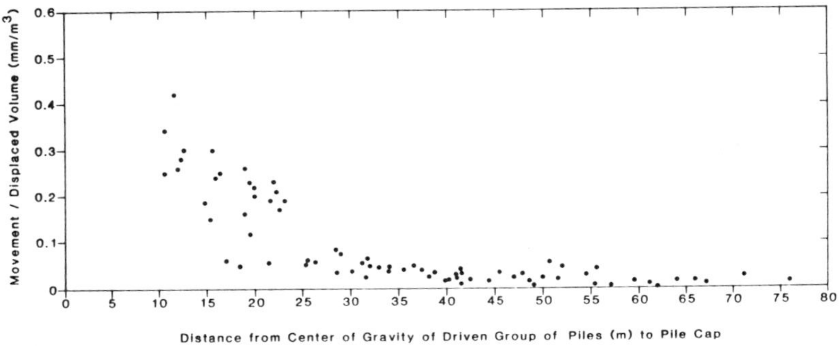

Figure 11: Distance/Movement Relationships

A hypothetical example for a simple geometric displacement model of lateral movements that would result from the installation of 10 cylinder piles each 24 inch (0.6m) and 130 foot (40m) long (total 3990 cubic feet (113m^3 displacement), is shown in Table 3 based on the observed movements plotted in Fig. 11. The middle column shows the approximate range between the mean of the plotted points in Fig. 11 and the maximum observed movements plotted in Fig. 11. The right-hand column in Table 3 gives the computed values of the difference in radii that would cause a volume of one-third of the driven pile volume 1340 cubic feet (38m^3) distributed over the surface of a right circular cone with nominal base radii equal to the distance from the 10 driven piles and height equal to the pile length. The remaining two-thirds of the volumetric displacement 2650 cubic feet (75m^3) is assumed to be in the vertical direction 1765 cubic feet (50m^3 within the 10 pile group and 880 cubic feet (25m^3) between the group and point of observed movement). This horizontal volumetric relationship correlates well with the observed movement up to one pile length away and then becomes larger than the observed movements. The assumed vertical ground displacements are similar to case histories reported in the literature (1, 3, 4, 5, 6).

These equivalent volumetric values are presented in Table 3 to illustrate that a simple geometric model can reasonably approximate the observed movements, even though it is realized that the actual displacement mechanism is much more complicated. The fact that the geometric model over-predicts movements beyond a radius of one pile length is reasonable, since this would be about the furthest expected zone of passive soil failure. More movements would be expected within the passive wedge (cone) than beyond.

TABLE 3

Mean Distance to 10-Pile Installation (m)	Observed Lateral Movement of Pile Caps (mm)	Computed Change in Base Radius of Cone (mm)
10	35-50	36
20	20-30	20
30	10-20	13
40	3-10	10
50	2-7	8
60	1-5	7
70	0-2	6
80	0-1	5

The effects of lateral displacements on the structural integrity of the piling was evaluated somewhat empirically by visual inspection of displaced cylinder piles throughout the project area, by static load testing, by retapping with the pile driving hammers, and by theoretical analyses of the bending capacity of the piles.

The verticality and curvature of the hollow cylinder piles were observed by lowering a lighted plumb-bob down the piles and measuring or estimating the curvature. The results of these surveys were consistently inconsistent. No pattern of tilting or bending could be established. Many of the most displaced piles were not bent and many were not out-of-plumb in directions consistent with the most probable pile top movements. The most severe bends in piles occurred at splices which were believed to have either been welded crooked or broke during or following driving. The only broken piles found were suspected of being knocked over by excavating equipment during pile cap excavation when excavated materials were stockpiled along the cut bank. The curvatures of the more gradually bent piles were in random directions and could not be related to driving sequence or the amount or direction from the driven pile locations. These observations lead to the conclusion that the displacements of piles caused by driving nearby piles did not cause significant bending and that some of the out-of-planned position and/or curvature probably occurred during driving of the out-of-position piles rather than from displacments caused by adjacent piles. The general absence of damage in the prestressed concrete cylinder piles plus the results of the above observations lead to the conclusion that the displaced steel pipe piles for the pipeway and the displaced cylinder piles in the 16-pile group illustrated in Fig. 8 were structurally capable of carrying their anticipated loads. New pedestals and anchor bolts were requred for the more severely displaced pipeway pile caps.

Measurements of ground heave were not taken, yet it was visually apparent that 3 to 7 feet (1 to 2m) of ground heave occurred within large groups of closely spaced piles. It was therefore estimated that average ground heave within the perimeter of the pile groups was approximately equivalent to one-half of the pile volumetric displacement divided by the area of the pile group, as has been reported in the previously referenced papers as being typical for several generally level sites.

Measurements of pile heave were taken in most of the surveyed areas. Maximum pile heave in the 300-pile circular group located in a deep soft soil area was about 36 inches (900 mm) (0.6m diameter prestressed concrete cylinder piles, 52m long). One of the heaved piles was load tested to 400 tonnes and displayed essentially the same load/deformation relationship as other load tested piles that had not heaved. Since these piles derived practically all of their support in friction, the ground heave and pile heave was judged to have negligible influence on their load bearing capacity.

Summary and Conclusions: Case B--The pile movement surveys and studies on this project showed that ground displacement from driven piles caused vertical pile heave of up to 36 inches (900 mm) and lateral pile movement of up to about 17 inches (440 mm). Measurable lateral movements occurred as far away as 245 feet (75m) from where piles were driven. Lateral pile movements of up to 4 inches (100 mm) occurred adjacent to excavations that were no more than 7 feet (2m) deep and during time periods when no piles were being driven. Construction traffic loads contributed to part of this movement. Piles within excavations 6 to 10 feet (2 to 3m) deep were moved laterally by the ground movements caused by the construction slopes and earth moving equipment loads. Piles within unstable slopes were broken from smaller lateral movements than the movements from ground displacements caused by nearby pile installations that did not break piles.

Excluding the unstable slope areas, combinations of nearby open excavations and installation of displacement piles on the high ground next to an excavation caused several times more movement than indicated in Fig. 11.

Previously installed piles within the zone of movement did not appear to influence the amount of movement caused by installing more distant piles.

Conclusions--The conditions at the Indonesia site suggest that the following factors contributed to the unusually large and distant lateral movements:

- No fractures occurred in the surface layers and the underlying soft soils were not extruded to the soil surface.

- The weakest soil stratum was confined below the relatively strong root zone and overlying relatively dense hydraulic fill. Displacements adjacent to pile installations would lift and spread these relatively strong layers.

- As vertical heave increased, the mounded surface caused larger radial components of force and movement.

- The displacements of the stronger near-surface layers would be carried laterally on the underlying soft layer as it was being squeezed laterally.

- o Any nearby low ground area, such as a pile cap excavation, would have significantly smaller confinement over the softest layer. The typically 6 to 10 foot (2 to 3m) deep excavations which removed all the sand fill and part of the root zone would remove essentially all of the lateral resistance within the strong and dense near-surface soils and would impose high shear stresses in the weak underlying layer. Thus, the low areas would offer significantly smaller resistance to laterally displaced soils and would attract ground movements caused by nearby pile installations.

The fact that nearly all inspected piles were not severly bent was partially attributed to the great stiffness of the cylinder piles (the only piles that could be inspected below ground level).

Appendix I - References

1. Bozozuk, M., Fellenius, B.H., and Samson, L., "Soil Disturbance from Pile Driving in Sensitive Clay", Canadian Geotechnical Journal, Vol. 15, No. 3, August 1978, pp. 346-341.

2. Cummings, A.E., Kerkhoff, G.O., and Peck, R.B., "Effect of Driving Piles into Soft Clay", Transactions, ASCE, Vol. 115, 1950, pp. 275-285.

3. Fellenius, B.H., and Samson, L., "Testing and Drivability of Concrete Piles and Disturbance to Sensitive Clay", Canadian Geo-technical Journal, Vol. 13, 1976, pp. 139-160.

4. Hagerty, A.M., and Peck, R.B., "Heave and Lateral Movements Due to Pile Driving", Journal of the Soil Mechanics and Foundations Division, ASCE, Vol. 98, No. SM11, November 1971, pp. 1513-1532.

5. Orrje, O., and Broms, B., "Effect of Pile Driving on Soil Properties", Proceedings, ASCE, Vol. 93, No. SM5, 1967, Proc. paper 5415, Sept. 1967, pp. 59-73.

6. Peck, R.B, Hanson, W.E., and Thornburn, T.H., Foundation Engineering, 2nd ed., John Wiley and Sons, Inc., New York, NY, 1974.

PILE FOUNDATION DESIGN USING PILE DRIVING ANALYZER

By

Virgil A. Baker,[1] A.M. ASCE
Neil A. Thomsen,[2] A.M. ASCE
Christopher R. Nardi,[3] A.M. ASCE
Michael J. Talbot,[4] A.M. ASCE

ABSTRACT

A case history of pile foundation design and construction for Pacific Bell's San Ramon Valley Administrative Complex is presented. Dynamic monitoring of driven piles with a pile driving analyzer and static load tests were used to estimate pile capacities during an extensive testing program. The pile test program is described, and the results and conclusions of the test program are presented. Following the completion of the pile test program, the pile driving analyzer was used during the indicator and production pile driving phases of foundation construction. A total of 3,020 piles were installed. Both the indicator and production pile driving phases are described. Conclusions regarding the application of pile driving analysis using dynamic monitoring methods for foundation design are summarized.

INTRODUCTION

Pacific Bell (formerly Pacific Telephone & Telegraph Company) is planning to move its present administrative center from San Francisco to San Ramon, California. To provide facilities for this move, the $180,000,000 San Ramon Valley Administrative Complex (SRVAC) is being constructed on a 97-acre (39 hectare) site in the Bishop Ranch Business Park in San Ramon, California. The new administrative complex will be located approximately 24 miles (40 km) east of downtown San Francisco.

---

[1]Vice President, InSituTech, 262 Grand Avenue, Oakland, CA, 94610
[2]Vice President, DCM/Thomsen Consultants, 1777 N. California Blvd., Walnut Creek, CA, 94596
[3]Staff Geotechnical Engineer, J. H. Kleinfelder & Associates, 1901 Olympic Blvd., Walnut Creek, CA, 94596
[4]Staff Geotechnical Engineer, J. H. Kleinfelder & Associates, 1901 Olympic Blvd., Walnut Creek, CA, 94596

The project involves construction of a three to four-story cross-shaped steel-frame building with a footprint area of approximately 570,000 square feet (53,000 square meters). Column loads will range from 750 to 1,200 kips (3,340 to 5,340 kN). A site plan showing the building layout is presented in Figure 1. Because of the building loads, varying foundation elevations and the presence of compressible soils at the site, it was necessary to use 55 to 80 foot (17 to 24 m) long, 12-inch (0.3 m) square, pre-cast, pre-stressed concrete friction piles for the structural foundation. The pile foundation design was accomplished by combining borehole information, laboratory test results, cone penetration test data, and the results of pile load tests consisting of static load tests and dynamic analysis of pile driving.

SUBSURFACE CONDITIONS

For the subsurface investigation, a total of 18 borings and 9 Cone Penetration Test (CPT) probes were advanced in the building footprint area. Borings were drilled to depths ranging between 30 and 100 feet (9 to 30.5 m). CPT probes were advanced to depths of 20 to 55 feet (6 to 16.5 m).

The upper three to five feet of soil at the site consist of black to dark brown, stiff to very stiff, expansive silty clay. The black silty clay is underlain by alluvial soil deposits of silty clay which are interfingered with layers and lenses of silt, sand, and gravel. Water well logs in the area indicate that these alluvial soil deposits are 500 to 700 feet deep.

The shear strength and compressibility of the subsurface soils were evaluated by Cone Penetration Tests, Standard Penetration Tests (SPT), laboratory triaxial compression tests, unconfined compression tests, and constant rate-of-strain consolidation tests. The results of all the undrained shear strength tests performed for this investigation are shown in Figure 2. Based on the results of the field and laboratory tests, the upper 5 feet (1.5 m) of soil has moderate shear strength and low compressibility. From 5 to 25 feet (1.5 to 7.5 m) below the ground surface, sporadic layers of moderately compressible soils with low shear strength are present. Below 25 feet (7.5 m), the soils are much less compressible with moderate to high shear strengths. The sporadic conditions encountered in the upper 25 feet indicated that if spread footings were used to support the building loads, unacceptable differential settlement of the structure would result. Since a bearing stratum of sufficient strength, thickness, and continuity was not found underlying the site, it was necessary to found the structure on friction piles.

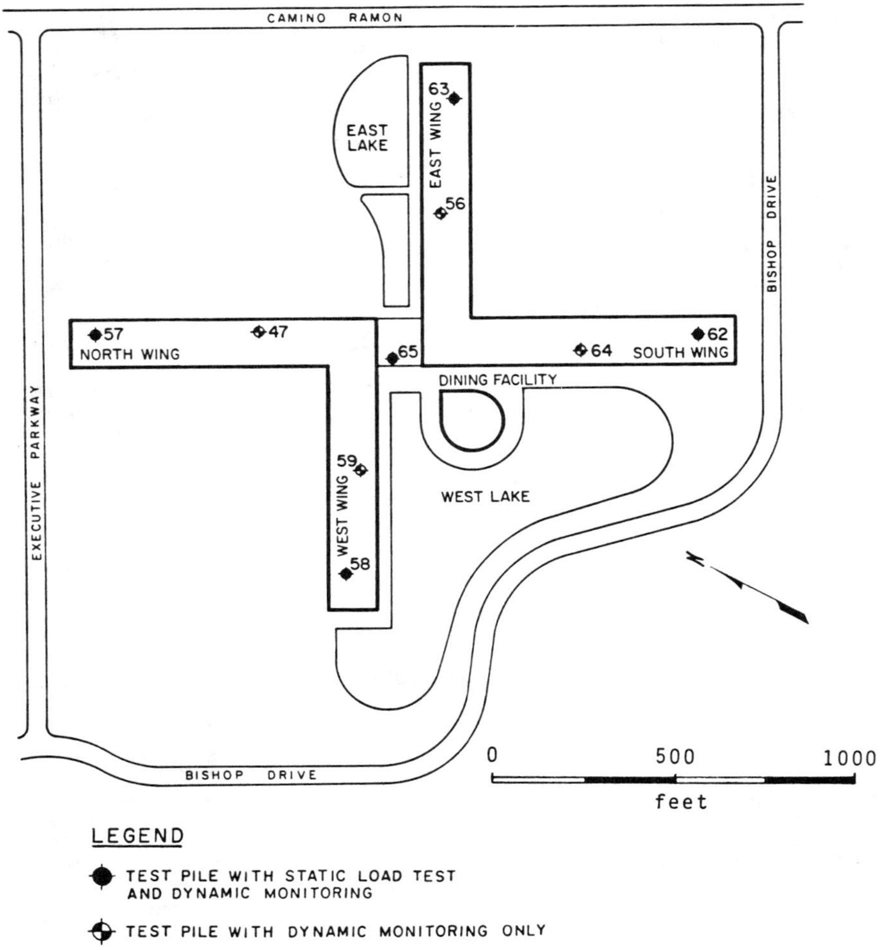

Fig. 1 - Site Plan with Pile Test Program

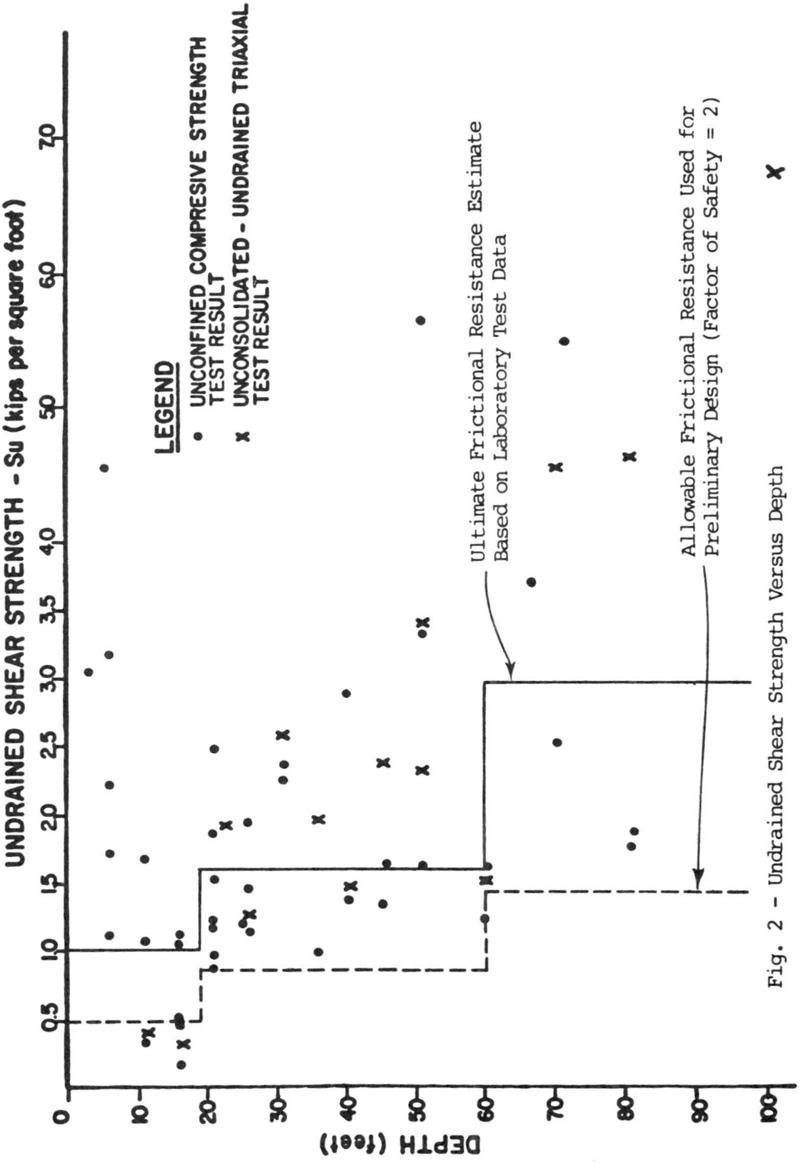

Fig. 2 - Undrained Shear Strength Versus Depth

PRELIMINARY FOUNDATION DESIGN

Based on the undrained shear strength data shown in Figure 2, frictional resistances for the soils to be penetrated were evaluated by various methods, including the Adhesion Factor Method, the "Lambda" method and the "Beta" Method (5). Based on the evaluations using these methods and engineering judgment, preliminary design frictional resistance values were recommended for 12-inch (0.3 m) square single driven concrete piles. These preliminary values are given in Table 1 and are shown schematically on Figure 2. The recommended frictional resistances given in Table 1 indicated that 12-inch (0.3 m) square driven concrete piles approximately 65 and 75 feet (20 and 22.5 m) long would be required to support structural loads of 100 and 125 tons (890 and 1,110 kN), respectively.

TABLE 1

Preliminary Recommended Design Allowable Frictional Resistance for 12-Inch Single Driven Concrete Piles

Depth Below Existing Ground Surface (ft)	Frictional Resistance (kips per sq ft)
0 to 20	0.5
20 to 60	0.8
60 to 100	1.5

Because of the absence of other large buildings with deep foundations in the area and the quantity of piles that would be required (3,020 piles), it was concluded that an extensive pile testing program should be carried out at the site prior to commencement of production pile driving. The objective of the pile test program was to provide information for verification and/or revision of the frictional resistances shown in Table 1. It was also hoped that the information obtained from the pile test program would result in shorter piles with a concomitant cost savings for the project.

PILE TEST PROGRAM

The pile test program was performed in five phases. The first phase consisted of a pre-field wave equation analysis using the computer program SWEAP. This analysis was performed prior to test pile driving and was based on hammer and drive assembly data supplied by the piling contractor (Kiewit Pacific Company). The second phase consisted of driving test piles at the nine locations shown in Figure 1. All test piles were dynamically monitored during driving with a Pile Driving Analyzer manufactured by Pile Dynamics Inc., of Warrensville Heights, Ohio. The third phase consisted of performing static load tests on five test piles. The static load tests were performed one to two weeks after the piles had been driven. The fourth phase consisted of restriking or retapping the test piles after the static load tests had been completed. Restrike behavior was also monitored with the Pile Driving Analyzer. The fifth phase consisted of analysis of all the data obtained. The computer program CAPWAP was used to analyze the dynamic analyzer data.

Pre-Field Pile Driving Analyses

Prior to the start of pile testing, the piling contractor furnished schematic representations and relevant hammer specifications for a Vulcan 80C differential-acting air hammer, and a FEC 3000 single-acting diesel hammer system. The FEC 3400 diesel hammer, used to drive test Pile No. 57 and later throughout the restrike, was added by the contractor during the field program and was not modelled in this pre-field study.

The SWEAP (small WEAP) wave equation computer program, developed by Goble & Associates and distributed by Pile Dynamics, Inc.[4], was used to analyze the pile driving behavior anticipated in the test program. This program rigorously models the thermodynamic characteristics of diesel hammer behavior, which is an essential part of the resulting solution. In brief, the program solves, in an iterative manner, for the penetration, stresses, and related parameters obtained for an assumed soil resistance distribution and prescribed input stroke. For the assumed resistance, the program determines a return stroke that matches the input stroke value. In this manner, the "variable stroke" feature of the diesel hammer is properly modeled by the program in developing the behavior predictions.

Results of the SWEAP analyses are illustrated in Figure 3a in terms of bearing capacity (R-UT), ram stroke, and maximum/minimum pile stresses versus blow count for the cases involving the FEC 3000 diesel hammer. Three hammer efficiency levels (50, 70 and 90 percent) were assumed in developing these curves, covering the expected range of

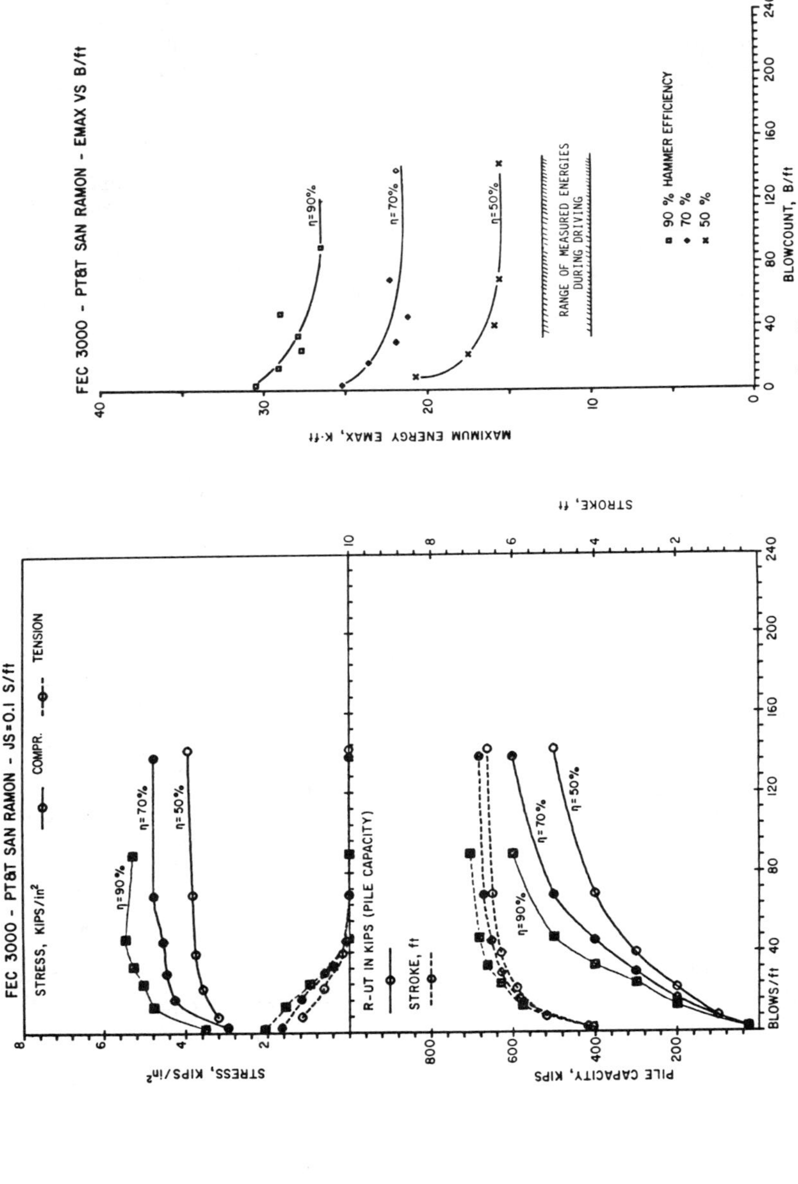

Fig. 3 - SWEAP analytical results for FEC 3000 hammer with J = 0.1 s/ft normal hammer performance. The assumed hammer efficiency levels are based on the mechanical performance of the hammer alone and excludes any consideration of energy losses that might occur due to the hammer cushions and pile cushions. Skin friction was assumed to comprise 90 percent of the total resistance, and a Smith damping constant of 0.1 sec/ft (0.33 sec/m) was used along the shaft. Note that under easy driving (low blowcount) conditions, SWEAP predicts very high tensile stresses for the proposed driving system. Companion plots of maximum transferred energy (E max) were generated from these analyses as shown in Figure 3b.

It is useful to mention that for hammer efficiencies of between 50 and 90 percent (see Figure 3b), the energy actually delivered to the pile (as predicted by wave equation analyses) ranges between 1/3 and 1/2 of the manufacturer's rated energy of the hammer. It is quite common to find such an amount of energy lost in transmission between the hammer and the pile through the helmet, capblock and cushion. This is a phenomenon usually verified by dynamic measurements. These analytical results were used in the field to compare measured data with predicted behavior, with some useful consequences that are described later.

Test Pile Installation and Monitoring

The pile test program consisted of driving piles at nine locations as shown in Figure 1. All test piles were dynamically monitored during driving as described in a later section. In addition, at five of the locations, reaction piles were driven so that static load tests could be performed. In total, 29 piles were driven during the pile test program consisting of 20 reaction piles and 9 test piles.

Three different hammers were used to drive the 29 piles. The hammer details are presented in Table 2. The FEC 3400 hammer was used for restrike since dynamic monitoring during initial driving indicated that the FEC 3000 hammer was performing poorly based on pre-field analysis results (see Figure 3b). The contractor had recognized that the hammer was in need of repair. Therefore, the dynamic monitoring indicated that the FEC 3000 hammer probably would not be capable of delivering sufficient energy to overcome the soil resistance expected during restrike and acheive additional penetration.

TABLE 2

Rated and Delivered Energies of Pile Driving Hammers
Used During Pile Test Program

Hammer	Manufacturer's Rated Energy (ft:lb)	Measured Delivered Energy (ft:lb)
Vulcan 80C* (differential-acting air)	24,950	10,000 - 13,000
FEC 3000** (diesel)	34,450-63,000	10,000 - 15,000
FEC 3400** (diesel)	42,900-71,500	18,000 - 26,000

*Pile cushion - 5" plywood: hammer cushion - 17-1/2" micarta and aluminum sheets.
**Pile cushion - 5" plywood: hammer cushion - 3" micarta and aluminum sheets.

At eight of the test locations, the piles were driven with relative ease to the intended depths of 65 to 78 feet (20 to 24 m) below ground surface. The reaction piles were driven to depths of 64 feet (19.5 m). However, at Pile Test No. 56, the test pile encountered refusal (greater than 120 blows/foot) at a depth of 59 feet (18 m), and driving was halted with the pile tip 6 feet (1.8m) above the intended tip elevation.

Static Pile Load Tests

Five axial compression static load tests were performed in accordance with ASTM D-1143 approximately one to two weeks after the piles were driven. The locations of the static load tests are shown in Figure 1. The piles were loaded to 500 kips (2,220 kN) using the Standard Loading Procedure described in ASTM D-1143. After completing the Standard Loading Procedure, the piles were loaded to 600 kips (2,670 kN) using the Quick Load Test Method described in ASTM D-1143. A calibrated pressure gauge and 1,000 kip (4,440 kN) hydraulic jack system were used to apply the compressive load. Vertical deflection of the pile butt was measured using three dial gauges attached to a pair of reference beams installed adjacent to two sides of the pile.

The piles were loaded to 200 percent of the design load, or 500 kips (2,220 kN). At this load, piles experienced settlements of 0.50 to 0.77 inches (1.3 to 2.0 cm) with the exception of Pile Test No. 62 which experienced a plunging failure when the 500 kip (2,220 kN) load was applied (i.e. pile settled at such a rapid rate that the load could not be maintained). At the design load of 250 kips (1,110 kN), piles experienced settlements of 0.15 to 0.27 inches (0.38 to 0.69 cm). An example load settlement curve (Pile No. 62) is presented in Figure 4. The results of the static load tests are given in Table 3. Of the five static load test piles, four piles had ultimate capacities greater than 600 kips (2,670 kN), i.e., the piles did not fail at 240 percent of the design load which was the capacity of the testing system. However, Test Pile No. 62 did fail when the 500 kip (2,220 kN) load was applied. The ultimate capacity of that pile was estimated to be between 450 and 500 kips (2,000 and 2,220 kN).

Dynamic Monitoring of Program

Pile driving monitoring applications for this project included: (1) monitoring of compression and tension stresses in piles during driving; (2) monitoring of pile integrity during driving; (3) hammer performance determination; and (4) estimation of pile bearing capacity.

Fourteen piles were dynamically monitored during the pile installation phase of the Pile Test Program. This included 9 test piles and 5 reaction piles. After the static pile load test phase of the program was completed, 8 test piles were restruck or retapped with the FEC 3400 hammer. Set up (soil freeze) of 200 to 400 kips (890 to 1780 kN) between initial driving and restrike driving made it difficult to move the test piles during restrike. The initial 14 dynamically monitored piles and the 8 restruck test piles were monitored using a Model GA Pile Driving Analyzer. Subsequently, 26 indicator piles were monitored as part of the indicator pile driving program using the Model GB Pile Driving Analyzer, an improved version of the GA analyzer.

Dr. George Goble and his colleagues developed the pile driving analyzer system at Case Western Reserve University.(1,2) The method involves attaching two strain gages and two accelerometers typically 2 to 3 pile diameters from the top of a pile. At each pile driving hammer impact, signals from the transducers are fed into a small field computer. A pile driving analyzer system is illustrated in Figure 5. The system consists of a field computer or pile driving analyzer, a tape recorder, an oscilloscope, strain gages, acceleration transducers, and cables with connectors.

The analyzer performs Case Method calculations of static pile capacity utilizing values of measured force and velocity obtained at an initial prescribed time, and 2L/c time later (2L/c is the time

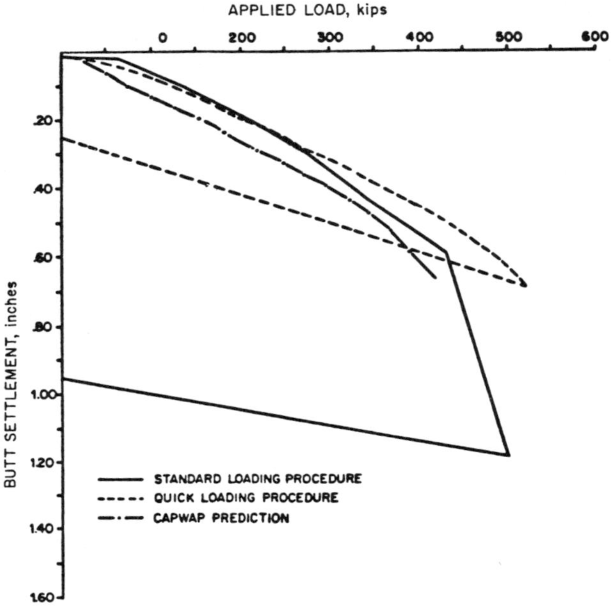

Fig. 4 - Static Pile Load Test Results for Test Pile No. 62

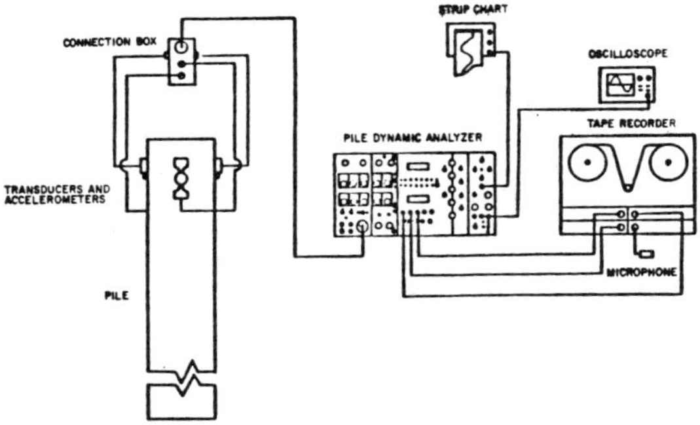

Fig. 5 - Schematic Diagram of GB Pile Driving Analyzer

TABLE 3: SUMMARY OF PILE LOAD TEST RESULTS

Pile Load Test Number and Location	Penetration Length Below Existing Ground Surface (ft)	ESTIMATED ULTIMATE AXIAL PILE CAPACITIES (Kips)				Ultimate Design Capacity Predicted from Table 1	Set-up Factor[4]	Time Duration between Case EOI and Case BOR (days)
		Static Load Test	Case EOI[1]	Case BOR[2]	CAPWAP[3]			
47 (North wing)	75.5	---	340	>565	---	522	1.7	20
56 (East wing)	59.0	---	>200	>510	---	330	2.6	3 hrs.
57 (North wing)	75.5	>600	265	>508	---	522	1.9	16
58 (West wing)	65.5	>600	200	>428	>460	402	2.1	20
59 (West wing)	73.2	---	415	>615	---	494	1.5	20
62 (South wing)	72.5	475	365	>425	>415	486	1.2	17
63 (East wing)	71.2	>600	370	>500	>535	470	1.4	16
64 (South wing)	74.5	---	210	430	400	510	2.0	21
65 (Central core)	65.5	>600	295	>656	>515	402	2.2	17

Notes:
1. Case EOI is Case Method Capacity at "end of initial" driving.
2. Case BOR is Case Method Capacity at "beginning of restrike" driving.
3. CAPWAP is CAPWAP Method Capacity determined from restrike.
4. Set-up Factor is CASE BOR capacity divided by CASE EOI capacity.

required for the stress wave to pass from the level of the instrumentation to the pile tip, distance L, and back at a velocity, c), in a closed form equation to determine static soil resistance(3). The standard capacity value, RSP, uses the first relative peak in the velocity record as the initial time in the computation, as illustrated in Figure 6. The maximum capacity calculation, RMX, varies the initial time point within reasonable limits to determine the maximum value using the same equation. The pile driving record and Case Method capacity estimates are shown versus depth in Figure 7 for Test Pile No. 62.

At the completion of the field program, the analyzer was brought back to the office in order to reprocess dynamic measurement data, perform CAPWAP analyses with the processed data, and correlate the data with static load test results.

The taped data was played back through the analyzer using varying J values (the Case damping factor) until there was agreement between Case method predictions and CAPWAP results. Reasonably good agreement was obtained using an average value of J = 0.5. This value of J was used to determine Case method capacities during the subsequent indicator pile program.

For all but Test Pile No. 64, the earliest restrike blows produced negligible penetration (refusal). Even with the larger (FEC 3400) hammer, it was not possible to deliver a large enough hammer blow to fully overcome the available soil resistance without risking structural damage to the pile butt and damage to the instrumentation. Under refusal conditions, the dynamic methods can only "detect" the resistance forces mobilized by the restrike blows. Therefore, the capacity estimates at the beginning of restrike presented in Table 3 should be considered lower-bound values.

CAPWAP Analysis Results

CAPWAP is a wave equation method which uses measured acceleration records as input to the discretized pile-soil model in order to predict the measured force at the pile butt(3). An interactive procedure allows the operator to modify the elemental soil resistance parameters (ultimate resistance, quake, and damping) iteratively to improve the match between measured and predicted butt forces as a function of time. Once a satisfactory match is obtained, the results include the ultimate pile resistance encoutered by the hammer blow, along with distribution of the resistance versus depth. Finally, using the elemental soil parameters in a static model, a load-settlement curve can be generated directly.

As described above, in the absence of full soil resistance mobilization due to an insufficient hammer blow, the CAPWAP results underpredict the ultimate pile capacity to some degree. The CAPWAP

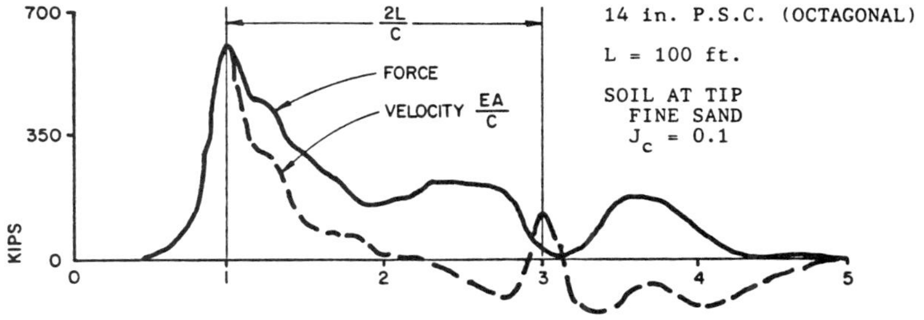

$$RTL = \frac{FT1 + FT2}{2} + \frac{Mc}{2L}(VT1 - VT2)$$

$$= (601 + 17)/2 + (589 - 111)/2$$

$$= 309 + 239$$

$$= 548 \text{ KIPS}$$

$$D = J_c \bullet (2F1 - RTL - STAT)$$

$$= J_c \bullet (1202 - 548 - 12)$$

$$= J_c \bullet 642$$

$$RSP + RTL - D$$

$$= 548 - 0.1 \bullet (642)$$

$$= 484 \text{ KIPS}$$

$$\text{LOAD TEST} = 470 \text{ KIPS}$$

$$R_u = RTL - D$$

$$470 = 548 - J_c(642)$$

$$J_c = 0.121 \text{ (corrected)}$$

Fig. 6 - Example of Case Method Determination of Pile Capacity

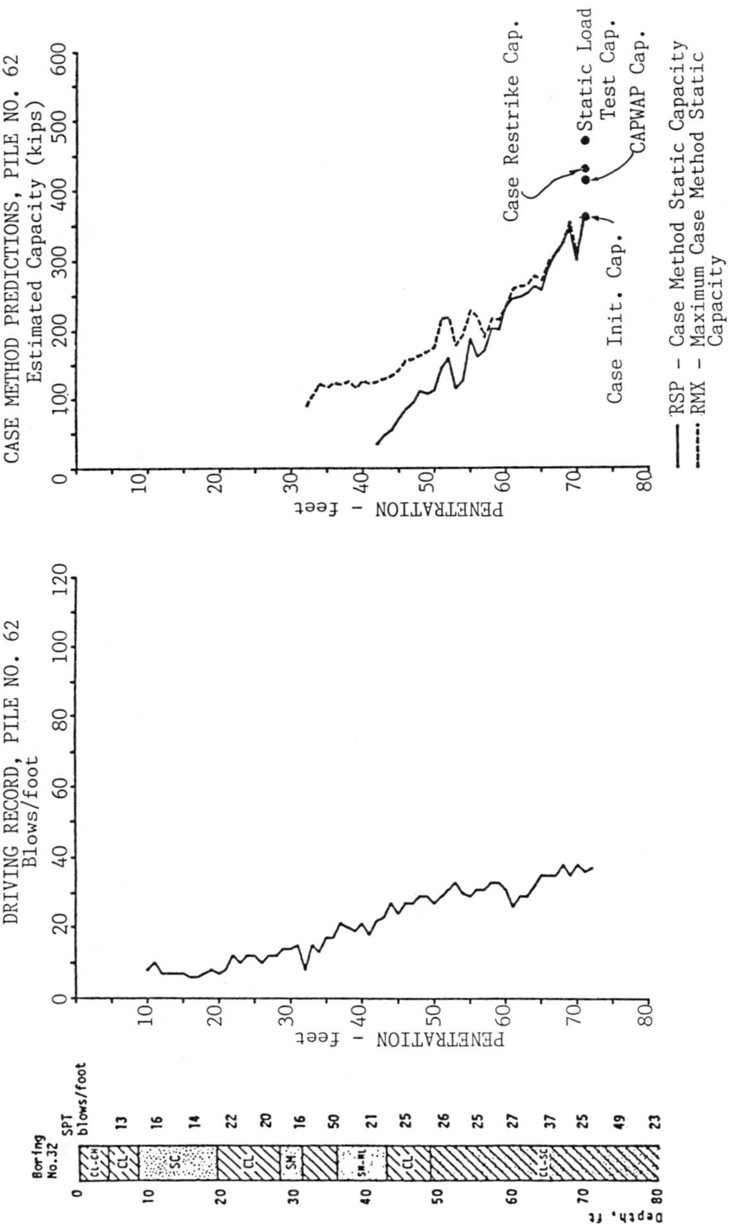

Fig. 7 - Installation Data for Test Pile No. 62 with Log of Nearby Boring results for the test piles are summarized in Table 3. The resistance distribution versus depth results for Test Pile No. 62 are shown in Figure 8. Negligible penetration for the early restrike blow in this case resulted in failure of the hammer blow to mobilize the full soil resistance available near the pile toe. As restrike driving proceeded, the soil resistance forces at shallow depths degraded sufficiently to permit further pile penetration.

The static load-settlement prediction based on the CAPWAP load transfer model obtained from the dynamic analyses is plotted with the field test results in Figure 4. The agreement between the predicted and measured settlements is generally good. As previously mentioned, the underestimate of ultimate capacity is probably caused by the limited hammer energy available.

RESULTS OF PILE TEST PROGRAM

The results of the 5 static load tests and 9 dynamically monitored tests are summarized on Table 3. The results of all the pile tests demonstrate that, except for Test Pile 62 and 64 (both in the South Wing of the building), the ultimate capacities predicted during the preliminary foundation design were below the capacities obtained during the testing program. Consequently, except for the South Wing, it was clear that frictional resistances recommended during the preliminary foundation design could be increased. The revised frictional resistances are presented in Table 4. Dynamic monitoring indicated the frictional resistances at depth were higher than originally predicted and that resistances near the surface were lower than predicted. The revised frictional resistances presented in Table 4 resulted in reduced pile lengths. The total reduction in pile length for the project was on the order of 7000 to 8000 feet (2100 to 2400 m) of pile.

The pile test program, especially the dynamic monitoring phase, yielded some other very useful information that was incorporated into the recommendations for the pile foundation design. One of the most important results of the dynamic monitoring was to indicate that driving criteria based on blow counts is often unreliable if hammer performance and pile capacities are not evaluated by an adequate testing program. Therefore, it was recommended that all piles be driven to the design pile tip elevation as determined utilizing frictional resistances given in Table 4, rather than driving to some prescribed blow count. The exception to this was piles that encountered very hard driving. Analysis of the dynamic monitoring data indicated that for the FEC 3400, potentially damaging compressive stresses could occur in the pile during very hard driving. Therefore, in order to avoid damaging the piles, it was recommended that driving with the FEC 3000 and 3400 be halted when the blow count exceeded 120 blows/foot (394 blows/meter). It was also clear that this criteria should be re-evaluated for different hammers the contractor might wish to use.

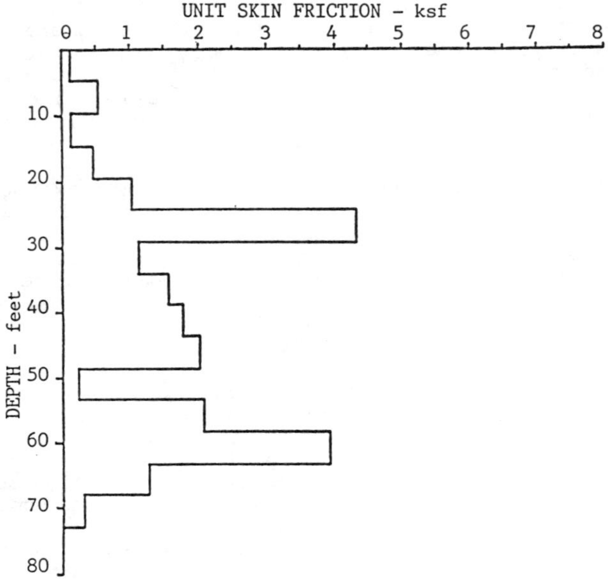

Fig. 8 - Frictional Resistance of Test Pile No. 62 as Estimated Using CAPWAP Analysis

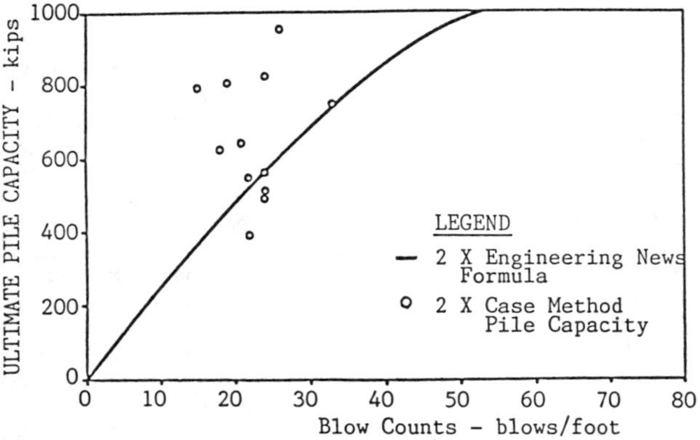

Fig. 9 - Ultimate Pile Capacities Versus Blow Counts for Delmag D-36/23 as Estimated by Dynamic Monitoring and Engineering News Formula

TABLE 4

Final Design Allowable Friction Resistance
for 12-inch Single Driven Concrete Piles

Depth Below Original Ground Surface (ft)	Frictional Resistance (kips per square foot)
0 to 20	0.3
20 to 60	(0.9 or 1.0)*
60 to 100	(1.5 or 1.7)*

* Use 0.9 and 1.5 kips per square foot for the South Wing, use 1.0 and 1.7 kps per square foot for the East, North and West Wings, Dining Facilities, and Central Facilities.

The subsurface investigation as well as the pile test program indicated the presence of thin discountinous dense gravel and sand layers. There were two reasons for avoiding early driving refusal in these layers well above the design tip elevation. First, in order to avoid potential differential settlement, it was desired to drive piles within the same group to the same tip elevation and to drive piles in adjacent groups to similar elevations. Therefore, it would be undesirable to have some piles "hang-up" while adjacent piles were driven to a different tip elevation. Also, the top 20 feet (6.1 m) of the piles had extra reinforcing steel. The extra steel had been added for seismic design purposes. As a result, the project structural engineer required that all piles be driven to within 5 feet (1.5 m) of the design tip elevation. Consequently, it was necessary to ensure that pile hammers used at the site delivered enough energy to the piles so that the piles could be driven through thin gravel layers. Thus, the piles would not encounter early refusal. Based on the SWEAP analysis and experience in the area, it was concluded that pile hammers used at the site should be capable of delivered energies, as measured with a Pile Driving Analyzer, of at least 15,000 ft-lb. (20,000 m-kN). This requirement was included in the project specifications.

CAPWAP results indicated the sporadic presence of a relatively soft soil layer at a depth of 50 to 60 feet (15 to 18 m) below the original ground surface. Consequently, it was recommended that all piles be driven to a minimum depth of 60 feet (18.3 m) below the original ground surface.

Dynamic monitoring also allowed measurement of compressive and tensile stresses developed in the pile during driving. Maximum compressive stresses of 5,800 psi (39 MPa) and maximum tensile stresses of 1,300 psi (9.0 MPa) were measured during driving. This information was given to the project structural engineer for consideration in structural design of the piles.

INDICATOR PILE PROGRAM

Because of the large numbers of piles to be driven and the significant lateral and vertical variability in subsurface conditions at the site, indicator piles were driven at regularly spaced intervals throughout the building footprint before the final foundation construction drawings were issued. The indicator piles were located at final production pile locations and were intended to be load bearing piles for the building. The purpose of the indicator piles was to provide information for either verifying or revising frictional resistances given in Table 4 on a section-by-section basis throughout the site.

A total of 84 indicator piles were driven across the site. Twenty-six of these piles were monitored using a Pile Driving Analyzer. A summary of the capacities of dynamically monitored indicator piles driven with the Delmag D-36-23 hammer are shown in Table 5. The capacities shown in Table 5 do not include set-up effects and are dependent on an assumed damping value of 0.5. Figure 9 illustrates the blow count versus capacity information contained in Table 5. The capacities plotted in Figure 9 are twice the capacities shown in Table 5 in order to take into account set-up effects which were assumed to double the initial capacities. The ultimate capacities as determined using twice the Engineering News formula value, which also takes into account the set-up effects, are shown for comparison (5). Examination of Figure 9 indicates that use of the Engineering News formula alone for this project would have resulted in a much more conservative design.

TABLE 5

Indicator Pile Data for Piles Driven
With Delmag D-36/23 Hammer

Pile No.	Length (ft)	Average Delivered Energy (ft.-kips)	Case Method Static Resistance* (kips)	Blowcount (blow/ft)
1	75	25.4	405	19
2	82	23.9	478	26
3	82	23.3	396	15
4	70	23.8	540	15
5	70	22.4	271	20
6	70	30.1	350	21
7	78	18.8	323	21
8	78	17.4	259	24
9	76	26.1	311	18
10	80	18.3	271	22
11	82	25.2	373	33
12	76	31.7	411	24
13	72	29.5	196	22
14	82	21.6	280	24
15	82	19.6	244	24

*J (assumed damping value) = 0.5 as evaluated during the pile test program.

The final frictional resistances presented in Table 4 were verified by the indicator pile program across most sections of the site. However, areas of softer driving were identified by dynamic monitoring in portions of the East Wing and in the Dining Facilities. Therefore, lower frictional resistances recommended for the South Wing were also used in these soft driving areas.

Based on the evaluation of all the dynamic analyzer data, it was concluded that piles with final blow counts of less than 15 blows/ft. (49 blows/meter) or less than 20 blows/ft. (66 blows/meter), if driven with a follower, might indicate piles with reduced capacities. Consequently, if such soft driving conditions were encountered during production pile driving, the driving records of adjacent piles were carefully examined. If necessary, the estimated pile capacities at a column location were reduced and extra piles were added, if required, by the project structural engineer.

Production Piles

Production pile driving operations were started on July 18, 1983 by Kiewit Pacific Co. of Concord, California. Complete driving records were obtained for each pile. The driving records were reviewed and evaluated on a daily basis by the project geotechnical engineers. Four pile driving hammers were used for production driving: Kobelco K-35, an ICE 1070, and two Delmag D-36/23. The Delmag hammers were used to drive the majority of piles. The Pile Driving Analyzer was used to measure hammer performance as well as identify any problems involving the hammer-pile system. Details of the hammer performance monitoring are given on Table 6. The four hammers used by the contractor did meet project specifications.

TABLE 6

Rated and Delivered Energies of Pile Driving Hammers Used During Production Pile Driving

Hammer	No. of Measurements	Manufacturer's Maximum Rated Energy (ft:lb)	Measured Delivered Energy (ft:lb)
ICE 1070 (Double-acting Diesel)	1	70,000	25,100
Kobelco K-35 (Diesel)	2	75,270	26,400-31,900
No. 1 Delmag D-36/23 (Diesel)	16	38,000-83,100	17,400-31,700
No. 2 Delmag D-36/23 (Diesel)	1	38,000-83,100	24,400

Analysis of the dynamic monitoring data for the four hammers presented in Table 6 indicated that potentially damaging compressive stresses could occur during hard driving at a lower blow count than for the FEC 3000 and 3400 hammers which were used during the pile test program. Accordingly, for the four hammers used during production driving, it was recommended that driving be halted when the blow count exceeded 60 blows/foot (200 blows/meter).

A total of 3,020 production piles were driven between July 18, 1983, and July 1, 1984. During this period, driving operations were not continuous due to scheduling with other construction operations and periods of inclement weather. At various times, the contractor used either one or two pile driving rigs. Average production pile length was approximately 65 feet (20 m) with minimum and maximum lengths of 55 and 82 feet (17 and 25 m). Average production per rig was approximately 20 piles per day. Production was dependent on such factors as local soil conditions, pile length, and the depth which piles were to be driven below grade. A steel pipe follower was used where pile cutoffs were below the existing ground surface.

Of the 3,020 piles driven at the site, 18 were broken during driving and were replaced. Thirty-eight piles spalled which resulted in cessation of driving before the design pile tip elevation was reached. One to two feet of the butt were removed to form an even striking surface before driving could begin again. Seventy piles encountered refusal above the design tip elevation. Of these, fifteen piles encountered refusal greater than five feet above the design tip elevation. Since the upper portion of the piles were reinforced to carry lateral building loads imposed by earthquake shaking, piles adjacent to those meeting refusal greater than 5 feet from design tip elevation and extra piles if required, were driven in 9 inch (23 cm) diameter pre-drilled holes, which extended to five feet above the design tip elevation.

Piles at 50 column locations encountered soft driving conditions. Based on evaluation of these driving records, as well as records for piles driven at adjacent column locations, the project geotechnical engineers informed the project structural engineers that the pile capacities at those locations should be reduced by an estimated amount. Then, the structural engineers re-evaluated the column loads. When the re-evaluation of the column loads indicated that the estimated reduced pile capacities were inadequate, the structural engineer required extra piles at these column locations.

CONCLUSIONS

The use of the Pile Dynamic Analyzer for pile foundation design at the Pacific Bell project resulted in many benefits when compared to design techniques which rely exclusively on laboratory tests of soil samples or driving formulas such as the Engineering News Formula. These benefits include:

- Reduction in the number of expensive static load tests that might otherwise be required.
- Improved understanding of subsurface conditions at the site and their effect upon pile capacities and drivability.
- More realistic pile foundation design parameters.

- o Evaluation of the performance of pile hammers.

- o An immediate assessment of compression and tension stress levels developed in the piles.

- o Development of more realistic installation criteria.

- o Assessment of potential pile installation problems on site.

The Pile Driving Analyzer and associated dynamic measurements are not substitutes for sound engineering judgement. However, when used to complement other information, it can be an invaluable tool providing better fundamental knowledge of hammer - pile - soil interaction while allowing increased efficiency and economy in pile foundation construction.

ACKNOWLEDGMENTS

We would like to thank Ron Tapper of Pacific Bell for his assistance and cooperation throughout the geotechnical investigation, pile test program, and foundation construction. In addition, we appreciate the cooperation and support of the management and staff of Skidmore, Owings & Merrill (project designer), Kiewit Pacific Company (piling contractor) and Swinerton, Walberg Company (general contractor). We would also like to thank Michael Holloway of InSituTech for assistance during the preparation of this paper. The preparation of this paper was generously funded by a professional development grant from the management of J. H. Kleinfelder & Associates.

REFERENCES

1. Goble, G.G., Rausche, F. and Moses, F., "Dynamic Studies on the Bearing Capacity of Piles, Phase III," Report No. 18, Case Western Reserve University, Cleveland, Ohio, August 1970.

2. Goble, G.G., Kovacs, W.D. and Rausche, F., "Field Demonstration: Response of Instrumented Piles to Driving and Load Testing," Proceedings of the Specialty Conference on the Performance of Earth Earth and Earth-Supported Structures, ASCE, Vol. 3, June 1972, pp. 3-38.

3. Goble, G.G., Likins, G. and Rausche, F., "Bearing Capacity of Piles from Dynamic Measurements," Ohio Department of Transportation Report No. OHIO-DOT-05-75, Columbus, Ohio, March 1975.

4. Goble, G.G. and Raushe, F., "Wave Equation Analysis of Pile Driving," FHWA Report No. IP-76-14, Federal Highway Administration, Washington, D.C., July 1976.

5. Poulos, H.G. and Davis, E.H., Pile Foundation Analysis and Design, John Wiley and Sons, New York, NY, 1980.

Role of a Standard in the Production of Drilled Shafts

Lymon C. Reese, F.ASCE*

Introduction

The preparation of drawing and written documents that describe the details of the installation of drilled shafts for a project requires the designer to have a considerable amount of information at hand. An up-to-date standard provides some of the information. But much more than a standard obviously is needed. The role of the standard in the design process is discussed. The discussion will not be comprehensive but will indicate important areas of consideration.

Literature Review

In spite of the fact that the Standard for drilled shafts that has been written is as independent as possible of new information, research and actual field experience are being reported regularly. New methods of construction and new information on concrete and other materials improve inspection techniques; these and a large number of other topics will undoubtedly be discussed in the literature. The users of the Standard will need to keep abreast of new developments as a matter of course.

Construction Methods

The Standard emphasizes methods of construction that employ rotary drilling with a power-driven kelly. Augers or drilling buckets are usually attached to the bottom of the kelly but the kelly and auger are sometimes replaced by a flight auger. Three specific methods of construction were identified: the dry method, the casing method, and the slurry displacement method. Some attention was given to details of each of these three methods.

However, several other methods of construction are being used, particularly outside the United States. A device called a hammer-grab is being used to construct drilled shafts. A heavy bucket of the clamshell type is fastened to the bottom of a heavy bar. The bucket and bar are lifted in a guide with power equipment and dropped. The heavy teeth of the open bucket are capable of penetrating soft rock and can extract boulders.

Another device that has had considerable use is the so-called "rodless drill." Electric motors that drive cutting tools are lowered by cables. The cuttings are removed by continual fluid flow and the

*The Nasser I. Al Rashid Chair, Dept. of Civil Engineering, The University of Texas at Austin.

excavation is guided by a cylinder that is situated above the cutters. The rodless drill has a definite advantage in some situations in that little headroom is required for the drilling operation.

While neither of the two methods of excavation are addressed specifically in the Standard, these and other methods of advancing a hole are covered in general terms. Over the next several years, there may be many new methods of construction introduced in the United States and many variations developed of the methods currently in use. Thus, writers of specifications may have to make use of basic principles in preparing the guidelines for construction. Excavations should be made with as little weakening of the in situ soil as possible, weak soil and sediments should be removed prior to placing concrete, and the concrete should be placed so as to produce the strongest bond between the soil and the concrete.

Subsurface Investigations

The presumption is implicit in the writing of the Standard that investigations are made with conventional methods. The diameter of the usual soil boring is from four to six inches, or perhaps a little larger, and samples are taken with thin-walled tubes or with driven tubes. Devices such as cones or vanes can be inserted in an open borehole to obtain additional information on the in situ properties of the soil. The use of a large diameter hole, made with drilled-shaft equipment, is recommended in the Standard.

Accurate information on stratigraphy and soil properties is essential in order that an appropriate construction technique be selected and in order for predictions to be made on bearing capacity. Current methods of subsurface investigation sometimes fail to reveal critical information. Rapid progress is currently being made on the use of indirect methods in obtaining a three-dimensional picture of subsurface conditions (7). If such methods are proven to be effective and if they come into general use, some jobs should require that seismic or other indirect methods be used to obtain subsurface data.

Because the nature of ground water is of critical importance in making excavations for drilled-shaft foundations, the storage and permeability of an aquifer are needed pieces of data. The Standard does not address directly the necessity of surveying ground-water conditions because frequently such a study is unnecessary and, when desirable, the survey would be too expensive except for the largest jobs. But advances in the ability to survey the ground water could lead to the desirability in specifying such surveys in a sizeable number of construction projects.

Concrete

The character of concrete as it is placed and hardens has a profound effect on the quality of a drilled shaft. Unfortunately, the technology of concrete as a material is insufficiently well known to allow the Standard to address adequately the properties of fresh concrete. When placed in most excavations for drilled shafts (shallow, large-diameter excavations are exceptions), concrete should have excellent flow characteristics. The numerical term that is given to the fluidity of concrete is "slump," but a high-slump concrete can be harsh, can experience bleeding, and can have other undesirable characteristics. It is not possible to give numerical values that specify the mix because materials vary so widely from place to place. Tests beyond the slump test need to be developed that are easy to use, are generally accepted, and give numerical values to the flow characteristics of fresh concrete.

Concrete that expands during hardening would be desirable in some cases. The expansion would increase the lateral stress at the interface of the concrete and soil, and, hence, would increase the ability of the drilled shaft to carry load. However, research is needed to show the benefits to be gained from the use of expansive concrete and to show whether or not there are detrimental effects. Such research is now underway (11). After some time a rewritten version of a standard for drilled shafts may be more specific than at present concerning the way in which concrete is to be made.

Design

The Standard is relatively silent on the design of drilled shafts. Only some general comments are given. A complete treatment of design is left to other documents (6,8,10).

With regard to design for axial loading, it is of interest to note the kinds of information that the engineer can find in the literature that will aid in design. The results of load tests, particularly on drilled shafts that are instrumented for the measurement of load along the length of the foundation, are of great importance. Several dozens of such tests have been reported in the technical literature. The tests generally show that the compressive load is sustained by skin friction and by end bearing. Correlations have been indicated between unit load transer, both in skin friction and in end bearing, and soil properties. Extreme caution, however, should be used in the selection of such correlations in making designs. A careful study of the results of such load tests have revealed that the behavior of drilled shafts under load is strongly dependent on the details of construction. Instrumented drilled shafts that were employed in field tests as a rule were constructed with great care; therefore, the performance of such foundations could well be more favorable than the average production shaft.

The load transfer in skin friction is likely to be more sensitive to the details of construction than is the load transfer in end bearing. Assuming that good contact exists between concrete and soil at the base of the foundation, the load in end bearing can be computed by using the same procedures as are used to compute the load on a spread footing. But the load in skin friction is strongly dependent on the lateral stress at the interface of the concrete and the soil as well as on the nature of the soil at the interface (3). Thus, the computation of the load in skin friction involves many more assumptions than the computation of the load in end bearing. Furthermore, while numerous load tests have been performed and the unit skin friction measured in a number of instances, the soil conditions at a site could be unique and respond in a different manner than have the soils at the test sites. Thus, the designer of a drilled shaft for a particular project may use the results from previous tests as a guide and use an appropriate factor of safety (perhaps three or more) or he could specify a load test at the site. In any case, careful inspection must be done of the production shafts to insure good construction.

If a load test is specified, consideration should be given to the use of internal instrumentation in the drilled shaft to enable the engineer to estimate the amount of load that is distributed in end bearing and in skin friction. Several types of instrumentation have been suggested (1,2) and other types will undoubtedly become available with time. Such instrumentation will yield more information than just the way the load is distributed; if the axial stiffness of the drilled shaft is taken into account, curves can be developed that show unit load transfer, t, as a function of shaft movement, z, for various points along the length of the foundation (9). If methods of predicting t-z curves can be worked out, it will then be possible to use a small computer to estimate a load-settlement curve for a drilled shaft under axial loading. Such a procedure will possibly become routine in the future and would then be addressed by a revised standard.

The design of drilled shafts to sustain lateral loads is mentioned in the Standard. Much progress has been made in the past several years in making such designs, but unfortunately the design procedures have not been included to any extent in textbooks. Some information that has been published in design guides on the lateral response of deep foundations is incomplete and misleading. Futhermore, research is continuing on the lateral response of piles and new information becomes available from time to time. Therefore, the engineer who designs drilled shafts to sustain lateral loads must become acquainted with the technical literature and become adept at the solution of the nonlinear, fourth-order differential equation that describes the reponse of a pile or drilled shaft when subjected to lateral loading.

Inspection

The Standard gives attention to the inspection of the construction of drilled shafts. Inspection is an important function of any project involving earthwork but it is of critical importance concerning the construction of drilled shafts. There are numerous examples in technical literature of how errors in construction can lead to defective foundations (4). A technique that has been employed routinely by some contractors and engineers is to place a transducer at the top of a completed shaft, to strike the top of the foundation, and to use an instrument to record the time it takes a wave to travel to the bottom of the foundation and to be reflected to the top (5). The technique has been worked out for performing a dynamic load test on a completed foundation (12). While the Standard points the way to making inspections of drilled shafts, advancements are being made on the development of techniques that can have a significant impact on the way inspections are made.

Conclusions

The Standard provides an up-to-date guide for the design and construction of drilled shafts. However, the designer must become acquainted with a wealth of material if a proper job is to be done. New information is being added regularly to the technical literature that will also influence actual practice. Therefore, a revision of the Standard will be required in due course.

References

1. Baguelin, F., Bustamante, M., Frank, R., and Jezequel, J. F., "Bearing Capacity of Piles," Annales de L'ITBTP, July-August, 1975, pp. 111-119.

2. Barker, W. R., "Instrumentation for Measurement of Axial Load in Drilled Shafts," M.S. Thesis, The University of Texas at Austin, May, 1970.

3. Bernal, Juan B., and Reese, Lymon C., "Study of the Lateral Pressure of Fresh Concrete as Related to the Design of Drilled Shafts," conducted for State Department of Highways and Public Transportation in cooperation with U. S. Department of Transportation, Federal Highway Administration, by the Center for Transportation Research, Bureau of Engineering Research, The University of Texas at Austin, Research Report 308-IF, May, 1984.

4. Fleming, W. K., and Sliwinski, Z. J., "The Use and Influence of Bentonite in Bored Pile Construction," Dept. of the Environment, Report No. P63, Sept., 1977.

5. Goble, G. G., and Rausche, Frank, "Pile Load Test by Impact Driving," Highway Research Record No. 333, 1970, pp. 123-129.

6. Meyerhoff, G. G., "Compaction of Sands and Bearing Capacity of Piles," Journal of the Soil Mechanics and Foundations Division, ASCE, Vol. 85, No. SM6, 1963, pp. 1-29.

7. Nazarian, S., Stokoe, K. H., and Hudson, W. R., "Use of Spectral-Analysis-of-Surface-Waves Method for Determination of Moduli and Thicknesses of Pavement Systems," Transportation Research Record 945, 1983.

8. Reese, Lymon C., "Design and Construction of Drilled Shafts," The Twelfth Terzaghi Lecture, Journal of the Geotechnical Engineering Division, ASCE, Vol. 104, No. GT1, Jan., 1978, pp. 91-116,

9. Reese, Lymon C., "Design and Evaluation of Load Tests in Deep Foundations," Behavior of Deep Foundations, ASTM STP670, a symposium sponsored by ASTM Committee D18 on Soil and Rock for Engineering Purposes, June 28, 1978.

10. Reese, Lymon C., and Wright, Stephen J., "Construction of Drilled Shafts and Design for Axial Loading," Drilled Shafts Design and Construction Guidelines Manual, Vol. One, July, 1977.

11. Sheikh, Shamim A., O'Neill, Michael W., and Mehrazarin, Mohammad, "Application of Expansive Cement in Drilled Shafts," Research Report No. UHCE 83-14, University of Houston, Aug., 1983, 170 pp.

12. TNO Institute for Building Materials and Building Structures, "Dynamic Pile Testing," Report No. BI-77-13, Delft, Netherlands, Jan., 1977.

GEOTECHNICAL PERSPECTIVE -- PROPOSED DRILLED SHAFT STANDARD

Clyde N. Baker, Jr., F. ASCE*

INTRODUCTION

In order to prepare a rational written review of the working draft standard, "Design and Construction of Drilled Shafts", it is necessary to have agreement on, "What is a standard?" and, "What is the primary purpose and objective of the standard?" In the writer's understanding, a standard is like an agreed upon manual of practice, and as such, sets the level of performance and attainment that is considered adequate. A standard is not a code nor is it a specification. It must be seen as part of an overall framework. The prime objective of the standard should be that, if followed (along with appropriate specifications), an adequately engineered and constructed end-product will result.

PROBLEMS CAUSED BY OUR ADVERSARIAL SYSTEM

In the United States, we have a society which tends to use the adversarial system. This is evident in labor-management-government conflicts and is evident in our justice system. It is also involved in our competitive bidding system where a buyer may try to buy a service or an end-product at the cheapest price and then hire somebody else to oversee the work to see that he gets what he paid for. If something goes wrong, somebody has to be responsible and blamed and somebody must pay to correct what went wrong. This makes the wording of any standard very critical.

DESIRABLE REQUIREMENTS OF STANDARD

Insofar as is possible, the standard must reconcile the different interests of the owner, designers, and contractors. While these interests can vary from project to project and from individual to individual, in general, the owner wants an end product that will serve his needs at minimum cost, and the designers want to accomplish the owners objective but have the additional interests of not wanting to be exposed to any more professional liability than absolutely necessary, and the contractor's interest is to construct the foundation as quickly and cheaply as possible within the specification requirements.

The standard also needs to reflect current successful practice as well as perhaps move in the direction of further improvements in the practice.

*Senior Principal Engineer, STS Consultants, Ltd., Northbrook, Illinois

QUESTIONS FOR THE STANDARD

Should construction procedures that affect ground movement be specified by the engineer, or should only the performance requirements be set and the procedures necessary to accomplish the performance requirements left to the contractor? The standard as proposed appears to be in conflict in different sections. In the section on responsibilities, the standard states, "The owner and his representatives shall generally accept responsibility for damage to adjacent structures if the job is built according to plans and specifications." However, in the section on construction methods, in 7.4.7, covering loss of ground, the standard states, "The contractor shall take such steps as necessary to prevent damage to structures or facilities on adjacent sites resulting from loss of ground during or subsequent to his construction." Perhaps this ambiguity is merely an indication that there is no one best way that the project can be successfully completed, but rather there are different possible approaches.

The project can be set up and successfully completed with specifications that require certain specific construction procedures designed to minimize ground loss on the one hand, and a project can also be successfully completed with specifications wherein the requirement of preventing ground loss is stated but the procedures are left to the contractor. In either case, the personnel specifying or deciding on the necessary construction procedure must be sufficiently knowledgeable and experienced in caisson construction in similar soils, or must have sufficient knowledge of soil mechanics to be able to predict the behavior of the soils in question (if lacking experience in caisson construction in the specific soils) so that the proper procedures are used.

What should be the role of the design geotechnical engineer during construction? Must he provide the foundation inspection? The standard as proposed says "All construction operations shall be observed and documented by the engineer or his designated representative." In Section 1.4 on definitions, under the description of the term "engineer", the standard states "A geotechnical engineer is best capable of carrying out most of the functions noted in this document that call for the participation of an engineer. The standard also states that "A Registered Professional Engineer or the equivalent is desirable." In the opinion of the writer, the standard has a weakness here in that it does not discuss continuity between design and construction. It does not imply recognition of the fact that there is normally both a Structural Engineer and a Geotechnical Engineer involved on most significant drilled shaft projects. Many architects and structural engineers feel that the Geotechnical Engineer responsible for the evaluation of the soil conditions, foundation analysis, and foundation design recommendations should be responsible for the foundation observation and documentation, and the writer strongly supports this view.

If a different Geotechnical Engineer does the foundation observation than performed the foundation analysis and design recommendations, there is a problem of divided responsibility. However, there is a much more significant problem in that the new geotechnical engineer may not be aware of everything that went into the design recommendations and may have a totally different set of experiences on which he will be basing his field observations and field decisions. This can lead to quite different decisions being made by the new firm as compared to those that would have been made by the firm responsible for the original soil evaluation and design recommendations. The problems could be reduced, perhaps, if the new firm were required to totally review the soil evaluation report and any assumptions involved in the design recommendations and then to agree with the design recommendations and in effect, assume comparable responsibility for the design recommendations as if they had made them themselves.

The writer is aware of several instances where a different firm did the foundation observation than the one that did the design recommendations, and because of different evaluation procedures, required the foundations to go much deeper than necessary at significant increase in foundation costs to the owner. It could be argued that since the owner probably had a role in the selection of the second firm on the basis of a cheaper price, he is reaping his just reward when he pays for his ultimate foundation costs. However, unfortunately, life usually is not that simple, and there can be other consequences, including disputes on changed conditions, and lawsuits involving the original soil engineer, the owner, and third parties.

The essence of the problem is that foundation observation and approval of bearing stratum in many cases involves considerable judgment based on experience and the judgment and experiences of both individuals and firms vary within a range wide enough so that significant differences in decisions can result. Particular geologic profiles where this is certainly the case in the writer's experience are residual silts, saprolites, and weathered schists in the Piedmont region of the eastern United States, the glaciated areas of the northern Midwest and Northeast, and seamy rock profiles anywhere. In these profiles, it is difficult to set up specific test procedures and required test results for all eventualities so that in the final analysis, considerable engineering judgment is involved.

As a final point on this subject, the engineer in overall charge of the foundation observation and documentation must be a Registered Engineer. It should be stated as "mandatory" and not just as desirable.

ADDITIONAL REQUIREMENTS DESIRABLE IN THE STANDARD

It would be very desirable if the standard required that a pre-job conference (prior to start of construction) among the contractor, architect (and owner's representative if different from the architect), structural engineer, and geotechnical engineer be held to discuss construction procedures, coordination, communication, report submittals, and who is to receive reports, and any questions different parties might have with regard to the project construction. Oftentimes, the project specifications call for a pre-job conference but it would be helpful if the desirability of the pre-job conference were explicitly stated in the standard. Specifications often require the contractor to submit his written procedures for construction of the shafts for review by the engineer (structural engineer and geotechnical engineer) and architect. This desired procedure could also be stated in the standard.

Since soil and groundwater conditions often vary between borings, and since changed conditions claims are the basis for many disputes and costly lawsuits in drilled shaft construction, a fair and expeditious procedure for settling disputes and deciding changed conditions claims is essential. The current draft standard when referring to changed conditions disputes states that, "The construction documents should allow such situations to be mediated or arbitrated without delay." This is certainly a positive statement and a step in the right direction. However, that statement would be even better if it were worded so that inclusion of the mediation-arbitration principal in drilled shaft construction documents was a "must" requirement rather than a "should" requirement. Also, the statement should not read "mediation or arbitration", but should read "mediation and arbitration" or "mediation-arbitration", since there are significant procedural differences between the two and what is desired is a system that involves both, to ensure the fastest, most fair resolution of the dispute at the lowest overall total cost.

ADDITIONAL COMMENTS, QUESTIONS, AND TECHNICAL DIFFERENCES

In reading the standard, it appears that there is a lack of consistency in the use of the terms shall, will, must, and should, and since each of these terms has different legal significance, the writer recommends that the document be reviewed very carefully by a person knowledgeable in the law to make certain that the intent is clear and the legal implications understood. The words "must" and "shall" should be used with care in a standard because of the possible legal implications if a project is undertaken wherein the required procedure is not followed (for whatever reason).

In the following paragraphs, specific recommendations are made for changes or additions to the standard:

1. In Paragraph 3.2.C, the statement is made that, "Descriptions of the foundations of those structures must be given if it can be reasonably expected that subsurface soil movements could occur at those foundations." The writer recommends that the word "given" be replaced by the word "determined" since the owner sometimes does not know the information and makes it part of the specifications for the contractor to find it out.
2. <u>Paragraph 3.3.3 -- Ground Water</u> This paragraph, as written, is too idealistic and too all encompassing. The implication after reading the paragraph is that if the contractor does not find the conditions with regard to water seepage, quantity, and rate to be exactly as predicted by the geotechnical engineer, he will have a claim for an extra. While the contractor should have an agreed-upon procedure for claiming an extra where there are changed conditions, it is unrealistic to take all the risk out of the bidding process and put it on the back of the geotechnical engineer. The writer is in agreement with the general objective of the paragraph, i.e., giving the caisson contractor the maximum amount of information that will affect his bid so that his risks are reduced and there is a concrete basis for changed conditions claims. However, as currently worded, it would

drastically increase the exploration programs for small to medium sized projects. As a minimum, the paragraph could be weakened by beginning it with the phrase "Insofar as is practical . . ." and changing the words "must" to "should".

3. Paragraph 6.2 -- Concrete In Table 6.1, the writer would change the recommended slump ranges slightly as follows: For shafts with close reinforcement or whenever casing is to be extracted, the recommended minimum slump should be 5 inches.

   For concrete placed underwater or under drilling mud, a distinction should be made between tremie concrete and pumped concrete. For tremie concrete, the slump should not be less than 7 inches. However, for pumped concrete, the slump can go as low as 5 inches because of the pumping ability to push the concrete through the pipe. If no distinction is to be made between pumped concrete and tremie concrete, the recommended range should be 7 to 9 inches, or at most 6 to 9 inches, rather than the 5 to 9 inches as currently indicated.

4. Under 6.4 and 6.4.1, more differentiation should be made between the use of drilling mud to hold a hole open long enough for casing insertion in the casing method on the one hand, and the use of drilling mud for the slurry method where the slurry stays in the hole until it is displaced by the concrete.

   Paragraphs 6.4.1a and 6.4.1b apply primarily to the former, whereas 6.4.1c applies primarily to the latter.

5. Under 6.4.4, the standard should state who determines the quality of the slurry and who corrects the caisson if defective slurry is used and creates problems. In the writer's opinion, the quality of the slurry is the responsibility of the contractor and he is also responsible if defective slurry is used which later creates problems in the caisson.

6. Paragraph 7.3 -- Tolerances The tolerances should agree with the ACI recommended tolerances which have just been revised slightly. The revised ACI recommended tolerances are shown below:

Proposed

Concrete shafts out of plumb - Use the following tolerances:

Category A. For unreinforced shafts extending through materials offering no or minimal lateral restraint; i.e., water, normally consolidated organic soils, and soils that might liquify during an earthquake - 12.5% of shaft diameter.

Category B. For unreinforced shafts extending through materials offering lateral restraint (soils other than those indicated in Category A) - not more than 1.5% of the shaft length.

Category C. For reinforced concrete shafts - out of plumb tolerance determined by the engineer based on the structural properties of the shaft and lateral restraint properties of the medium penetrated. See project specifications.

The Proposed Standard out of plumb tolerance of 2% for the total length of the shaft would be excessive for a number of possible soil profile situations, but is also probably acceptable for a large number of soil profile situations, and particularly would be acceptable in those firm soil situations where the shafts are designed primarily in friction or combination friction and end bearing.

The recommended ACI tolerance specification separates shafts into different categories of unreinforced and reinforced and categories of soils offering virtually no lateral restraint and other soils. Since the behavior of caissons or shafts in these different categories can differ significantly, this breakdown is an attempt at a rational approach to the problem of allowable tolerances.

7. Paragraph 7.5.5 -- Inspection of Concrete The standard calls for minimizing time at the site making test cylinders, performing slump tests, and measuring air entrainment, and instead calls for as much inspection as possible at the plant site. This is contrary to ACI recommended procedures which require the testing of concrete be performed at the point of discharge. This should be modified appropriately.

8. In Sections 9, 10, and 11 on design, the terms "appropriate theory", "semi-empirical methods", and "state-of-the-art procedures" are used. These are subject to wide interpretation which may be good or bad, depending upon one's point of view. Since the phrases lack specificity, they leave the geotechnical engineer with considerable flexibility which the writer believes is desirable. At the same time, they can be a source of disagreement in disputes.

CONCLUSIONS

The authors of the proposed standard are to be commended for their effort at pulling together a standard representing a variety of points of view (contractor, design engineer, and geotechnical engineer). The writer believes that much of the standard is excellent and will provide useful reference for the design and construction profession. The questions, comments, and recommendations presented in this paper are intended to be helpful in the discussion and eventual adoption of the standard as modified and approved.

A Structural Engineering Perspective for Drilled Shafts

Walter E. Hanson, F.ASCE*

INTRODUCTION

Substructure and superstructure designs are usually apportioned between geotechnical and structural engineers. When this occurs, satisfactory results can be obtained only by full communication and cooperation among the specialists. Unfortunately, foundation and superstructure designers do not always possess good working knowledge and appreciation of the other's concerns in the design process. That is, design philosophies and methods of determining load and resistance parameters, which are the basic ingredients in the determination of safety and settlements of the foundation and of the performance of the whole structure, are not always mutually understood, and communication is difficult.

This commentary is an attempt to close the gap between the Building Code Requirements for Reinforced Concrete (ACI 318-83) (1,2) and the design of drilled shafts. The context of design includes the selection of loads and resistances for the proportioning of the shaft as well as for the structural design which follows the proportioning step.

BUILDING CODE REQUIREMENTS (ACI 318-83)

Strength and Serviceability

Section 9.1.1 of the ACI Building Code states: "Structures and structural members shall be designed to have design strengths at all sections at least equal to the required strengths calculated for the factored loads and forces in such combinations as are stipulated in this code." Factored loads are usually the service loads specified in building codes multiplied by appropriate load factors which vary according to type of load--for example, 1.4 for dead load and 1.7 for live load.

---

*Chairman, Hanson Engineers, Inc., Springfield, Illinois.

It is important to elaborate on the terms design strength, required strength, and their relationship to normal strength, because all three terms appear in the ACI code. The required strength is that necessary to resist factored loads or related internal moments and forces. As stated above, the design strength must be equal to or greater than the required strength. The design strength is calculated by multiplying the nominal strength by a strength reduction factor less than one. Theoretically, the reduction accounts for uncertainties in all the assumed parameters involved in design and construction which may combine to result in understrength. In the ACI code, the strength reduction factor, designated Ø, varies from 0.70 to 0.90 depending upon the type of structural member. (Because of the soil-shear connotation of Ø to geotechnical engineers, it is unfortunate that the same symbol was chosen by the ACI 318 Committee.) It would seem appropriate to use a reduction factor of 0.70 to obtain the "design strength" in the structural design of drilled shafts by the ACI strength procedures.

Thus, the margin of structural safety in the strength design procedure is provided in two ways: (1) the required strength is computed by multiplying service loads by load factors to provide for excess load effects, and (2) the design strength is computed by multiplying normal strength by a reduction factor to account for uncertainties in the assumed parameters, computations, and construction. Accordingly, the basic relationship of the ACI code may be expressed as

$$\text{Required Strength} \leq \text{Design Strength}$$

or

$$1.4D + 1.7L \leq \emptyset\ [\text{Normal Strength}]$$

The reader is referred to Chapter 9 of the ACI Code and to the corresponding chapter of the Commentary for further information and discussion of loads, load factors, and strength requirements.

In addition to the above load and strength considerations (sometimes called the ultimate strength method), Section 9.1.2 of the ACI code states: "Members also shall meet all other requirements of this code to insure adequate performance at service load levels." This requirement is primarily concerned with the deflections and cracks which occur in reinforced concrete structures and which must be held within tolerable limits regardless of safety from the standpoint of strength.

ACI Code References to Foundations

Chapter 15 of the ACI code applies mainly to spread footings and pile caps. Nevertheless, it may be assumed that the basic provisions would apply also to drilled shafts. Section 15.2.1 requires that the structural design of footings conform to the strength method to resist factored loads. The subsequent Section 15.2.2 states that the base area of footings shall be determined from unfactored (service) loads using a permissible soil pressure "selected through principles of soil mechanics." The choice of loads and the selection of the permissible soil pressure for design are discussed from the geotechnical viewpoint in the following section.

GEOTECHNICAL CONSIDERATIONS

General Comments

As mentioned above, the ACI code recommends that the area of foundation support be determined from service loads and permissible soil pressures, whereupon the structural analysis and design is made using factored loads and other strength procedures. It is therefore necessary that the geotechnical engineer have a good knowledge of the character of the loads and load combinations as well as of the superstructure type prior to selection of soil parameters for design. It is also necessary that the structural engineer recognize the loading conditions and combinations to which the soil parameters apply. Thus, the need exists for full communication and cooperation between the specialists.

Like all other types of foundations and, indeed, all structural elements, a drilled shaft must be safe against outright failure. At the same time, settlement must be within tolerable limits to prevent damage (maintain serviceability) of the superstructure. Outright failure is a function of the shear strength of the soil or rock and of the shaft itself, while settlements depend on deformation characteristics of all materials. Therefore, the choices of design loads and the associated soil resistance parameters are intimately related to the strength and deformation characteristics of the material which support the shaft and to the characteristics of the loading conditions (3). As only one example of many, live loads of short duration may cause little settlement of clay due to the long time required for consolidation, whereas the same load and duration may cause many times more settlement if the shaft is supported by a granular deposit. In both cases, the shaft may possess approximately equal and adequate safety against outright failure.

Soil Parameters for Service Loads

Since service loads are generally used to determine the dimensions of the shaft (in some cases, the area of the base; in others, the depth of the shaft to obtain sufficient side shear), the "permissible" soil parameters should possess appropriate intrinsic factors of safety. It would seem reasonable that the factor of safety for foundation resistance be at least equal to that for superstructure design. Using a 1.4 load factor for dead load and 1.7 for live load, and if the strength reduction factor is taken as 0.7, the factor of safety against outright failure would vary between 2.0 and 2.4, depending upon the distributive character of the loading. These are values which assume the same degree of risk and reliability in the soil data and in the design and construction of the substructure as in a column of the superstructure. On the other hand, a greater intrinsic factor of safety for the foundation is generally selected for design. A value of 3.0 is common where service loads are involved for determining the dimensions of a foundation, including drilled shafts. In terms of the ACI code, if the total design service load is equally divided between dead load and live load, a factor of safety of 3.0 corresponds to a strength reduction factor of 0.52.

On certain projects, it may be worthwhile to use probability theories of reliability and risk in the selection of factors of safety. Such theories may be applied to determine load factors as well as resistance parameters of the subsoil, and results may serve as guides to engineering judgment. As explained previously, whether the soil parameters selected for safety against outright failure are appropriate for settlement and serviceability is an entirely different matter.

CONCLUSIONS

The structural designs of drilled shafts (as well as other types of foundations) usually conform to the strength provisions of the ACI code using factored loads and strength reduction factors. However, such structural designs usually follow proportioning steps based on service (unfactored) loads and on soil resistance parameters which may also involve settlement considerations in addition to safety against outright failure. Therefore, a mutual understanding of all aspects of superstructure and substructure design must exist between the specialists (structural and geotechnical engineers) usually involved.

An intrinsic factor of safety equal to, but not less than, 2.0 appears justified for permissible soil parameters used in conjunction with service loads, assuming conditions are as favorable as those commonly associated with the structural design of building columns, particularly those which carry primarily fixed loads. On the other hand, a safety factor of 3.0 or even greater may be justified based upon the character of the loads, the degree of risk and reliability associated with the soil data, and with the design and construction of the foundation.

The permissible soil parameters chosen to meet certain safety requirements against failure may not be appropriate to meet settlement and serviceability requirements. This is, of course, analogous to a concrete beam which possesses adequate strength, but which does not meet deflection and serviceability requirements.

REFERENCES

1. ACI Committee 318, Building Code Requirements for Reinforced Concrete, American Concrete Institute, Detroit, Michigan, Nov. 1983, Chapters 8, 9 and 15.

2. ACI Committee 318, Commentary on Building Code Requirements for Reinforced Concrete, American Concrete Institute, Detroit, Michigan, Nov. 1983, Chapters 8, 9 and 15.

3. Peck, Hanson & Thornburn, Foundation Engineering, 2nd Ed., John Wiley & Sons, N.Y., 1974, Sect. 17.3, pp. 264 and Sect. 23.1, pp. 375.

ASCE DRILLED SHAFT STANDARD: UNIVERSITY PERSPECTIVE

Fred H. Kulhawy, F. ASCE*

A critique is presented of the proposed ASCE standard on drilled shafts. Conceptually, the version reviewed addresses the proper issues. However, this version can only be considered a first draft. There are numerous technical issues of concern, but most can be addressed editorially. Areas where major work is necessary relate primarily to the shaft design sections, establishment of loads, and selection of a factor of safety.

INTRODUCTION

This brief paper has been prepared by invitation to provide a perspective on the proposed ASCE drilled shaft standard (April 1984 version) by someone within a university, who teaches foundation engineering, conducts research on drilled shafts, and consults on drilled shaft design and construction. In preparing this paper, emphasis has been placed on discussing major issues of concern to the writer. Detailed comments will be transmitted separately to the Sub-Committee on Drilled Shafts.

OVERALL SCOPE AND CONCEPT

The proposed standard generally addresses the appropriate issues required to provide a guideline document for the design and construction of drilled shafts. In the writer's opinion, the choice of the word "standard" is unfortunate because of the formal connotations it conveys and the likelihood of it becoming a legal reference very rapidly.

Consider the definitions of standard as found in Webster's Unabridged Dictionary:

1. standard (noun). Of eighteen major definitions, the following are appropriate in our context: (3a) something that is established by authority, custom, or general consent as a model or example to be followed, (3b) a definite level or degree of quality that is proper and adequate for a specific purpose, (4) something that is set up and established by authority as a rule for the measure of quantity, weight, extent, value, or quality, (7a) a carefully thought-out method of performing a task.

2. standard (adjective). Of eight major definitions, the following are appropriate in our context: (1) constituting or afford-

*Professor, School of Civil and Environmental Engineering, Cornell University, Ithaca, NY 14853 ing a standard for comparison, measurement, or judgment, (2a) having qualities or attributes required by law or established by custom, (4) having recognized and permanent value, (5) substantially uniform and well-established by usage in the speech and writing of the educated and widely recognized as acceptable and authoritative.

From these definitions, it is clear that a "standard" can be thought of in different contexts, ranging from a more formal authoritative or legal rule to a generally accepted approach. The writer does not believe that a geotechnical system, involving exploration, analysis, design, and construction variables, can be standardized in the formal sense. The term "guidelines", implying a recommended approach, would be more correct and palatable, and would not create an albatross that could be regretted in the future.

A second overall point is that the proposed standard apparently has been written from the viewpoint and philosophy of a designer whose main interest is in building, and perhaps bridge, structures. There is repeated emphasis on loads being obtained from building codes, there is extrapolation from building structural design to foundation structural design, and the exploration section appears to focus on the detailed exploration typically conducted for buildings and bridges. However, drilled shafts are used for other purposes as well. Two of the more prominent applications are: (1) foundations for tower and pole structures used for the transmission of electric power and communication signals, and (2) earth retaining structures. The entire exploration, analysis, design, and perhaps construction philosophy will vary with the application, as will the applicable codes or design criteria. These points must be brought out in the proposed standard to make it sufficiently general.

A third overall point is the repeated emphasis that a geotechnical engineer is "best capable" of carrying out the necessary functions. Simply stated, this is not correct! Engineering geologists often are better in the exploration phase, and similarly structural engineers and sometimes architects are better in analysis and design. And in construction, none of the above can match a competent and conscientious contractor. What should be brought out is that the overall exploration-analysis-design-construction process should be under the direction of a licensed professional engineer with experience in drilled shafts. Any more specificity is bound to create problems.

DEFINITIONS AND NOTATION

The definitions provided are a good starting point for discussion. Some need clarification, and others need to be added. For instance, bearing strata implies that shafts derive their support principally from tip resistance; this is not generally true. The definition of casing does not differentiate between casing used to advance a hole, to provide safety for an inspector, or to be left in place as part of the foundation. Others could be mentioned as well.

Terms which need to be defined include the following: construction methods (dry, casing, slurry), failure, excessive movement, collapse, outright failure. The last several are ambiguous to many engineers. Perhaps alternative terms would be more precise.

One final point to make in this section is the description of probability concepts as alternate theory, while the traditional factor of safety concept is implied to be the more definitive approach. Consciously or subconciously, we all use probability in the evaluation of likely foundation loads and potential strength or other resistance parameters. Loads are estimated liberally while resistance parameters are estimated conservatively. Probability-based design concepts represent just a consistent and formal approach to what we already do. These concepts should be treated as such.

MATERIAL AND IN-SITU PROPERTY EVALUATION

Within the several sections dealing with materials and property assessment, a number of points must be made. First, under subsurface conditions, a great deal of text is devoted to evaluating rock characteristics, yet very little is said about evaluating soil characteristics. This is a major problem because it implies that soil evaluation is well-known and straightforward, while rock evaluation is just the opposite. Neither is true and the text should reflect this in a balanced perspective.

When addressing questions of concrete, a considerable amount of text is devoted to slump. Up to 9 inch (229 mm) slump concrete is recommended as a matter of course, as if this was a subject which is established definitively in the literature. From what the writer has seen, and has discussed with concrete specialists, concrete with slumps above 6 to 7 inches (152 to 178 mm) have not been researched and documented adequately to recommend such a sweeping endorsement.

A second point regarding concrete is that the only equation in the proposed standard is for the modulus of concrete as a function of cylinder strength. This equation should be deleted and simply referenced by designation to the appropriate ACI specification. For this equation, it is especialy true because there is a definite chance of it being modified in the new code edition.

DESIGN

The writer believes that the sections on design are inadequate and require substantial re-thinking. There really are no guidelines given and there is much too much reference made to using the state-of-the-art, appropriate theory, or special design approaches. The term state-of-the-art is particularly inappropriate in this regard. This term is defined in the Concise Oxford Dictionary as the current stage of development or knowledge of a subject. This definition poses some interesting questions. First, who makes the definitive judgment on what the state-of-the-art is? Second, what form of literature (theses, reports, papers, texts) is included? Third, are all written languages included?

The point to be made is that state-of-the-art is a term which should not be used in a proposed standard. It is inappropriate, and probably impossible, to expect every drilled shaft designer to be right at the cutting edge of knowledge development on the subject.

The proposed standard should stress sound, modern design concepts. Qualitative and conceptual descriptions should be included which illustrate how a drilled shaft works and how it develops its resistance to load. However, the details of selecting specific design equations should be the realm of the designer.

For instance, consider the axial uplift of a straight drilled shaft. An example description might be as follows. The equilibrium equation is:

$$Q_u = W + Q_{su} + Q_{tu} \tag{1}$$

in which Q_u = uplift capacity, W = weight, Q_{su} = side resistance, and Q_{tu} = tip resistance. The side resistance represents the available shearing resistance summed over the side of the shaft, taking into account applicable freeze-thaw, swell-shrink, and other local phenomena which may lead to a design decision to discount a portion of the shaft depth. The tip resistance can develop over the tip area from suction during undrained loading and from bonding of the shaft concrete to the soil or rock at the tip, resulting in a tip tension. Tip suction dissipates with time and therefore is only applicable during transient live loading. Tip tension in soil is small and requires extraordinary clean-out procedures to develop; therefore it should be disregarded in normal design. Tip tension in rock can be as large as the lower tensile strength of the rock or concrete, but can only develop after careful clean-out. Only limited field conditions will allow this special clean-out.

An approach such as given above defines the general design components, but allows the designer the flexibility to select their own design equations and design concepts.

A final point to be made is that the proposed standard only considers axial and lateral loads, and the axial mode is heavily oriented toward compression. The axial uplift mode needs to be addressed more fully, and overturning moment, torsion, and combined modes need to be included. The moment and torsion modes are particularly important for the pole structures noted previously.

LOAD TESTING

In the writer's opinion, a particularly heavy emphasis has been placed on load testing as a design tool. A great deal of information has accumulated in the literature on the behavior of drilled shafts, particularly under axial loads, and our knowledge of the subject has improved greatly. This knowledge has demonstrated clearly that it is no longer necessary to "prove" every design for axial load, especially in well-known ground conditions. However, the knowledge base is much less for lateral load and overturning moment, and is even less for torsion and combined loading. For these loading modes, perhaps the emphasis is warranted when the project economics dictate.

A second point with respect to load testing is the strong endorsement of relatively detailed instrumentation for all load tests. This simply is not warranted for all projects and this decision should be up to the discretion and need of the designer.

DESIGN LOADS AND FACTORS OF SAFETY

The discussion of design loads and factors of safety does not focus on the key issues for geotechnical foundation design. From a foundation standpoint, it is necessary to know the following:

1. dead, sustained live, and transient live loads, along with the duration and frequency of the live (particularly transient) loads

2. function of the structure, and limiting serviceability criteria for it

3. geologic variability of the site

4. drainage conditions at the site, to evaluate undrained vs. drained response

5. soil properties for the appropriate loading and drainage conditions

6. reliability and quality of the soil property data

7. reliability of the foundation design model used

8. method of analysis, considering total vs. effective stress methods

Without knowing these factors with some degree of confidence, a conservative design must be developed.

Factors of safety, or levels of reliability, are developed using all of the above criteria, and can vary significantly depending on the input data quality. For example, with well-defined loads, no very stringent serviceability criteria, little site variability, either undrained or drained response evaluated in terms of effective stresses, appropriate soil properties determined directly with proper equipment, and a proven design model, a high factor of safety is inappropriate. However, the reverse would be true if the conditions were opposite.

Blanket statements recommending a factor of safety of three as a general rule, and not less than two when all factors are known extremely well, are a disservice to the profession because the above eight factors are not addressed specifically. Each factor and its infuence must be examined carefully and individually.

Consider a simple example of a straight-sided drilled shaft in clay, 5 feet in diameter, 5 feet deep, with average operative undrained shear strength (adhesion) along the shaft equal to 750 psf, and a potential tip suction of 1/2 atmosphere acting over the tip area during undrained transient live loading. For these conditions, using the notation in Equation 1, the weight (W) = 14.7 kips, the side resistance (Q_{su}) = 58.9 kips, and the tip resistance (Q_{tu}) = 41.5 kips, giving an uplift capacity (Q_u) = 115.1 kips for an undrained transient live load.

The question now presents itself as to how designers might approach this problem to compute a design capacity. Table 1 outlines five approaches the writer has seen used within the past two years, giving the design equation, computed design capacity, and the ratio of the available uplift capacity to the computed design capacity. For illustration only, a FS = 3 was used. It is clear that there are gross inconsistencies. Designers 1, 3, and 5 used FS = 3 on the weight, which is quite conservative, while designers 2 and 4 used FS = 1 on the weight, which is perhaps a bit unconservative. Designers 3 and 4 neglected the tip suction, and designer 5 neglected the tip suction and side resistance, reflecting extreme conservatism.

Table 1. Design Capacity Example

Designer	Design Equation	Design Capacity, Q_{ud}, kips (using FS = 3)	Q_u/Q_{ud}
1	$Q_{ud} = (W + Q_{su} + Q_{tu})/FS$	38.4	3.0
2	$Q_{ud} - W = (Q_{su} + Q_{tu})/FS$	48.2	2.4
3	$Q_{ud} = (W + Q_{su})/FS$	24.5	4.7
4	$Q_{ud} - W = Q_{su}/FS$	34.3	3.4
5	$Q_{ud} = W/FS$	4.9	23.5

1 kip = 4.45 kN

The point to be made by this example is that recommendation of a particular factor of safety, without specifying the design equation and all of the eight factors noted previously, should not be done. Different factors of safety should be used for different assumptions, loading conditions, etc. Without this specificity, the number is meaningless.

SUMMARY AND CONCLUSIONS

The proposed ASCE drilled shaft standard conceptually addresses the proper issues of concern, but can only be considered as a first draft. There are many positive aspects of this draft, as will be seen by anyone reading it. However, there are also a number of significant shortcomings. This brief paper has focused on some of the major shortcomings, and has offered a number of suggestions for future consideration.

Standards and the Contractor

James L. Parks*

INTRODUCTION

Contractors engaged in the construction of drilled shafts have long been interested in the development of improved standards and specifications. The International Association of Drilled Shaft Contractors have written their own standards and have cooperated fully with other groups in similar activities. There are several reasons for our interest in better standards, as outlined in the following paragraphs.

In a general sense, the writing of up-to-date standards is an educational process. Contractors have a role in preparing standards, of course. The construction of drilled foundations where there exists a variety of purposes, terrains, environments, and subsurface conditions leads to knowledge that is important to pass on. In addition to participating in cooperative endeavors with engineers and owners, as described herein, the ADSC has carried out an educational program consisting of seminars, workshops, conferences, and the distribution of literature. As a group, we welcome the opportunity to participate in the San Francisco conference of ASCE.

INTERESTS OF THE CONTRACTOR

At the outset, the interests of the Contractor in a particular job, and in general, are set forth. Good standards and specifications figure strongly in each of these interests.

1. We are interested in construction of good quality. The failure of a caisson or drilled shaft, for any reason, is bad for the industry. Almost without exception, Contractors will go "an extra mile" to do whatever is necessary to get a foundation of good quality. We strongly recommend free and open discussions between the owner, engineer, and Contractor so that any possibility of poor construction can be eliminated.

An example concerns the minimum diameter of a drilled shaft that can be installed and the slump of concrete. It is impossible to get good construction with 2-inch slump if the shaft is to be 14 inches in diameter and 50 feet deep.

2. Contractors are interested in efficient methods of construction. If the cost of a job is as small as possible, drilled shafts are more competitive with other forms of deep foundations and owners may be able to build a project when otherwise they could not. The owner benefits because he gets his structure sooner than otherwise and with a lower cost.

---

*President, Parks Drilling Company, Columbus, Ohio.

An example of more efficient methods of construction concerns the specification of casing, and consequently the size of the drilled hole. Because pipe whose diameter is given by its outside dimension is readily available and is much cheaper than other pipe, Contractors strongly recommend the use of "OD" pipe. Thus, if the casing method of construction is used, the diameter of the drilled hole must be a couple of inches less than the casing size. Furthermore, Contractors recommend that casing sizes (and hole sizes) be specified in 6-inch increments, reducing the number of sizes of casing and tools that must be brought to a job.

3. Contractors are anxious to eliminate disputes that arise during construction. Someone has said that Mother Nature never went to the university and did not learn about soil conditions from textbooks! With no intention to criticize the work of the geotechnical engineer, there are occasions when the soil borings do not reflect the subsurface conditions. The geology may be complex, the small-diameter borings used by the engineer may not encounter boulders that are present, or the owner may set up an inadequate budget for the soil investigation. Then there arises the bane of a foundation job, "change-of-conditions." Many examples can be given of difficulties that develop over such circumstances: large cost over-runs, drilling Contractors going bankrupt, lawsuits over periods of years, and the like. Preliminary agreements should be worked out to deal with change-of-conditions and other unforeseen difficulties so that the job can proceed in a fair and orderly way.

4. Contractors must make a profit. The system of competititve bidding should assure the owner of getting a fair price. After an agreement is reached, the Contractors should be able to proceed with the job in the way that was planned. An example of a situation where a Contractor had his profit margin threatened occurred when the contractual documents required the base of the drilled shafts to be founded on shale "with a bearing value of 25 tons/sq ft." The original soils consultant indicated informally the depth and kind of material where the required bearing value could be found but another soils engineer was employed for inspection. Which incidently is in my opinion a very poor practice. The owner hired the original soils engineer but made the cost of inspection a part of the General Contractor's bid. The General Contractor then hired the soils engineer, who was low bidder, to do the inspection. The second engineer had a completely different idea of what constituted proper bearing. The result was that the Contractor was required to drill much deeper than he had anticipated, requiring more machine and operator time and extra-depth cost to owner. It is my strong opinion that the cost of inspection and selecting an engineer must be under control of the owner or his representative. In no case should the General Contractor or Caisson Contractor be hiring and paying the inspectors that work for him.

DESIRABLE CHARACTERISTICS OF STANDARDS

In addition to contributing toward construction that is of high quality, efficient, and without disputes, standards should have other characteristics from the standpoint of the Contractor.

Standards should allow innovation. The strength of any industry lies in the development of better methods of doing things. The drilled-shaft industry is no exception. Over the years improved machines, drilling tools, and construction methods have appeared and these improvements must continue. The "hammer-grab" is an example of the type of machine and construction that may become important in the U.S. A heavy, clam-shell bucket is fixed to the bottom of a heavy, steel bar. The bucket and bar are raised and dropped and can penetrate strong formations. Holes of various cross sections can be cut. The hammer-grab has found limited use in the U.S. to date but may become important in the future. The standards as drafted by the committee are believed to be able to accommodate new ideas.

The language of standards should be clear. The Committee worked hard to eliminate ambiguities and expect the final document to be quite clear and free from the possibility of various interpretations.

The standards should receive acceptance and come into general use. If the standards are employed by specification writers in all cases, design and construction of drilled shafts should improve. Too often Contractors are given a set of specifications that are poorly written and out-of-date. These specifications add cost to a job and can result in difficulties for all concerned.

The standards should not discourage research. There has been a number of research projects over the past one or two decades that have influenced considerably the design and construction of drilled shafts. More research needs to be done but sometimes ideas get "set in concrete" and tend to freeze practice. It is hoped that the new standards will encourage rather than discourage research. An example of useful research involves the Mustran cell which was developed at the University of Texas. Mustran cells can be used to measure the load distribution over the full length of a drilled shaft. Full-scale load tests which use the results to lower cost of construction have become commonplace and have made our product very competitive with piling. Whole new territories, especially in the coastal south, have opened up for economical foundations.

Standards should be rewritten periodically. While efforts were made by the Committee to make the drilled-shaft standard as flexible as possible, changes in design practice or construction procedures will no doubt occur over the next several years that require the standards to be revised. It is hoped that ASCE will continue to monitor the standards and take steps to make revisions at the appropriate time.

EXAMPLES OF POOR PRACTICE

A few examples are given where good standards have had or could have a strong influence on the performance of the work. The examples that are cited are from the point of view of the Contractor and it is agreed that engineers that are designing and inspecting drilled shafts would cite different examples or view these examples differently.

There are many specifications that still require no free fall of concrete over 5 feet. Most of the time this requirement is not enforced in the field but it can become very costly to the Contractor who makes that assumption. If appropriate concrete is used, the specification is unnecessary and can lead to defective foundations.

Some specifications require 1% plumbness. I can remember a job where the engineer, after 1/2 hour with the use of plumb bobs and strings on a 10-foot diameter, 12-foot deep hole, told us we were out of plumb by 3/16th of an inch. But if we tapped the temporary casing on one side with the kelly bar, the hole would be plumb enough!

Some specifications require a dry hole with no leeway to pour if you cannot get a seal. This had led to many lawsuits because the Contractor has had no method to finish the job.

Many other examples can be cited but these few should serve to indicate the strong interest in Contractors getting the adoption of the standards that are being considered.

CONCLUDING COMMENTS

Improved standards, if put into general use, will be of benefit to owners, engineers, and Contractors. In addition to the new standards, however, there needs to be a continued program of information exchange so that the best methods of design and construction will become widely known. As new methods for design and construction become available, they should become known throughout the industry and specifications can be modified to reflect the better methods.

Government Agency Perspective on Standard for Drilled Shafts

Samuel D. Stone, M.ASCE*

ABSTRACT

Relative to the proposed ASCE standard for Design and Construction of Drilled Shafts, the Tennessee Valley Authority (TVA) would primarily be considered an "owner". However, in past installations of drilled shafts, TVA has variously performed the functions of geotechnical engineer, structural engineer, driller-constructor, inspector, and owner. Most drilled shaft installations have been performed through design-construct contracts. TVA's past drilled shaft work has equalled or exceeded the criteria in the proposed standard. Therefore, the proposed standard will probably have little impact upon the agency's operation relative to drilled shaft foundations.

The Tennessee Valley Authority (TVA) is an independent corporate agency of the federal government established by the U.S. Congress in May 1933. The agency is involved with flood control, navigation, production of electric power, agriculture and industrial development, conservation and development of natural resources in the Tennessee River basin.

TVA is somewhat unique among federal agencies. One unique feature is that TVA performs most of its construction with its own employees rather than using construction contracts with commercial firms. However, TVA does contract for "specialty" construction where it is not cost effective to maintain full-time construction capability.

TVA has installed a few thousand drilled shafts in the past 10 years. In most cases, the drilled shafts were installed under contract by companies specializing in drilled shaft construction. Typical installations were for the foundations of a large office building complex, hyperbolic natural draft cooling towers, turbine buildings, electrostatic precipitators, and various other power plant features. Since the soil overburden is relatively shallow in most of the TVA region, these drilled shafts were typically socketed into structurally

---

*Principal Civil Engineer, Tennessee Valley Authority, Knoxville, Tennessee.

adequate bedrock. In many instances, the shafts penetrated weathered shales or solution features in limestone to reach competent rock. Shaft sizes have ranged up to 78-inch diameter in the overburden and 72-inch diameter in rock. Drilled shafts founded entirely in overburden are generally not appropriate at TVA's facilities due to typically large loads, settlement requirements, and the structural capacity of the soil.

Relative to the proposed ASCE standard for Design and Construction of Drilled Shafts, TVA would primarily be considered an "owner". However, in past installations of drilled shafts, TVA has on occasions worn nearly all of the available "hats", i.e., geotechnical engineer, structural engineer, driller-constructor, inspector, and owner. The variation in roles has been dictated by differing case-by-case circumstances.

TVA has employed the following options for the installation of drilled shafts.

a - Design-Construct by TVA.
b - Design by TVA, Construction by contract.
c - Design-Construct by contract.

In most cases, the option of design-construct by contract has been utilized.

TVA has also used different options in supplying geotechnical data for contract work. Initially, bedrock was investigated entirely by core drilling and conventional logging by trained geologists. Later, core drilling was supplemented by percussion borings that were geophysically logged, i.e., sonics, gamma-gamma density, mechanical caliper, and natural gamma ray. Favorable experiences led to a change in the ratio of core borings to percussion borings with geophysical logging to where percussion boring with geophysical logging is now the primary investigative technique. In recent jobs, TVA has, as a minimum, drilled and geophysically logged one percussion boring at each drilled shaft location. These borings extend well below the estimated bearing elevation to ensure that no adverse defects exist beneath the shaft. This thorough exploration has been well received by all parties, since many unknowns are removed allowing potential contractors to bid with some confidence. In addition, TVA has the information to reasonably assure that the drilled shaft is located at the optimum bearing elevation. Occasionally, additional hand drilled percussion borings are made during construction if TVA's inspector detects or suspects an anomaly while inspecting the completed excavation.

With the previous material serving as background information, I would like to discuss the role of the proposed ASCE Standard For Design and Construction of Drilled Shafts and other external codes and standards upon TVA's internal operation.

TVA manages its design and construction processes through a series of controlled documents called Design Criteria, Design Standards, Design Guides, General Construction Specifications, Project Construction Specifications, and technical specifications for purchase requisitions (contract work). Design Criteria specify the engineering requirements which provide the basis for conceptual and detailed design. Design Standards are documents that present specific methods that must be used in the design process. Design Guides are documents presenting acceptable methods that may be employed in the design process. Design Guides must be considered, but the degree to which they are used is determined by the individual engineer. General Construction Specifications cover a general subject area, e.g., "Plain and Reinforced Concrete," and apply to all TVA projects. Project Construction Specifications apply to one project only. The latter may supplement or modify information contained in a general construction specification or cover a subject unique to one project which is not addressed in a general construction specification, e.g., "Earth and Rock Foundations and Fills, Hartsville Nuclear Plant." Technical specifications for purchase requisitions define the design and/or construction requirements and are site-job specific. It should be emphasized that TVA is not obligated to adopt and use consensus codes and standards published by external agencies. However, it is quite common to see codes, standards, test methods, procedures, etc., published by others listed in TVA's design criteria documents, construction specifications, and technical specifications for purchase requisitions. A partial listing of external agencies whose codes, standards, test methods and procedures are often used at TVA are:

American Concrete Institute (ACI)
American Institute of Steel Construction (AISC)
American National Standards Institute (ANSI)
American Society of Civil Engineers (ASCE)
American Society of Mechanical Engineers (ASME)
American Society for Testing and Materials (ASTM)
American Welding Society (AWS)
American Water Works Association (AWWA)
American Association of State Highway and Transportation Officials (AASHTO)

Obviously TVA has not made a commitment on the adoption of the proposed ASCE Standard for Design and Construction of Drilled Shafts which is nearing completion and balloting among members of the Foundation and Excavation Standards Committee. TVA has had considerable input in formulating the proposed drilled shaft standard. The present draft has been reasonably well received by TVA engineers associated with drilled shaft design and construction. Typically, TVA's past drilled shaft work has equalled or exceeded the criteria in the proposed standard. Although I have no authority to establish TVA policy or the ability to forecast the future, I personally feel that TVA would have no serious problems with the standard as presently drafted.

I believe one of the greatest services that the proposed drilled shaft standard may provide will be to small government agencies and bodies, i.e., small municipalities and counties. Many of these government agencies cannot justify maintaining large engineering staffs or full-time foundation engineering expertise. The proposed standard should be of assistance from the planning stage through construction. Additionally, I suspect that many larger local, state, and federal agencies will adopt the standard.

I believe the standard merely reflects what would be considered by most practicing engineers as good engineering practices with no overkills, excessive constraints, or gold plating. Additionally, the proposed standard is well balanced in considerations relative to the engineer, the contractor, and the owner without imposing excessive restraints on any of the parties.

SUBJECT INDEX

Page number refers to first page of paper.

AUTHOR INDEX

Page number refers to first page of paper.